TOPICS IN
APPLIED
ABSTRACT
ALGEBRA

The Brooks/Cole Series in Advanced Mathematics
Paul J. Sally, Jr., Editor

Series editor Paul J. Sally, Jr. and Brooks/Cole have developed this prestigious list of books for classroom use. Written for post-calculus to first year graduate courses, these books maintain the highest standards of scholarship from authors who are leaders in their mathematical fields.

Titles in this prestigious series include:

Probability: The Science of Uncertainty with Applications to Investments, Insurance, and Engineering (0-534-36603-1)
Michael A. Bean

A Course in Approximation Theory (0-534-36224-9)
Ward Cheney and Will Light

Advanced Calculus: A Course in Mathematical Analysis (0-534-92612-6)
Patrick M. Fitzpatrick

Fourier Analysis and Its Applications (0-534-17094-3)
Gerald B. Folland

Introduction to Analysis, Fifth Edition (0-534-35177-8)
Edward Gaughan

The Mathematics of Finance: Modeling and Hedging (0-534-37776-9)
Joseph Stampli and Victor Goodman

Algebra: A Graduate Course (0-534-19002-2) and
Geometry for College Students (0-534-35179-4)
I. Martin Isaacs

Numerical Analysis: Mathematics of Scientific Computing, Second Edition (0-534-33892-5)
David Kincaid and Ward Cheney

Ordinary Differential Equations (0-534-36552-3)
Norman Lebovitz

Introduction to Mathematical Modeling (0-534-38478-1)
Daniel Maki and Maynard Thompson

Introduction to Fourier Analysis and Wavelets (0-534-37660-6)
Mark A. Pinsky

Real Analysis (0-534-35181-6)
Paul Sally, Jr.

Beginning Topology (0-534-42426-0)
Sue Goodman

TOPICS IN APPLIED ABSTRACT ALGEBRA

S.R. Nagpaul
Ohio University

S.K. Jain
Ohio University

THOMSON
™
BROOKS/COLE

Australia • Canada • Mexico • Singapore • Spain
United Kingdom • United States

Publisher: *Bob Pirtle*
Assistant Editor: *Stacy Green*
Editorial Assistant: *Katherine Cook*
Marketing Manager: *Tom Ziolkowski*
Marketing Assistant: *Erin Mitchell*
Project Manager, Editorial Production:
Cheryll Linthicum
Print Buyer: *Rebecca Cross*
Permissions Editor: *Stephanie Lee*
Production Service: *Cecile Joyner/*
The Cooper Company

Text Designer: *John Edeen*
Art Dirctor: *Vernon Boes*
Copy Editor: *Carol Reitz*
Illustrator: *International Typesetting*
and Composition
Cover Designer: *Vernon Boes, Jennifer*
Mackres
Cover Printer: *Coral Graphic Services*
Compositor: *International Typesetting*
and Composition
Printer: *Courier Corporation/Westford*

For more information about our products,
contact us at:
Thomson Learning Academic Resource Center
1-800-423-0563

For permission to use material from this text,
contact us by:
Phone: 1-800-730-2214
Fax: 1-800-730-2215
Web: http://www.thomsonrights.com

Library of Congress Control Number: 2004107658

ISBN 0-534-41911-9

Thomson Higher Education
10 Davis Drive
Belmont, CA 94002-3098
USA

Asia
Thomson Learning
5 Shenton Way #01-01
UIC Building
Singapore 068808

Australia/New Zealand
Thomson Learning
102 Dodds Street
Southbank, Victoria 3006
Australia

Canada
Nelson
1120 Birchmount Road
Toronto, Ontario M1K 5G4
Canada

Europe/Middle East/Africa
Thomson Learning
High Holborn House
50/51 Bedford Row
London WC1R 4LR
United Kingdom

Latin America
Thomson Learning
Seneca, 53
Colonia Polanco
11560 Maxico D.F.
Mexico

Spain/Portugal
Paraninfo
Calle Magallanes, 25
28015 Madrid, Spain

CONTENTS

1

BOOLEAN ALGEBRAS AND SWITCHING CIRCUITS

2

BALANCED INCOMPLETE BLOCK DESIGNS

3

ALGEBRAIC CRYPTOGRAPHY   128

4

CODING THEORY   160

5

SYMMETRY GROUPS AND COLOR PATTERNS   208

6

WALLPAPER PATTERN GROUPS   249

PREFACE

This book presents some interesting applications of abstract algebra to practical real-world problems. Whereas many applications of calculus are presented in undergraduate courses, usually no such applications are given in courses on abstract algebra. The object of this book is to fill this lacuna. It is hoped that this will make the study of abstract algebra more interesting and meaningful, especially for those whose interest in algebra is not confined to mere abstract theory. Among the applications discussed in the book are designing and simplifying switching circuits and gate networks used in computer science; designing block designs to conduct statistical experiments for unbiased study of samples; designing secret-key cryptosystems and public-key cryptosystems for secure transmission of sensitive or secret data; designing error-correcting codes for transmission of data through noisy channels; computing the number of color patterns of a given design; and classifying the symmetry groups of wallpaper patterns.

The book may be used as a text for senior-level and beginning graduate-level students for a course in *applied* abstract algebra. It is addressed to two categories of students: (1) those who are majoring in mathematics and are interested in knowing about applications of what they have learned in an abstract algebra course, and (2) those who are majoring in other disciplines like physics, computer science, and engineering, and deal with these applications in their area of study but often do not have sufficient understanding of the mathematics involved. Of course, the book may also be used profitably as a supplementary text for a regular course in abstract algebra or as a reference book for all scientists in general. This book may indeed be looked upon as a companion volume to all books in abstract algebra.

The book is self-contained and does not assume a prior knowledge of abstract algebra, although a knowledge of elementary linear algebra is taken for granted. The algebraic theory used in the applications is fully developed here. However, this book is not meant to be used as a text for a course in abstract algebra per se. The amount of algebraic theory presented here is just what is required for the applications discussed in the book.

The opening chapter, called Chapter 0, gives a somewhat-condensed account of the basic algebraic systems—namely, groups, rings, and fields, and their salient properties. We have omitted the proofs of several theorems in this chapter that are quite easy to prove. A student who has already done a course in abstract algebra may skip this chapter initially and refer to it subsequently if and when necessary. The remaining six chapters that contain the applications are independent of one another and may be read in any order. However, the chapters have been arranged in ascending order of mathematical complexity. The first chapter is the simplest and requires very little beyond the basic notions of sets and mappings. The last two chapters are relatively difficult and use some fairly complex group theory.

We are very grateful to those who have reviewed the text and offered valuable and helpful comments: John Dauns, Tulane University, New Orleans; Edward Formanek, Pennsylvania State University; Sarah Spence, Olin College, Massachusetts; and Jeffrey D. Valer, The University of Texas at Austin.

S.R. Nagpaul
S.K. Jain

CHAPTER

0

PRELIMINARY ALGEBRAIC CONCEPTS

In this preliminary chapter, we give a somewhat condensed account of certain basic algebraic structures and their important properties that will be used in the practical applications to be presented in the subsequent chapters. Proofs of several theorems that are quite easy to prove have been omitted. A student who has already completed a course in abstract algebra may skip this chapter initially and may refer to it subsequently if and when necessary.

0.1 SETS, MAPPINGS, RELATIONS, AND BINARY OPERATIONS

0.1.1 Sets

A *set* is, informally speaking, a collection of objects. The objects in the collection are called *elements* (or *members*) of the set. We ordinarily use capital letters to denote sets and lowercase letters to denote elements of a set. If a is an element of a set S, we say a *belongs* to S (or a is in S), and we write $a \in S$. If a is not an element of S, we write $a \notin S$.

Some letters are reserved for particular sets and are always used with that meaning. We list some of them here. (They are usually written in **boldface** or, as shown here, in blackboard bold type.)

$\mathbb{N}$ denotes the set of all natural numbers 1, 2, 3, ...

$\mathbb{Z}$ denotes the set of all integers 0, ± 1, ± 2, ...

$\mathbb{Q}$ denotes the set of all rational numbers.

$\mathbb{R}$ denotes the set of all real numbers.

$\mathbb{C}$ denotes the set of all complex numbers.

Two sets A, B are said to be *equal,* written $A = B$, if they consist of the same elements; that is, every element of A belongs to B, and every element of B belongs to A.

A set A is called a *subset* of a set B, written $A \subset B$, if every element of A is an element of B. Trivially, $A \subset A$. If $A \subset B$ and $A \neq B$, then A is called a *proper* subset of B, written $A \subsetneq B$.

From these two definitions it follows that two sets A and B are equal if and only if A is a subset of B and B is a subset of A.

We often use the notation $\Rightarrow$ to denote implication, and we use the two-headed arrow $\Leftrightarrow$ to mean "if and only if." With this notation, the foregoing definitions can be stated as follows:

$A = B$ means $x \in A \Leftrightarrow x \in B$.

$A \subset B$ means $x \in A \Rightarrow x \in B$.

If a set has no elements, it is called *empty* and denoted by $\varnothing$. The empty set is a subset of every set. Because there is no element in the set $\varnothing$, the condition $x \in \varnothing \Rightarrow x \in A$ is satisfied *vacuously.*

A set may be described in several ways. A set that contains a finite number of elements may be exhibited by a list of all its elements, separated by commas, within braces. Thus $\{1, 2, 3\}$ denotes the set consisting of elements 1, 2, and 3. The order in which the elements are written is immaterial. Thus $\{1, 2, 3\} = \{2, 1, 3\}$. If the set has a large number of elements, we may write only a few representative elements and use ellipsis points to stand for the rest, provided the meaning is clear. Thus $\{1, 2, \ldots, 100\}$ denotes the set consisting of all integers from 1 to 100. We may use this method even for infinite sets, provided the meaning is clear. The set of all integers can be written as $\mathbb{Z} = \{0, \pm 1, \pm 2, \ldots\}$.

More generally, we use the notation

$$S = \{x \in A \mid p(x)\}$$

to denote the set S consisting of all those elements x in a given set A for which the statement $p(x)$ is true. For example, $S = \{x \in \mathbb{Z} \mid -2 < x \leq 2\}$ means $S = \{-1, 0, 1, 2\}$. Sometimes, when the meaning is unambiguous and reference to the set A is not necessary, we simply write $S = \{x \mid p(x)\}$.

Definition 0.1.1 Let X be a set. The set of all subsets of X is called the *power set* of X and denoted by $\mathcal{P}(X)$.

For any set X, the empty set $\emptyset$ and X itself are subsets of X, so $\emptyset$, $X \in \mathcal{P}(X)$. For example, let $X = \{a, b, c\}$. Then

$$\mathcal{P}(X) = \{\emptyset, \{a\}, \{b\}, \{c\}, \{a, b\}, \{a, c\}, \{b, c\}, \{a, b, c\}\}$$

A set X is said to be *finite* or *infinite* according to whether it has a finite or an infinite number of elements. It is clear that if X is an infinite set ($\mathbb{Z}$, for example), then X has infinitely many subsets. The following theorem gives the number of subsets of a finite set.

THEOREM 0.1.2 Let X be a finite set that has n elements. Then X has 2^n distinct subsets, so $\mathcal{P}(X)$ has 2^n elements.

Proof: Let us count the number of ways in which we can construct an arbitrary subset A of X. For each element $x \in X$, we have two choices—namely, to include or not to include x in A. Since there are n elements in X, we have $2 \times 2 \times \cdots \times 2 = 2^n$ distinct subsets of X.   ■

The number of elements in a finite set X is called the *order* of X and denoted by $|X|$. Theorem 0.1.2 thus says that $|\mathcal{P}(X)| = 2^{|X|}$. (This justifies the nomenclature of power set.)

We now define three operations by which we obtain new sets from any two given sets.

Definition 0.1.3 Let A and B be sets.

(a)   The *union* of A and B, written $A \cup B$, is the set

$$A \cup B = \{x \mid x \in A \text{ or } x \in B\}$$

(b)   The *intersection* of A and B, written $A \cap B$, is the set

$$A \cap B = \{x \mid x \in A \text{ and } x \in B\}$$

(c)   The *difference* $A - B$ is the set

$$A - B = \{x \mid x \in A \text{ and } x \notin B\}$$

Two sets are said to be *disjoint* if their intersection is empty—that is, if they have no common elements. If $A \subset B$, then $B - A$ is called the *complement* of A in B.

If A and B are both subsets of a set X, then clearly $A \cup B$, $A \cap B$, $A - B$, and $B - A$ are all subsets of X. If all sets under consideration in a given context are subsets of a set U, then U may be referred to as the *universal set*. In that case, $U - A$ is called the *complement* of A and written A'. For example, if we are dealing with the power set of a given set X, then X itself is the universal set in relation to the members of $\mathcal{P}(X)$.

By an *algebra of sets* we mean the power set $P(X)$ (for any given set X) together with the operations of union, intersection, and complement.

We now state some important properties of these operations of union, intersection, and complement. The proofs are left as an exercise.

THEOREM 0.1.4 Let A, B, C be sets. Then

(a) $A \cup B = B \cup A$; $A \cap B = B \cap A$ (*commutative laws*)
(b) $(A \cup B) \cup C = A \cup (B \cup C)$;
    $(A \cap B) \cap C = A \cap (B \cap C)$ (*associative laws*)
(c) $A \cup (B \cap C) = (A \cup B) \cap (A \cup C)$;
    $A \cap (B \cup C) = (A \cap B) \cup (A \cap C)$ (*distributive laws*)

In view of the associative laws, we can omit parentheses and simply write $A \cup B \cup C$ and $A \cap B \cap C$ for the union and intersection of A, B, and C, respectively.

THEOREM 0.1.5 Let X be a nonempty set, and for any $S \in P(X)$ let S' denote the complement of S in X. Let $A, B \in P(X)$. Then

(a) $(A')' = A$; $A \cup A' = X$; $A \cap A' = \emptyset$
(b) $(A \cup B)' = A' \cap B'$; $(A \cap B)' = A' \cup B'$ (*DeMorgan's laws*)

THEOREM 0.1.6 Let A and B be finite sets. Then
$$|A \cup B| = |A| + |B| - |A \cap B|$$

The definition of union and intersection can be generalized to more than two sets. Let $\mathcal{F}$ be a set whose members are themselves sets. (We refer to $\mathcal{F}$ as a *family* of sets.) The union and intersection of the family $\mathcal{F}$ are defined, respectively, as
$$\bigcup \mathcal{F} = \{x \mid x \in A \text{ for some member } A \text{ of } \mathcal{F}\}$$
$$\bigcap \mathcal{F} = \{x \mid x \in A \text{ for each member } A \text{ of } \mathcal{F}\}$$
If $\mathcal{F}$ consists of a finite number of sets, say $\mathcal{F} = \{A_1, \ldots, A_n\}$, then $\bigcup \mathcal{F} = A_1 \cup \cdots \cup A_n$ and $\bigcap \mathcal{F} = A_1 \cap \cdots \cap A_n$.

0.1.2 Mappings

Let A, B be sets. A *mapping* (or a function) from A to B is a rule that associates with each element in A exactly one element in B. We write
$$f : A \to B$$
to mean that f is a mapping from A to B. The sets A and B are called the *domain* and the *codomain* of the mapping f, respectively. For any $x \in A$, the unique element

in B that f associates with x (or *assigns* to x) is called the *image* of x and written $f(x)$. If $f(x) = y$, we call x a *preimage* of y and say that f takes x to y. Further, if reference to f is not necessary, we simply say that x goes to y and write $x \mapsto y$. The *range* of f, written Im f, is defined to be the set of those elements in B that are images of elements in A; that is,

$$\text{Im } f = \{y \in B \mid y = f(x) \text{ for some } x \in A\}$$

Two mappings $f : A \to B$ and $g : C \to D$ are said to be *equal*, written $f = g$, if $A = C$, $B = D$, and $f(x) = g(x)$ for all $x \in A$.

Let $f : A \to B$. If distinct elements in A have distinct images under f, then f is said to be *injective* (or one-to-one). In other words, f is injective if for all $x, x' \in A$, $f(x) = f(x') \Rightarrow x = x'$. If every element in B is the image of some element in A, then f is called a *surjective* (or onto) mapping. In other words, f is surjective if Im $f = B$. If f is injective and surjective, then it is called a *bijective* mapping or a *one-to-one correspondence*. A bijective mapping from a set A to itself is called a *permutation* on the set A.

Given two mappings $f : A \to B$ and $g : B \to C$, we define their *composite*, written $g \circ f$, to be the mapping

$$g \circ f : A \to C$$

given by the assignment rule

$$g \circ f(x) = g(f(x)) \text{ for all } x \in A$$

Given a set A, we define the *identity mapping* on A, written id_A or simply i, to be the mapping

$$id_A : A \to A$$

given by the rule

$$id_A(x) = x \text{ for all } x \in A$$

Given a mapping $f : A \to B$, if there exists a mapping $g : B \to A$ such that

$$g \circ f = id_A \quad \text{and} \quad f \circ g = id_B$$

then the mapping f is said to be *invertible* and g is called an *inverse* of f. In other words, g is an inverse of f if for all $x \in A$, $y \in B$, $f(x) = y \Leftrightarrow g(y) = x$. It is easy to show that f is invertible if and only if it is bijective. Further, if f is invertible, it has a unique inverse, which is written f^{-1}.

0.1.3 Relations

Let A, B be sets. The set of all ordered pairs (x, y), where $x \in A$, $y \in B$, is called the *Cartesian product* of A and B and written $A \times B$. In symbols,

$$A \times B = \{(x, y) \mid x \in A, \ y \in B\}$$

For example, let $A = \{1, 2\}$, $B = \{a, b, c\}$. Then

$$A \times B = \{(1, a), (1, b), (1, c), (2, a), (2, b), (2, c)\}$$

The notion of a relation (in ordinary life and in mathematics) is intuitively familiar to us. We give below a formal definition for it.

Definition 0.1.7 Let A and B be sets. A mapping $R : A \times B \to \{0, 1\}$ is called a *relation* from A to B. Given $a \in A$, $b \in B$, if $R(a, b) = 1$, we say that a is in relation R to b, and we write $a R b$. If $R(a, b) = 0$, then a is not in relation R to b, and we write $a \not R b$.

In the case $A = B$, we say R is a relation on A (or in A).

This formal definition of a relation is very general but abstract, and at first sight it may not appear to correspond to our intuitive understanding of a relation. In practice, we define a particular relation R by stating a rule that determines, for any given ordered pair (a, b), whether or not a is in relation R to b. For example, we define the relation $<$ of one real number being less than another as follows: Given $a, b \in \mathbb{R}$, we say a is less than b and write $a < b$, if there exists a positive real number c such that $a + c = b$. To understand how this manner of defining the relation $<$ corresponds with the formal definition of a relation we provided, let us confine our attention to the set $A = \{1, 2, 3\}$. In this set, we have $1 < 2$, $1 < 3$, $2 < 3$. We define the mapping $R : A \times A \to \{0, 1\}$ as follows:

$$R(1, 1) = 0, \quad R(1, 2) = 1, \quad R(1, 3) = 1,$$
$$R(2, 1) = 0, \quad R(2, 2) = 0, \quad R(2, 3) = 1$$

The mapping R defines the same relation as the one given in Definition 0.1.7.

We most often deal with relations that have some special property. The following definition states some of the most important properties.

Definition 0.1.8 Let R be a relation on a set A.

(a) R is said to be *reflexive* if $a R a$ for every $a \in A$.
(b) R is said to be *symmetric* if for all $a, b \in A$, $a R b \Rightarrow b R a$.
(c) R is said to be *antisymmetric* if for all $a, b \in A$, $a R b, b R a \Rightarrow a = b$.
(d) R is said to be *transitive* if for all $a, b, c \in A$, $a R b, b R c \Rightarrow a R c$.

Further, if R is reflexive, symmetric, and transitive, then R is called an *equivalence relation*. If R is reflexive, antisymmetric, and transitive, then R is called a *partial order.*

EXAMPLES

1. A trivial example of an equivalence relation is the relation of equality in any set.

2. For any complex numbers a, b, let $a R b$ mean $|a| = |b|$. Then R is an equivalence relation on $\mathbb{C}$.
3. Let A denote the set of all lines in a plane. For any $a, b \in A$, let $a R b$ mean that a is parallel to b. Further, let us agree to consider any line as being parallel to itself. Then R is an equivalence relation on A. Similarly, the relation of one triangle being congruent to another is an equivalence relation.
4. The relation $\leq$ in $\mathbb{R}$ is reflexive, antisymmetric, and transitive; hence it is a partial order. The relation $<$ is antisymmetric and transitive but not reflexive.
5. The relation $\subset$ is a partial order in the power set $\mathcal{P}(X)$ for any set X.

The most important property of an equivalence relation on a set is that it gives rise to a partition of the set, as explained below.

Definition 0.1.9 Let X be a set. Let P be a set consisting of some nonempty subsets of X. We say P is a *partition* of the set X if:

(a) $\bigcup P = X$; that is, the union of all members of P is equal to X.
(b) Every two members of P are disjoint.

Clearly, the two conditions stated in the definition are equivalent to saying that each element of the set X lies in one and only one member of P. For example, let $X = \{1, 2, 3, 4, 5\}$. Then the set $\{\{1, 3\}, \{2\}, \{4, 5\}\}$ is a partition of X.

Definition 0.1.10 Let E be an equivalence relation on the set X. Given $a \in X$, the set of all elements in X that are in relation E to a is called the *equivalence class* of E determined by a (or the equivalence class of a under E) and denoted by $Cl_E(a)$; that is,

$$Cl_E(a) = \{x \in X \mid x E a\}$$

The set of all equivalence classes of E in the set X is denoted by X/E and is called the *quotient set* of X by E.

For the sake of convenience, we usually write the equivalence class $Cl_E(a)$ as simply $[a]$ or $\bar{a}$ when explicit reference to E is not necessary.

We now prove that the equivalence classes form a partition.

THEOREM 0.1.11 Let E be an equivalence relation on the set X. Then the quotient set X/E is a partition of X.

Proof: Let $a \in X$. Then $a E a$ and hence $a \in [a]$. Thus every equivalence class $[a]$ is a nonempty subset of X, and the union of all equivalence classes is equal to X. Now let $a, b \in X$, and suppose the equivalence classes $[a]$ and $[b]$ have a common element c. Then cEa and cEb, so for every $x \in [a]$, xEa, aEc, cEb, and hence xEb. Therefore $[a] \subset [b]$. Similarly $[b] \subset [a]$ and hence $[a] = [b]$. Thus, for all $a, b \in X$, the equivalence classes $[a]$, $[b]$ are either equal or disjoint. This proves that X/E is a partition of X. ∎

We now describe an equivalence relation in $\mathbb{Z}$ that plays an important role in number theory and other applications. Given any integers a, b, we say a *divides* b and write $a \mid b$ if there exists an integer c such that $ac = b$.

Definition 0.1.12 Let n be a fixed positive integer. Given $a, b \in \mathbb{Z}$, we say a is *congruent* to b *modulo* n and write $a \equiv b \pmod{n}$ if n divides $a - b$.

For example, $11 \equiv 5 \pmod{3}$, since 3 divides $11 - 5 = 6$.
We show in the following theorem that this relation, called *congruence modulo n* and written $\equiv_n$, is an equivalence relation on $\mathbb{Z}$.

THEOREM 0.1.13 For any positive integer n, congruence modulo n is an equivalence relation on $\mathbb{Z}$. The equivalence class of an element $a \in \mathbb{Z}$ under $\equiv_n$ is given by

$$[a] = \{a + tn \mid t \in \mathbb{Z}\}$$

Moreover, the quotient set of $\mathbb{Z}/\equiv_n$ consists of exactly n equivalence classes, given by

$$\mathbb{Z}/\equiv_n = \{[0], [1], \ldots, [n-1]\}$$

Proof: For every $a \in \mathbb{Z}$, $a - a = 0$ is divisible by n and hence $a \equiv a \pmod{n}$. If $a \equiv b \pmod{n}$, then n divides $a - b$, so n also divides $b - a$. Hence $b \equiv a \pmod{n}$. If $a \equiv b \pmod{n}$ and $b \equiv c \pmod{n}$, then n divides $a - b$ and also $b - c$. Therefore n divides $(a - b) + (b - c) = a - c$. Hence $a \equiv c \pmod{n}$. This proves that $\equiv_n$ is an equivalence relation.

Let $a \in \mathbb{Z}$. If $x \equiv a \pmod{n}$, then n divides $x - a$, so $x - a = tn$ for some integer t. Hence the equivalence class determined by a is $[a] = \{a + tn \mid t \in \mathbb{Z}\}$.

By Euclidean algorithm, there exist unique integers q, r such that $a = qn + r$ and $0 \leq r \leq n - 1$. Then $a - r = qn$, so $a \equiv r \pmod{n}$ and hence $[a] = [r]$. Moreover, no two integers in the set $\{0, 1, \ldots, n - 1\}$ are congruent modulo n. Hence there are exactly n equivalence classes—namely, $[0]$, $[1]$, $\ldots$, $[n - 1]$. ∎

The following theorem gives a useful property of the congruence relation. The proof is left as an exercise.

THEOREM 0.1.14 Let n be a positive integer, and let $a, b, c, d \in \mathbb{Z}$. If $a \equiv b \pmod{n}$ and $c \equiv d \pmod{n}$, then $a + c \equiv b + d \pmod{n}$ and $ac \equiv bd \pmod{n}$.

0.1.4 Binary Operations

A *binary operation* on a set S is a rule that associates with each ordered pair of elements in S a unique element of S. For example, the usual addition and multiplication of integers are binary operations on the set $\mathbb{Z}$. The set of all ordered pairs of elements in the set S is the Cartesian product $S \times S = \{(x, y) \mid x, y \in S\}$. A binary operation on S is therefore a mapping from $S \times S$ to S. It is customary to denote binary operations by symbols like $*$, $\cdot$, $+$, $\circ$, and so on. The image of an ordered pair $(x, y) \in S$ under a binary operation $*$ is written $x * y$ [instead of $*(x, y)$], as is the usual practice in ordinary arithmetic.

Let $*$ be a binary operation on S. We say $*$ is *commutative* if $x * y = y * x$ for all $x, y \in S$. We say $*$ is *associative* if $(x * y) * z = x * (y * z)$ for all $x, y, z \in S$. An element $e \in S$ is called an *identity* for $*$ if $e * x = x = x * e$ for all $x \in S$. For example, addition and multiplication of integers are both commutative and associative; the integer 0 is an identity for addition, and 1 is an identity for multiplication. It is easily shown that an identity, if it exists, is unique. Suppose e, e' are both identities for the binary operation $*$. Then $e = e * e' = e'$.

Let $*$ and $\circ$ both be binary operations on S. We say $*$ is *distributive* over $\circ$ if

$$x * (y \circ z) = (x * y) \circ (x * z) \quad \text{and} \quad (y \circ z) * x = (y * x) \circ (z * x)$$

for all $x, y, z \in S$. For example, multiplication in integers is distributive over addition (but addition is not distributive over multiplication).

A *unary operation* on S is simply a mapping from S to S. In general, an *n-ary operation* on S is a mapping from the n-fold Cartesian product $S \times \cdots \times S$ to S.

By an *algebraic system* we mean a nonempty set S together with one or more binary operations on S that satisfy some specified properties.

EXERCISES 0.1

1. Let A be a set of n elements. Show that the number of subsets of A of r elements is $\binom{n}{r} = \dfrac{n!}{r!(n-r)!}$. Hence deduce the total number of subsets of A.

2. Let A, B, C be finite sets. Prove that

$$|A \cup B \cup C| = |A| + |B| + |C| - |A \cap B| - |B \cap C| - |C \cap A| \\ + |A \cap B \cap C|$$

3. *(Generalization of Exercise 2.)* Let $A_1, \ldots, A_n$ be finite sets. Prove that

$$|A_1 \cup \cdots \cup A_n| = \sum_{r=1}^{n} (-1)^{r+1} N_r$$

where

$$N_r = \sum |A_{i_1} \cap \cdots \cap A_{i_r}|$$

the summation being over all combinations of r sets out of the given n sets.

4. The *symmetric difference* of two sets A and B, written $A \triangle B$, is defined as

$$A \triangle B = (A - B) \cup (B - A)$$

Prove that $A \triangle B = (A \cup B) - (A \cap B)$. Further, show that for any sets A, B, C, (a) $(A \triangle B) \triangle C = A \triangle (B \triangle C)$ and (b) $A \cap (B \triangle C) = (A \cap B) \triangle (A \cap C)$.

5. Let $\mathcal{F}$ be the family of all open intervals of the form $(0, a)$, where a is any positive real number. Find $\bigcup \mathcal{F}$ and $\bigcap \mathcal{F}$.

6. Let A and B be sets that have m and n elements, respectively. Show that there are n^m distinct mappings from A to B. How many among these are injective? What can you say about the existence of a surjective mapping from A to B?

7. Let A and B be sets with n elements each. Show that a mapping $f : A \rightarrow B$ is surjective if and only if it is injective.

8. Let $A = [0, 1] = \{x \in \mathbb{R} \mid 0 \leq x \leq 1\}$, $B = \{0, 1, 2\}$. Draw the graph of $A \times B$ in the xy-plane.

9. If A or B is an empty set, what is $A \times B$? Conversely, if $A \times B$ is an empty set, what conclusion can we draw about the sets A and B?

10. Show that the relation $\mid$ (where $a \mid b$ means a divides b) is a partial order in the set $\mathbb{N}$. Show further that $\mid$ is not a partial order in $\mathbb{Z}$.

11. Define a relation $\sim$ on the set $\mathbb{N} \times \mathbb{N}$ by the rule that $(a, b) \sim (c, d)$ means $a + d = b + c$. Show that $\sim$ is an equivalence relation. Write the equivalence classes of $(1, 1)$ and $(1, 2)$.

12. Let α be a fixed positive real number. For any $a, b \in \mathbb{R}$, let aEb mean that $a - b = n\alpha$ for some integer n. Show that E is an equivalence relation on $\mathbb{R}$. Describe the quotient set $\mathbb{R}/E$.

13. Let P be a partition of $\mathbb{Z}$ consisting of the subsets $\{\ldots, -7, 0, 7, 14, \ldots\}$, $\{\ldots, -6, 1, 8, 15, \ldots\}$, $\{\ldots, -5, 2, 9, 16, \ldots\}$, and so on. Describe the equivalence relation that gives rise to the partition P.

14. Let $p(n)$ denote the number of ways in which a set of n elements can be partitioned, with $p(0) = 1$. Prove the following recursion formula:

$$p(n + 1) = \sum_{r=0}^{n} \binom{n}{r} p(r)$$

Use this formula to compute $p(10)$.

15. Define a binary operation $*$ on the set $\mathbb{Z}$ by the rule $a * b = a^2 b$. Determine whether $*$ is (a) commutative, (b) associative, (c) left (right) distributive over $+$.

0.2 GROUPS AND SEMIGROUPS

Definition 0.2.1 A *group* is an algebraic system $(G, *)$, where G is a nonempty set and $*$ is a binary operation on G that satisfies the following conditions:

(a) $(a * b) * c = a * (b * c)$ for all $a, b, c \in G$ (*associative law*).

(b) There exists an element e, called an *identity,* such that $e * a = a = a * e$ for all $a \in G$.

(c) For every $a \in G$, there exists a corresponding element $a' \in G$, called an *inverse* of a, such that $a * a' = e = a' * a$.

Further, a group $(G, *)$ is said to be *abelian* if $a * b = b * a$ for all $a, b \in G$.

It can be shown that a group has a unique identity and every element in it has a unique inverse.

EXAMPLES

1. $(\mathbb{Z}, +)$, $(\mathbb{Q}, +)$, $(\mathbb{R}, +)$, and $(\mathbb{C}, +)$ are abelian groups.

2. Let $\mathbb{Q}^*$, $\mathbb{R}^*$, and $\mathbb{C}^*$ denote the sets of all nonzero elements in $\mathbb{Q}, \mathbb{R}$, and $\mathbb{C}$, respectively. Then $(\mathbb{Q}^*, \cdot)$, $(\mathbb{R}^*, \cdot)$, and $(\mathbb{C}^*, \cdot)$ are abelian groups.

If $(G, *)$ is a group, we say that G is a group under the operation $*$. A group $(G, *)$ is commonly referred to as group G when it is not necessary to specify the operation $*$.

As a matter of convention, the binary operation in a group is usually written $\cdot$ or $+$. Whereas $\cdot$ may be used for an abelian or a nonabelian group, $+$ is used only in the abelian case. When the operation is written $\cdot$, the group is called *multiplicative,* its identity is written e (or 1), and the inverse of an element a is written a^{-1}. When the operation is $+$, the group is called *additive,* the identity is written 0 (called *zero*), and the inverse of a is written $-a$ (called the *negative* of a). We write $b - a$ to denote $b + (-a)$.

As in ordinary algebra, it is normal practice to represent the operation $\cdot$ (multiplication) by juxtaposition; that is, $a \cdot b$ is written simply ab.

We shall normally use the multiplicative notation in stating theorems for groups in general. The corresponding results for an additive group can be easily written.

Before considering more examples and discussing the properties of a group, we define a more general algebraic system called a *semigroup.* The foregoing remarks on the notation for groups also apply to semigroups.

Definition 0.2.2 A *semigroup* is an algebraic system $(S, \cdot)$, where S is a nonempty set and $\cdot$ is an associative binary operation on S. Further, if the operation $\cdot$ is commutative, the semigroup S is said to be *abelian.*

Obviously, every group is a semigroup. An example of a semigroup that is not a group is $(\mathbb{Z}, \cdot)$.

Definition 0.2.3 Let S be a semigroup. An element e in S is called an *identity* if $ea = a = ae$ for all $a \in S$.

For example, $(\mathbb{N}, \cdot)$, where $\mathbb{N}$ is the set of all positive integers, is a semigroup with identity—namely, the integer 1. On the other hand, $(\mathbb{N}, +)$ is a semigroup without identity.

Definition 0.2.4 Let S be a semigroup with identity e. Let $a \in S$. If there exists an element $a' \in S$ such that $aa' = e = a'a$, then a is said to be *invertible* and b is called an *inverse* of a.

It is an easy exercise to show that if there exists an identity in a semigroup, it is unique, and that the inverse of an element, if it exists, is unique. The following theorem shows that the invertible elements in a semigroup with identity form a group. The proof is left as an exercise.

THEOREM 0.2.5 Let $(S, \cdot)$ be a semigroup, and let S^* denote the set of all invertible elements in S.

(a)   If $a \in S^*$, then $a^{-1} \in S^*$ and $(a^{-1})^{-1} = a$.
(b)   If $a, b \in S^*$, then $ab \in S^*$ and $(ab)^{-1} = b^{-1}a^{-1}$.

Consequently, S^* is a group under the operation $\cdot$ (restricted to S^*).

The following simple result is useful in working with elements in a group. The proof is easy.

THEOREM 0.2.6 Let G be a group, and let $a, b, c \in G$. Then

(a) $ab = ac \Rightarrow b = c$ (*left cancellation law*)
(b) $ba = ca \Rightarrow b = c$ (*right cancellation law*)

A group G is called a *finite group* if the set G has a finite number of elements. The number of elements in the set G is called the *order* of the group G and denoted by $|G|$. The examples of a group that have been mentioned so far are all infinite groups. We now present several examples of finite groups.

0.2.1 Groups of Integers Modulo *n*

Let n be a fixed positive integer. Recall that congruence modulo n is an equivalence relation in $\mathbb{Z}$, and the equivalence class of an element a, written $[a]$ or $\bar{a}$, is given by

$$[a] = \bar{a} = \{a + tn \mid t \in \mathbb{Z}\}$$

Let $G = \mathbb{Z}/ \equiv_n$ be the set of equivalence classes under congruence modulo n; that is, $G = \{[0], [1], \ldots, [n-1]\}$. Addition in $\mathbb{Z}$ induces a corresponding operation (again written $+$) in G. For any $[a], [b] \in G$, we define

$$[a] + [b] = [a + b]$$

This binary operation in G is *well defined*, by which we mean that the result of the operation is independent of the integers chosen to represent the equivalence classes. Suppose $[a] = [a']$ and $[b] = [b']$. Then $a \equiv a' \pmod{n}$ and $b \equiv b' \pmod{n}$. Therefore, by Theorem 0.1.14, $a + b \equiv a' + b' \pmod{n}$ and hence $[a + b] \equiv [a' + b']$. It is now easily verified that $(G, +)$ is an abelian group in which the identity is $[0]$, and for every $[a] \in G$, $-[a] = [n-a]$. This group is called the *additive group of integers modulo n*. The operation $+$ in G is called *addition modulo n*.

We similarly define multiplication in G (called *multiplication modulo n*) by the rule

$$[a] \cdot [b] = [ab]$$

It is easily verified that $(G, \cdot)$ is an abelian semigroup with identity $[1]$. It is called the *multiplicative semigroup of integers modulo n*. We have thus shown that $(\mathbb{Z}/ \equiv_n, +)$ is a group and $(\mathbb{Z}/ \equiv_n, \cdot)$ is a semigroup.

We now describe an alternative method for constructing a group under addition modulo n and a semigroup under multiplication modulo n.

Again let n be a fixed positive integer. By Euclidean division algorithm, given any integer a, we can find unique integers q, r such that $a = qn + r$ and $0 \leq r \leq n - 1$. The integer r is called the *remainder* or *residue* of a modulo n and denoted by $res_n(a)$ or $a \bmod n$. Clearly, $res_n(a) \equiv a \pmod{n}$. [Note the distinction between the congruence $b \equiv a \pmod{n}$ and the equality $b = a \bmod n$.]

Let $\mathbb{Z}_n$ denote the set $\{0, 1, \ldots, n-1\}$. We refer to $\mathbb{Z}_n$ as the *set of residues modulo n*. Addition and multiplication in $\mathbb{Z}$ induce corresponding binary operations on $\mathbb{Z}_n$, called *addition modulo n* and *multiplication modulo n*, which we temporarily denote by the symbols $+_n$ and $\cdot_n$, respectively. For all $a, b \in \mathbb{Z}_n$, we define

$$a +_n b = res_n(a + b) \quad \text{and} \quad a \cdot_n b = res_n(ab)$$

The operations $+_n$ and $\cdot_n$ are obviously commutative. We show that they are also associative. Let $a, b, c \in \mathbb{Z}_n$. Then, by the definition of $+_n$ and Theorem 0.1.14, we have $(a +_n b) +_n c \equiv a + b + c \equiv a +_n (b +_n c) \pmod{n}$, from which it follows that $(a +_n b) +_n c = res_n(a + b + c) = a +_n (b +_n c)$. A similar proof holds for the operation $\cdot_n$.

It is easily verified now that $(\mathbb{Z}_n, +_n)$ is an abelian group in which 0 is the identity, and for every nonzero element a, $-a = n - a$. Similarly, $(\mathbb{Z}_n, \cdot_n)$ is a semigroup with identity 1.

The two groups $(\mathbb{Z}/\equiv_n, +)$ and $(\mathbb{Z}_n, +_n)$ have the same algebraic structure in the following sense: For any $a, b, c \in \mathbb{Z}_n$, the equality $a +_n b = c$ holds in the group $(\mathbb{Z}_n, +_n)$ if and only if the equality $[a] + [b] = [c]$ holds in the group $(\mathbb{Z}/\equiv_n, +)$. A similar property holds for the semigroups $(\mathbb{Z}/\equiv_n, \cdot)$ and $(\mathbb{Z}_n, \cdot_n)$.

Two such groups (or semigroups) that have an identical structure are said to be *isomorphic*. (The term *isomorphic* means "having the same form.") We will give a formal definition of isomorphism and study its properties in a subsequent section.

Because of the isomorphism between the groups $(\mathbb{Z}/\equiv_n, +)$ and $(\mathbb{Z}_n, +_n)$, we refer to each as the group $\mathbb{Z}_n$. One may, according to one's taste or as required by the context, interpret the elements of the set $\mathbb{Z}_n$ to be the integers $0, 1, \ldots, n-1$ (as written) or the corresponding equivalence classes $[0], [1], \ldots, [n-1]$. Further, we omit the subscript n in the notation for addition and multiplication in $\mathbb{Z}_n$ (except when it is needed for the sake of clarity). We record the result of the foregoing discussion in the following theorem.

THEOREM 0.2.7 Let n be a positive integer, and let $\mathbb{Z}_n = \{0, 1, \ldots, n-1\}$. Then

(a) $(\mathbb{Z}_n, +)$ is an abelian group under addition modulo n.

(b) $(\mathbb{Z}_n, \cdot)$ is an abelian semigroup with identity under multiplication modulo n.

The group $(\mathbb{Z}_n, +)$ is called the *group of integers modulo n*.

We have seen (Theorem 0.2.5) that the invertible elements in a semigroup with identity form a group. We show that the invertible elements in the semigroup $(\mathbb{Z}_n, \cdot)$ are the integers relatively prime to n. To do so, we make use of the following result in number theory: The greatest common divisor of integers a and b, written $\gcd(a, b)$, is equal to d if and only if there exist integers s and t such that $sa + tb = d$.

Let $a \in \mathbb{Z}_n$. Suppose a is an invertible element in the semigroup $(\mathbb{Z}_n, \cdot)$ and $b = a^{-1}$. Then $ab \equiv 1 \pmod{n}$; hence $1 = ab - qn$ for some integer q and $\gcd(a, n) = 1$. Conversely, suppose $\gcd(a, n) = 1$. Then there exist integers s, t such that $1 = sa + tn$. Let $b = res_n(s)$. Then $b \in \mathbb{Z}_n$ and $ba \equiv sa \equiv 1 \pmod{n}$; hence $b = a^{-1}$. This proves that a is invertible if and only if $\gcd(a, n) = 1$; that is, a is relatively prime to n.

For any positive integer n, the number of positive integers not exceeding n and relatively prime to n is denoted by $\varphi(n)$. (The function φ is called *Euler's phi-function*.) It is obvious that $\varphi(p) = p - 1$ for any prime p. Thus we have the following result:

THEOREM 0.2.8 Let n be a positive integer. Let $\mathbb{Z}_n^*$ denote the set of invertible elements in the semigroup $(\mathbb{Z}_n, \cdot)$. Then $\mathbb{Z}_n^*$ consists of all positive integers not exceeding n and relatively prime to n. Consequently, $(\mathbb{Z}_n^*, \cdot)$ is an abelian group of order $\varphi(n)$.

In particular, for any prime p, $\mathbb{Z}_p^* = \{1, 2, \ldots, p-1\}$ is a group of order $p-1$ under multiplication modulo p.

The group $(\mathbb{Z}_n^*, \cdot)$ is called the *multiplicative group of integers modulo n*.

Example 0.2.9 Obtain the multiplicative group $\mathbb{Z}_{12}^*$, and find the product of every two elements in it.

Solution. The positive integers less than 12 and relatively prime to 12 are $1, 5, 7, 11$. Hence $\mathbb{Z}_{12}^* = \{1, 5, 7, 11\}$.

Using multiplication modulo 12 and writing a^2 for $a \cdot a$, we have

$$5^2 = 1, \quad 7^2 = 1, \quad 11^2 = 1, \quad 5 \cdot 7 = 11, \quad 5 \cdot 11 = 7, \quad 7 \cdot 11 = 5$$

The group $\mathbb{Z}_{12}^*$ is an example of a *Klein's 4-group*, which is defined as an abelian group of four elements, say $K = \{e, a, b, c\}$, in which

$$a^2 = e = b^2 = c^2, \quad ab = c, \quad bc = a, \quad ca = b$$

(In the additive notation, $2a = 0 = 2b = 2c$, $a + b = c$, $b + c = a$, and $c + a = b$.) We shall come across several examples of a Klein's 4-group.

If G is a finite group of m elements, we can explicitly show all the m^2 products in it in the form of a table, referred to as the *Cayley table*. The product ab is found at the intersection of the row starting with a and the column headed by b. For example, here is the Cayley table for the group $\mathbb{Z}_{12}^*$:

$\cdot$	1	5	7	11
1	1	5	7	11
5	5	1	11	7
7	7	11	1	5
11	11	7	5	1

We observe that each row and each column in the Cayley table is a permutation of the set G.

Let us now consider the practical problem of finding the inverse of an element a in the multiplicative group $\mathbb{Z}_n^*$. (If n is small, we can compute the product $a \cdot x$ for each x in $\mathbb{Z}_n^*$ until we find an element b such that $a \cdot b = 1$. But this is tedious for large n.) Because a is relatively prime to n, we know that there exist integers s, t such that $sa + tn = 1$, and then $a^{-1} = s \bmod n$.

To find such integers s, t, we use the *generalized Euclidean algorithm* (taught in high school algebra for finding the greatest common divisor). Since $\gcd(a, n) = 1$, we get a set of (say, k) equations of the following form:

$$n = q_1 a + r_1$$
$$a = q_2 r_1 + r_2$$
$$r_1 = q_3 r_2 + r_3$$

$$\vdots$$

$$r_{k-3} = q_{k-1}r_{k-2} + r_{k-1}$$
$$r_{k-2} = q_k r_{k-1} + 1$$

Substituting for r_{k-1} in the last equation from the equation immediately above it, we get

$$1 = r_{k-2} - q_k r_{k-1}$$
$$= r_{k-2} - q_k(r_{k-3} - q_{k-1}r_{k-2})$$
$$= (1 + q_k q_{k-1})r_{k-2} - q_k r_{k-3}$$

We repeat this process and, substituting for $r_{k-2}, r_{k-3}, \ldots, r_1$ successively from the preceding equations, we eventually arrive at an equation of the form $1 = sa + tn$. The following example illustrates this procedure.

Example 0.2.10 Find the inverse of 21 in the multiplicative group $\mathbb{Z}_{100}^*$.

Solution. By iterative division algorithm, we have

$$100 = 4(21) + 16$$
$$21 = 1(16) + 5$$
$$16 = 3(5) + 1$$

Now starting from the last equation and moving upward, we get

$$1 = 16 - 3(5)$$
$$= 16 - 3(21 - 1(16))$$
$$= 4(16) - 3(21)$$
$$= 4(100 - 4(21)) - 3(21)$$
$$= 4(100) - 19(21)$$

So the inverse of 21 in $\mathbb{Z}_{100}^*$ is equal to $(-19) \bmod 100 = 81$.

0.2.2 Groups of Matrices and Permutations

The groups that we have so far considered are all abelian. We now give some examples of nonabelian groups.

Let F denote any of the sets $\mathbb{Q}, \mathbb{R},$ or $\mathbb{C}$. Given positive integers m, n, let $F^{m \times n}$ denote the set of all $m \times n$ matrices with entries in the set F. Then $F^{m \times n}$ is an abelian group under addition, with the zero matrix as its identity.

The set $F^{n \times n}$ of all $n \times n$ matrices is a semigroup under multiplication, with the identity matrix I_n as its identity. Hence the invertible matrices in $F^{n \times n}$ form a group, which is called the *general linear group* in dimension n over F and denoted by $GL(n, F)$. If $n > 1$, the group is nonabelian.

Let X be a set. Let S_X denote the set of all permutations on X (that is, bijective mappings from X to X). For any $f, g \in S_X$, we define $f \cdot g = f \circ g$ (composite of the mappings f, g). Then $(S_X, \cdot)$ is a group and is called the *symmetric group on the set X*. If $X = \{1, 2, \ldots, n\}$, then S_X is written as S_n and is called the symmetric group of *degree n*. The order of the group S_n is $n!$. If $n > 2$, then S_n is a nonabelian group.

We provide a detailed treatment of permutation groups in Chapter 4.

EXERCISES 0.2

1. Suppose G is a group in which $a^2 = e$ for all $a \in G$. Show that G is abelian.

2. Suppose G is a group of four elements, $G = \{e, a, b, c\}$, such that $a^2 = b^2 = e$. Show that G is a Klein's 4-group.

3. Show that $\mathbb{Z}_8^*$, the multiplicative group of integers modulo 8, is a Klein's 4-group.

4. Show that the set

$$K = \left\{ \begin{bmatrix} 1 & 0 \\ 0 & 1 \end{bmatrix}, \begin{bmatrix} 1 & 0 \\ 0 & -1 \end{bmatrix}, \begin{bmatrix} -1 & 0 \\ 0 & 1 \end{bmatrix}, \begin{bmatrix} -1 & 0 \\ 0 & -1 \end{bmatrix} \right\}$$

is a group under multiplication of matrices. Further, show that K is a Klein's 4-group.

5. Show that the set Q consisting of the following matrices (where i denotes $\sqrt{-1}$) is a nonabelian group under multiplication:

$$\begin{bmatrix} 1 & 0 \\ 0 & 1 \end{bmatrix}, \quad \begin{bmatrix} -1 & 0 \\ 0 & -1 \end{bmatrix}, \quad \begin{bmatrix} 0 & 1 \\ -1 & 0 \end{bmatrix}, \quad \begin{bmatrix} 0 & -1 \\ 1 & 0 \end{bmatrix}$$

$$\begin{bmatrix} i & 0 \\ 0 & -i \end{bmatrix}, \quad \begin{bmatrix} -i & 0 \\ 0 & i \end{bmatrix}, \quad \begin{bmatrix} 0 & i \\ i & 0 \end{bmatrix}, \quad \begin{bmatrix} 0 & -i \\ -i & 0 \end{bmatrix}$$

(Q is called the *quaternion group*.)

6. Find the inverse of 11 in $\mathbb{Z}_{81}^*$.

7. Show that for each element $\alpha \in S_n$ ($n > 2$), there exists an element $\beta \in S_n$ such that $\alpha\beta \neq \beta\alpha$.

0.3 CYCLIC GROUPS, ORDER OF AN ELEMENT, AND DIRECT PRODUCT

0.3.1 Order of an Element

Let S be a semigroup, and let $a, b, c \in S$. Because of the associative property $(ab)c = a(bc)$, we can omit the parentheses and write abc without any ambiguity. This property can be extended to the product of more than three elements.

We define the *powers* of an element a in a semigroup S as follows: For any positive integer n, the nth power of a is defined as

$$a^n = aa \cdots a \quad (n \text{ factors})$$

More formally, we can define a^n recursively as

$$a^1 = a, \quad a^n = a^{n-1}a \qquad \text{for any positive interger } n > 1$$

The following formulas can be proved easily by induction: For all positive integers m, n,

$$a^m a^n = a^{m+n}$$
$$(a^m)^n = a^{mn} \tag{1}$$

If the semigroup S has an identity e, we define $a^0 = e$. Further, if a is an invertible element in S, we define $a^{-n} = (a^{-1})^n$ for any positive integer n. In this case, formulas (1) hold for all integers m, n. It follows further that $(a^n)^{-1} = a^{-n} = (a^{-1})^n$.

If S is an additive semigroup, then instead of powers of an element a, we define *multiples* of a as follows: For any positive integer n,

$$na = a + a + \cdots + a \quad (n \text{ terms})$$

If S has identity 0, we define $0a = 0$. If a has a negative, we define $(-n)a = n(-a)$ for all positive integers n. Formulas (1) take the following form:

$$ma + na = (m + n)a$$
$$n(ma) = (nm)a \tag{2}$$

Definition 0.3.1 Let G be a group. The *order* of an element $a \in G$, written $o(a)$, is the least positive integer m such that $a^m = e$. If no such integer m exists, then a is said to be of *infinite order*.

If G is additive, then $o(a)$ is the least positive integer m such that $ma = 0$.

It is obvious from the definition that, in any group, the identity is the only element of order 1.

In the additive group $\mathbb{Z}$, every nonzero element is of infinite order. For an example of an element of finite order, consider the additive group $\mathbb{Z}_{12}$. Under addition modulo 12, $8 + 8 = 4$ and $8 + 8 + 8 = 0$. Hence $o(8) = 3$.

The next two theorems give some elementary properties of the order of an element.

THEOREM 0.3.2 Let G be a group. An element a in G is of finite order if and only if $a^n = e$ for some nonzero integer n. If a is of infinite order, then $a^i \neq a^j$ for any two distinct integers i, j. Consequently, every element in a finite group is of finite order.

Proof: Suppose $a^n = e$ for some nonzero integer n. Then $a^{-n} = (a^n)^{-1} = e$. So $a^n = e$ for some positive integer n. Hence, by the well-ordering property of positive integers, there exists a least positive integer m such that $a^m = e$. So $o(a) = m$. The converse is obvious.

If a is of infinite order, then there is no nonzero integer n such that $a^n = e$. Suppose $a^i = a^j$ for some distinct integers i, j. Then $a^{i-j} = a^i(a^j)^{-1} = e$, a contradiction. Therefore, if a is of infinite order, $a^i \neq a^j$ for any two distinct integers i, j. So G contains infinitely many elements $a, a^2, \ldots$ Consequently, if G is finite, it cannot have an element of infinite order. ∎

The converse of the last statement in the theorem does not hold. If every element in a group G is of finite order, it doesn't necessarily follow that G is finite. (Construct an example!)

THEOREM 0.3.3 Let G be a group. Let $a \in G$ and suppose $o(a) = m$. Then

(a) $e, a, \ldots, a^{m-1}$ are m distinct elements.
(b) For any integer n, $a^n = a^{r(n)}$, where $r(n) = res_m(n)$ is the residue of n modulo m.
(c) $a^n = e$ if and only if $m \mid n$.

Proof:

(a) Suppose $a^i = a^j$ for some integers i, j, where $0 \leq i < j \leq m - 1$. Then $j - i$ is a positive integer less than m, and $a^{j-i} = e$, which contradicts the fact that $o(a) = m$. Hence $e, a, \ldots, a^{m-1}$ are all distinct elements.
(b) By Euclidean algorithm, given any integer n, there exist integers q, r such that $n = qm + r$ and $0 \leq r = res_m(n) < m$. So $a^n = a^{qm+r} = (a^m)^q a^r = e^q a^r = a^r$.
(c) Suppose $a^n = e$. Then $a^r = e$, where $0 \leq r = res_m(n) < m$. Since m is the least positive integer such that $a^m = e$, it follows that $r = 0$. Hence $n = qm$, so $m \mid n$. Conversely, if $m \mid n$, then $a^n = (a^m)^{n/m} = e^{n/m} = e$. ∎

It is obvious that if a is of infinite order, then a^n is also of infinite order for any nonzero integer n. The following theorem gives the relation between the orders of a and a^n when a is an element of finite order.

THEOREM 0.3.4 Let G be a group, $a \in G$. Let $o(a) = m$. Then for any positive integer n,

$$o(a^n) = \frac{m}{\gcd(m, n)}$$

Proof: Write $d = \gcd(m, n)$, $b = a^n$. Then $(b)^{m/d} = a^{nm/d} = (a^m)^{n/d} = e^{n/d} = e$. So b is of finite order, say $o(b) = k$. Then, by Theorem 0.3.3, $k \mid (m/d)$. Now $b^k = e$, so $a^{nk} = e$ and hence $m \mid nk$. Therefore $(m/d) \mid (n/d)k$. Since $d = \gcd(m, n)$, it follows that m/d and n/d are relatively prime. Hence $(m/d) \mid k$, so $k = m/d$. ∎

We now define a special type of a group.

0.3.2 Cyclic Groups

Definition 0.3.5 A group G is said to be *cyclic* if there is an element a in G (called a *generator* of G) such that each element in G is some power of a.

In the case of an additive cyclic group G with a generator a, each element in G is a multiple of a. For example, the additive group $\mathbb{Z}$ is cyclic, with the integer 1 as its generator, because every integer k is a multiple of 1—that is, $k = k1$. Likewise, for every positive integer n, the additive group $\mathbb{Z}_n$ is cyclic.

Let G be a (multiplicative) cyclic group with a generator a. Then $a^i \in G$ for all $i \in \mathbb{Z}$. On the other hand, each element of G is equal to some power a^i and hence $G = \{a^i \mid i \in \mathbb{Z}\}$. This, of course, does not mean that G is an infinite group because the powers a^i are not necessarily all distinct. In fact, we have seen that $\mathbb{Z}_n$ is a finite cyclic group of order n.

The following theorem shows that a finite group of order n is cyclic if and only if it has an element of order n.

THEOREM 0.3.6 Let G be a group of order n. Let $a \in G$. If $o(a) = n$, then G is a cyclic group with generator a and $G = \{e, a, \ldots, a^{n-1}\}$. Conversely, if G is cyclic and a is a generator of G, then $o(a) = n$.

Proof: Suppose $o(a) = n$. Then, by Theorem 0.3.3, $e, a, \ldots, a^{n-1}$ are n distinct elements in G. Since G has only n elements, it follows that $G = \{e, a, \ldots, a^{n-1}\}$. Hence G is cyclic and a is a generator of G.

Conversely, suppose G is cyclic and a is a generator, so $G = \{a^i \mid i \in \mathbb{Z}\}$. Since G has a finite number of elements, a must be of finite order. Let $o(a) = m$, and let b be any element in G. Then $b = a^i$ for some integer i. By Theorem 0.3.3, $a^i = a^r$, where r is the remainder on dividing i by m. Hence $G = \{e, a, \ldots, a^{m-1}\}$. Now $e, a, \ldots, a^{m-1}$ are m distinct elements. Since G has n elements, it follows that $m = n$ and hence $o(a) = n$. ∎

It follows from the theorem that if G is an infinite cyclic group, then its generator a is an element of infinite order. Thus the order of a generator of a cyclic group G (whether finite or infinite) is equal to the order of G.

A cyclic group G can have several generators; that is, there may be several elements in G each of which is a generator. For example, 1 and -1 are both generators of the additive group $\mathbb{Z}$.

The next two theorems give the precise number of generators of a cyclic group. We first consider the case of a finite group.

THEOREM 0.3.7 Let G be a cyclic group of order n. Then

(a)   The order of every element in G is a divisor of n; hence $x^n = e$ for all $x \in G$.
(b)   Given any positive divisor r of n, there are exactly $\varphi(r)$ elements of order r in G.

In particular, there are exactly $\varphi(n)$ elements of order n in G; hence G has $\varphi(n)$ generators. Moreover, if a is a generator of G, then for every positive integer j less than n and relatively prime to n, a^j is a generator of G.

Proof:

(a) Let a be a generator of G. By Theorem 0.3.6, $o(a) = n$ and $G = \{e, a, \ldots, a^{n-1}\}$. By Theorem 0.3.4, $o(a^i) = \dfrac{n}{\gcd(i, n)}$. Hence the order of each element in G is a divisor of n. Therefore, by Theorem 0.3.3, $x^n = e$ for all $x \in G$.

(b) Let r be any positive divisor of n, and suppose $o(a^i) = r$. Then $\gcd(i, n) = n/r$, so n/r divides i. We write $j = ir/n$. Then $\gcd(j, r) = 1$, so j is relatively prime to r. Thus the elements of order r in G are given by $a^{jn/r}$, where j is a positive integer less than r and relatively prime to r. Hence there are exactly $\varphi(r)$ elements of order r in G.

In particular, when $r = n$, it follows that there are exactly $\varphi(n)$ elements of order n in G, given by a^j, where j is relatively prime to n. By Theorem 0.3.6, these are the generators of G. ∎

THEOREM 0.3.8 An infinite cyclic group has exactly two generators.

Proof: Let G be an infinite cyclic group, and let a be a generator of G. So a is an element of infinite order, and $G = \{a^i \mid i \in \mathbb{Z}\}$. Suppose $b = a^m$ is also a generator of G. Then every element in G is a power of b. In particular, $a = b^k$ for some integer k, so $a = a^{mk}$ and hence $a^{mk-1} = e$. Since a is of infinite order, it follows that $mk - 1 = 0$, so $m = k = \pm 1$. Hence a and a^{-1} are the only generators of G. ∎

Example 0.3.9 Show that $\mathbb{Z}_{11}^*$, the multiplicative group of integers modulo 11, is cyclic, and find all its generators.

Solution. Consider the element 2 in the group $\mathbb{Z}_{11}^*$. By direct computation under multiplication modulo 11, we find $2^1 = 2$, $2^2 = 4$, $2^3 = 8$, $2^4 = 5$, $2^5 = 10$, $2^6 = 9$, $2^7 = 7$, $2^8 = 3$, $2^9 = 6$, $2^{10} = 1$. Thus every element in $\mathbb{Z}_{11}^*$ is a power of 2. Hence $\mathbb{Z}_{11}^*$ is cyclic, and 2 is a generator.

Now $\mathbb{Z}_{11}^*$ is a group of order 10 and $\varphi(10) = 4$. The integers relatively prime to 10 are 1, 3, 7, 9. Hence, by Theorem 0.3.7, the group $\mathbb{Z}_{11}^*$ has four generators. These are $2^1, 2^3, 2^7, 2^9$—that is, 2, 8, 7, 6.

(We shall prove subsequently that $\mathbb{Z}_p^*$ is a cyclic group for every prime p.)

0.3.3 Direct Product of Two Groups

Recall that the Cartesian product of two sets G, G' is defined to be the set $G \times G' = \{(x, x') \mid x \in G, x' \in G'\}$. If G, G' are groups, then the operations in G, G' induce a binary operation $*$ on the set $G \times G'$ as follows: For all (x, x'), $(y, y') \in G \times G'$, we define

$$(x, x') * (y, y') = (xy, x'y')$$

It is easily verified that $(G \times G', *)$ is a group with identity (e, e'), where e, e' are the identities of G, G', respectively. It is called the *direct product* of the groups G and G'.

Note that the operations in the groups G and G' are not necessarily the same. In keeping with our general practice, we have used the multiplicative notation for both G and G'. In case G, G' are both additive, we use the additive notation also for the direct product $G \times G'$.

Is the direct product of two cyclic groups again a cyclic group? The following theorem provides the answer.

THEOREM 0.3.10 Let G, H be cyclic groups of order m, n, respectively. Then their direct product $G \times H$ is cyclic if and only if m, n are relatively prime.

Moreover, if this condition holds, then (a, b) is a generator of $G \times H$ if and only if a and b are generators of G and H, respectively.

Proof: Suppose m, n are relatively prime. Let a and b be generators of G and H, respectively. Then $o(a) = m$ and $o(b) = n$. So $a^m = e$ and $b^n = e'$, where e, e' are the identities in G, H, respectively.

The direct product $G \times H$ is a group of order mn. Hence, by Theorem 0.3.6, (a, b) is a generator of $G \times H$ if and only if the order of (a, b) is equal to mn. Let $o(a, b) = k$. Then $(a^k, b^k) = (a, b)^k = (e, e')$, so $a^k = e$ and $b^k = e'$. Hence, by Theorem 0.3.3, $m \mid k$ and $n \mid k$. Since m, n are relatively prime, it follows that $mn \mid k$. On the other hand, $(a, b)^{mn} = (a^{mn}, b^{mn}) = ((a^m)^n, (b^n)^m) = (e^n, e'^m) = (e, e')$, and hence $k \mid mn$. This proves that $k = mn$, and hence (a, b) is a generator of $G \times H$.

Conversely, suppose m, n are not relatively prime. Then $d = \gcd(m, n) > 1$. We write $q = mn/d$. Let (x, y) be any element in $G \times H$. By Theorem 0.3.7, $x^m = e$ and hence $x^q = (x^m)^{n/d} = e$. Likewise, $y^q = (y^n)^{m/d} = e'$ and hence $(x, y)^q = (x^q, y^q) = (e, e')$. Therefore $o(x, y) \mid q$. Since $q < mn$, it follows that there is no element in $G \times H$ of order mn. Hence $G \times H$ is not cyclic. This proves that $G \times H$ is cyclic if and only if m, n are relatively prime.

Now we prove the second part of the theorem. We have already shown that if a and b are generators of G and H, respectively, then (a, b) is a generator of $G \times H$. Conversely, suppose (a, b) is a generator of $G \times H$, so $o(a, b) = mn$. Let $o(a) = s$ and $o(b) = t$. Then $(a, b)^{st} = (a^{st}, b^{st}) = (e, e')$ and hence $mn \mid st$.

By Theorem 0.3.7, $s \mid m$ and $t \mid n$. Therefore $s = m$ and $t = n$. Hence a and b are generators of G and H, respectively.   ■

By induction, the theorem can be generalized to the direct product of more than two groups.

EXERCISES 0.3

(Unless indicated otherwise, G denotes an arbitrary group.)

1. Show that every cyclic group is abelian.
2. Let $a, b \in G$. If $ab = ba$, show that $(ab)^n = a^n b^n$ for all n.
3. Find the order of every element in $\mathbb{Z}_{12}$.
4. Show that $o(a^{-1}) = o(a)$ for each $a \in G$.
5. Show that for all $a, b \in G$, $o(ab) = o(ba)$ and $o(aba^{-1}) = o(b)$.
6. Show that for every positive integer n, the nth roots of unity form a cyclic group under multiplication.
7. Show that the multiplicative group $\mathbb{Z}_{10}^*$ is cyclic. Find all its generators.
8. Show that the multiplicative group $\mathbb{Z}_9^*$ is cyclic. Find all its generators.
9. Find all generators of the multiplicative groups $\mathbb{Z}_{13}^*$ and $\mathbb{Z}_{17}^*$.
10. Show that a group of even order has at least one element of order 2.
11. Show that for every prime p, the multiplicative group $\mathbb{Z}_p^*$ has exactly one element of order 2.
12. Show that the direct product $\mathbb{Z}_2 \times \mathbb{Z}_2$ is a Klein's 4-group.

0.4 SUBGROUPS OF A GROUP

Definition 0.4.1 Let $(G, \cdot)$ be a group. A subset H of G is called a *subgroup* of G if H is itself a group under the operation $\cdot$ (restricted to H).

It is clear that G itself is a subgroup of G. The set $\{e\}$, consisting of just the identity in G, is also a subgroup. These two are called *trivial* subgroups of G. Any subgroup of G other than G and $\{e\}$ is called a *proper* subgroup. For example, the set of all even integers is a proper subgroup of $\mathbb{Z}$. This is a particular case of the following general result.

THEOREM 0.4.2 Let G be a group, $a \in G$. Let H be the set consisting of all powers of a; that is,

$$H = \{a^n \mid n \in \mathbb{Z}\}$$

Then H is a subgroup of G. Moreover, the order of H is equal to $o(a)$.

Proof: The identity e in G is an element of H, since $e = a^0 \in H$. For any elements $a^i, a^j \in H$, $a^i a^j = a^{i+j} \in H$. For any element $b = a^i \in H$, $b^{-1} = (a^i)^{-1} = a^{-i} \in H$. This proves that H is a group under the operation $\cdot$ in G. Hence H is a subgroup of G.

Since every element in H is a power of a, H is a cyclic group with generator a. Hence, by Theorem 0.3.6, the order of the group H is equal to the order of the element a. ∎

The subgroup H in Theorem 0.4.2 is called the *cyclic subgroup generated by* a and denoted by $[a]$. Thus each element in a group G generates a cyclic subgroup. In particular, $[e]$ is the trivial subgroup $\{e\}$. If $G = [a]$ for some $a \in G$, then G is itself a cyclic group.

If G is additive, then $[a]$ is the set of all multiples of a. For example, let $G = \mathbb{Z}$. For any integer m, let $m\mathbb{Z}$ denote the set of all multiples of m; that is, $m\mathbb{Z} = \{nm \mid n \in \mathbb{Z}\} = \{0, \pm m, \pm 2m, \ldots\}$. Then $m\mathbb{Z}$ is the cyclic subgroup of $\mathbb{Z}$ generated by m.

In the following theorem, we obtain necessary and sufficient conditions for a subset H of a group G to be a subgroup.

THEOREM 0.4.3 Let G be a group, and let H be a nonempty subset of G.

(a) H is a subgroup of G if and only if the following conditions hold:
   (1) $ab \in H$ for all $a, b \in H$.
   (2) $a^{-1} \in H$ for all $a \in H$.
(b) H is a subgroup of G if and only if the following condition holds:
   (3) $a^{-1}b \in H$ for all $a, b \in H$.

Proof:

(a) If H is a subgroup, then it is obvious that conditions (1) and (2) hold. Conversely, suppose the conditions hold. Since H is nonempty, we let $a \in H$. Then by (2), $a^{-1} \in H$. Now $a, a^{-1} \in H$ and hence by (1), $aa^{-1} \in H$, so $e \in H$. The associative property holds for all elements in G, so it holds for all elements in H. Thus H satisfies all the conditions for being a group under $\cdot$. Hence H is a subgroup of G.
(b) Suppose H is a subgroup, and let $a, b \in H$. Then $a^{-1} \in H$ and hence $a^{-1}b \in H$. Conversely, suppose condition (3) holds. Let $a \in H$. Then $a, a^{-1} \in H$ and hence $e = a^{-1}a \in H$. Now $a, e \in H$ and hence $a^{-1} = a^{-1}e \in H$. Thus for each $a \in H$, $a^{-1} \in H$. Now let $a, b \in H$. Then $a^{-1}, b \in H$ and hence $ab = (a^{-1})^{-1}b \in H$. Therefore, by (a), H is a subgroup of G. ∎

The proof of the following theorem is left as an exercise.

THEOREM 0.4.4 If H and K are subgroups of G, then $H \cap K$ is a subgroup of G.

This result can be easily generalized to any number of subgroups.

Subgroups of a cyclic group have some special properties that we prove in the next two theorems.

THEOREM 0.4.5 Every subgroup of a cyclic group is cyclic.

Proof: Let H be a subgroup of a cyclic group $G = [a]$. If $H = \{e\}$, then H is obviously cyclic because $H = [e]$. So let us suppose that $H \neq \{e\}$. Then H has some element other than e, say a^i, where i is some nonzero integer. Then $a^{-i} = (a^i)^{-1} \in H$, so $a^i \in H$ for some positive integer i. Hence, by the well-ordering property, there exists a least positive integer m such that $a^m \in H$. We write $b = a^m$ and claim that $H = [b]$.

Let a^n be any element in H. By Euclidean algorithm, $n = qm + r$, $0 \leq r < m$. Hence $a^r = a^{n-qm} = a^n (b)^{-q}$. Now $a^n, b \in H$ and hence $a^r \in H$. Since m is the least positive integer such that $a^m \in H$, it follows that $r = 0$. So $n = qm$ and hence $a^n = b^q$. Thus every element in H is some power of b. This proves $H = [b]$. ∎

THEOREM 0.4.6 Let G be a finite cyclic group of order n. Then the order of every subgroup of G is a divisor of n. Conversely, given any positive divisor d of n, there is a unique subgroup H of G of order d. Moreover, if $G = [a]$, then $H = [a^{n/d}]$.

Proof: Let H be a subgroup of G. By Theorem 0.4.5, H is cyclic. Let $H = [b]$. By Theorem 0.3.6, the order of H is equal to $o(b)$, and by Theorem 0.3.7, $o(b)$ divides n. This proves the first part of the theorem.

Let d be a positive divisor of n. Let $G = [a]$, so $o(a) = n$. We write $m = n/d$, $b = a^m$. Let $H = [b]$ be the subgroup generated by b. Now m divides n and hence $\gcd(m, n) = m$. By Theorem 0.3.4, $o(b) = o(a^m) = \dfrac{n}{\gcd(m, n)} = \dfrac{n}{m} = d$. Hence, by Theorem 0.3.6, H is a subgroup of order d.

To prove that $H = [a^{n/d}]$ is the unique subgroup of order d, let K be any subgroup of order d. By Theorem 0.4.5, K is cyclic. So $K = [c]$ for some $c \in G$, and $o(c) = d$. Let $c = a^s$. Then $a^{sd} = c^d = e$ and hence, by Theorem 0.3.3, $n \mid sd$. So $sd = qn$ for some integer q and hence $c = a^{qn/d} = b^q$. So every element in K is some power of b; hence $K \subset H$. But H and K have the same number of elements; hence $K = H$. This completes the proof of the theorem. ∎

Definition 0.4.7 Let G be a group, and let H be a subgroup of G. For any $a \in G$, the *left coset* of H determined by a is defined to be the set

$$aH = \{ah \mid h \in H\}$$

Likewise, the *right coset* Ha is defined as $Ha = \{ha \mid h \in H\}$.

In the sequel, we consider the left cosets of H. The corresponding results for right cosets are obtained similarly.

Let H be a subgroup of G. For any $a, b \in G$, we say a is *left congruent to* b modulo H, written $a \equiv b \pmod{H}$, if $a^{-1}b \in H$. It is easily verified that this is an equivalence relation on the set G. The equivalence class of an element a under this relation is the left coset aH. Since the equivalence classes form a partition (Theorem 0.1.11), we have the following result:

THEOREM 0.4.8 Let H be a subgroup of G. Then the left cosets of H form a partition of the set G.

For a finite group G, we get the following result, which is of fundamental importance in group theory:

THEOREM 0.4.9 (Lagrange theorem) Let H be a subgroup of a finite group G. Then the order of H divides the order of G.

Proof: Let $|G| = n$ and $|H| = m$. Let $H = \{h_1, \ldots, h_m\}$. Then, for any $a \in G$, $aH = \{ah_1, \ldots, ah_m\}$. If $ah_i = ah_j$, we get $h_i = h_j$; hence $h_1, \ldots, h_m$ are all distinct. So every left coset of H has m elements. Let k be the number of distinct left cosets of H. Since they form a partition of the set G, it follows that $n = km$. Hence m divides n. ∎

The following theorem follows immediately from Lagrange's theorem. The proof is left as an exercise.

THEOREM 0.4.10 Let G be a group of order n. Then for every element a in G, $o(a) \mid n$ and hence $a^n = e$.

Consequently, for every prime p, a group of order p is cyclic.

Let H be a subgroup of G. The set of all left cosets of H is written G/H. The order of the set G/H—that is, the number of left cosets of H in G—is called the *index* of H in G and denoted by $[G : H]$. For example, $[\mathbb{Z} : n\mathbb{Z}] = n$.

If G is finite, then $[G : H] = |G/H| = \dfrac{|G|}{|H|}$.

EXERCISES 0.4

(G denotes an arbitrary group unless stated otherwise.)

1. Show that every group of order less than 6 is abelian.
2. Find all subgroups of the symmetric group S_3.
3. Find all subgroups of $\mathbb{Z}_{12}$.
4. Show that for any positive integers m and n, $m\mathbb{Z} \cap n\mathbb{Z} = q\mathbb{Z}$, where $q = \mathrm{lcm}(m, n)$.

5. Let H and K be finite subgroups of G. If the orders of H and K are relatively prime, show that $H \cap K = \{e\}$.

6. Let G be the multiplicative group of nonzero real numbers, and let H be the set of all positive real numbers. Show that H is a subgroup of index 2 in G.

7. Let H and K be subgroups of finite index in a group G. Show that $H \cap K$ is again a subgroup of finite index.

8. If a group G has no proper subgroups, show that G is a group of order p for some prime p.

0.5 QUOTIENT GROUPS AND HOMOMORPHISMS

0.5.1 Quotient Groups

If H is a subgroup of an abelian group G, then $aH = Ha$ for all $a \in G$. Even in a nonabelian group, this property may hold for some special subgroups. Such subgroups play an important role in group theory and are called *normal subgroups*.

Definition 0.5.1 Let G be a group. A subgroup N of G is called a *normal subgroup*, written $N \triangleleft G$, if $aN = Na$ for all $a \in N$.

If N is a normal subgroup, then every left coset of N is also a right coset, so we need not distinguish between the left and the right cosets of N. An important example of a normal subgroup is the center of the group, defined next.

Definition 0.5.2 The *center* of a group G, written $Z(G)$, is defined as

$$Z(G) = \{a \in G \mid ax = xa \text{ for all } x \in G\}$$

In other words, the center consists of those elements that commute with every element in the group. It is easily verified that $Z(G)$ is a normal subgroup of G. If G is abelian, then obviously $Z(G) = G$.

The following theorem gives a necessary and sufficient condition for a subgroup to be normal. The proof is left as an exercise.

THEOREM 0.5.3 Let G be a group. A subgroup N of G is normal if and only if $a^{-1}Na \subset N$ for all $a \in G$.

The most important property of a normal subgroup N is that it determines another group, as stated in the following theorem. If A and B are subsets of G, we define $AB = \{xy \mid x \in A, y \in B\}$. It is easily shown that if H is a subgroup of G, then $HH = H$.

THEOREM 0.5.4 Let G be a group and N a normal subgroup of G. Then G/N is a group under multiplication.

Proof: Let $a, b \in G$. Then $(aN)(bN) = a(Nb)N = a(bN)N = ab(NN) = abN$. This multiplicative operation can be shown to be well-defined. Furthermore, for all $a, b, c \in G$, $[(aN)(bN)](cN) = abcN = (aN)[(bN)(cN)]$. For every $a \in G$, $(eN)(aN) = eaN = aN$. Hence $eN = N$ is the identity in G/N. Further, $(aN)(a^{-1}N) = aa^{-1}N = N$. Hence $a^{-1}N$ is the inverse of aN. This proves that G/N is a group.   ∎

The group G/N is called the *quotient group of G by N*. For example, let G be the additive group $\mathbb{Z}$ and $H = n\mathbb{Z}$. Then $\mathbb{Z}/(n\mathbb{Z}) = \{n\mathbb{Z} + i \mid i = 0, 1, \ldots, n - 1\}$ is a group of order n.

0.5.2 Homomorphisms of Groups

Definition 0.5.5 Let G and H be groups. A *homomorphism* from G to H is a mapping $f : G \to H$ such that

$$f(ab) = f(a)f(b) \text{ for all } a, b \in G$$

Moreover, an injective homomorphism is called a *monomorphism,* and a bijective homomorphism is called an *isomorphism.*

In keeping with our usual practice, we have written the definition of a homomorphism for multiplicative groups. If one or both of the groups are additive, the definition is suitably modified. For example, if G is multiplicative and H is additive, then $f(ab) = f(a) + f(b)$ for all $a, b \in G$.

As an example, let $G = \mathbb{R}^+$ be the multiplicative group of positive real numbers, and let H be the additive group $\mathbb{R}$. We define $f : G \to H$ by the rule $f(x) = \ln x$ for all $x \in G$. Then for all x, y, $f(xy) = \ln(xy) = \ln x + \ln y = f(x) + f(y)$. Hence f is a homomorphism. Moreover, f is surjective.

Definition 0.5.6 Let $f : G \to H$ be a homomorphism of groups. The *kernel* of f is defined to be the set

$$\ker f = \{x \in G \mid f(x) = e'\}$$

where e' is the identity in H.

In the following theorem, we prove some basic properties of a homomorphism. Recall that for any mapping $f : G \to H$, the range of f is the set $\mathrm{Im}\, f = \{y \in H \mid y = f(x) \text{ for some } x \in G\}$.

THEOREM 0.5.7 Let G and H be groups with identities e and e', respectively. Let $f : G \to H$ be a homomorphism. Then

(a) $f(e) = e'$.
(b) $f(a^{-1}) = (f(a))^{-1}$ for all $a \in G$.
(c) Im f is a subgroup of H.
(d) ker f is a normal subgroup of G.
(e) f is injective if and only if ker $f = \{e\}$.

Proof:

(a) $f(e)f(e) = f(ee) = f(e) = e'f(e)$. Hence, by the cancellation law in H, $f(e) = e'$.

(b) $f(a)f(a^{-1}) = f(aa^{-1}) = f(e) = e'$. Hence $f(a^{-1})$ is the inverse of $f(a)$.

(c) Clearly, Im f is a subset of H and $e' = f(e) \in$ Im f. Let $x, y \in$ Im f. Then $x = f(a)$, $y = f(b)$ for some $a, b \in G$. So $x^{-1}y = (f(a))^{-1}f(b) = f(a^{-1})f(b) = f(a^{-1}b) \in$ Im f. This proves that Im f is a subgroup of H.

(d) We write $K = \ker f$. Because $f(e) = e' \in H$, $e \in K$. Let $a, b \in K$. Then $f(a^{-1}b) = (f(a))^{-1}f(b) = e'e' = e'$, so $a^{-1}b \in K$. Hence K is a subgroup of G. To show that K is normal, let $a \in G$ and $k \in K$. Then $f(aka^{-1}) = f(a)f(k)f(a^{-1}) = f(a)e'f(a^{-1}) = e'$, so $aka^{-1} \in K$. Hence $aKa^{-1} \subset K$ for all $a \in G$. Therefore, by Theorem 0.5.3, $K \triangleleft G$.

(e) Suppose f is injective, and let $k \in K$. Then $f(k) = e' = f(e)$ and hence $k = e$. So $K = \{e\}$. Conversely, suppose $K = \{e\}$. Let $a, b \in G$ and suppose $f(a) = f(b)$. Then $f(a^{-1}b) = (f(a))^{-1}f(b) = e'$ and hence $a^{-1}b \in K$. So $a^{-1}b = e$, which implies $a = b$. Hence f is injective. ∎

Theorem 0.5.7 shows that every homomorphism $f : G \to H$ determines a normal subgroup of G—namely, ker f. The following theorem proves the converse; that is, every normal subgroup of G determines a homomorphism.

THEOREM 0.5.8 Let G be a group, and let $N \triangleleft G$. Then the mapping $f : G \to G/N$, given by $f(a) = aN$, is a surjective homomorphism and ker $f = N$.

Proof: For all $a, b \in G$, $(ab)N = (aN)(bN)$ and hence f is a homomorphism. Moreover, f is clearly surjective. Now the identity of the group G/N is $eN = N$, so ker $f = \{a \in G \mid aN = N\}$. But $aN = N$ if and only if $a \in N$. Hence ker $f = N$. ∎

The mapping f in Theorem 0.5.8 is called the *natural mapping* from G to the quotient group G/N.

Let us now consider isomorphisms of groups. Given groups G and H, if there exists an isomorphism $f : G \to H$, we say G is *isomorphic* to H and we write

$G \simeq H$. It can be shown that the relation $\simeq$ is an equivalence relation. An equivalence class under the relation $\simeq$ is called an *abstract group*. Two groups are said to be abstractly the same if they are isomorphic. For example, the Klein's 4-group is an abstract group. Any two Klein's 4-groups are isomorphic.

In the following theorem we consider isomorphisms between cyclic groups.

THEOREM 0.5.9 Every infinite cyclic group is isomorphic with the additive group $\mathbb{Z}$. Every cyclic group of order n is isomorphic with $\mathbb{Z}_n$, the additive group of integers modulo n.

Proof: Let G be an infinite cyclic group with a generator a. Consider the mapping $f : \mathbb{Z} \to G$ given by $f(i) = a^i$. For all $i, j \in \mathbb{Z}$, $f(i+j) = a^{i+j} = a^i a^j = f(i) f(j)$. Hence f is a homomorphism. The mapping f is obviously surjective. The element a is of infinite order; therefore, for any two distinct integers i, j, $a^i \neq a^j$. Hence f is injective. This proves that f is an isomorphism.

Now suppose G is a cyclic group of order n generated by a. Then $o(a) = n$ and $G = \{e, a, \ldots, a^{n-1}\}$. So $a^i = a^{res_n(i)}$ for every integer i. Consider the mapping $f : \mathbb{Z}_n \to G$ given by $f(i) = a^i$. The mapping f is obviously bijective. Let $i, j \in \mathbb{Z}_n$. Then

$$f(i)\, f(j) = a^i a^j = a^{i+j} = a^{res(i+j)} = f(i+j)$$

This proves that f is an isomorphism.   ∎

We write C_n to denote the abstract cyclic group of order n.

Let $f : G \to H$ be an injective homomorphism. Then, on restricting the codomain of the mapping f to Im f, we obtain an isomorphism from G to Im f. Hence $G \simeq \text{Im } f$.

The following theorem, known as the *fundamental theorem of homomorphisms of groups,* shows that every homomorphism determines an isomorphism.

THEOREM 0.5.10 Let $f : G \to H$ be a homomorphism of groups. Then $G/\ker f \simeq \text{Im } f$.

In particular, if f is surjective, then $G/\ker f \simeq H$.

Proof: By Theorem 0.5.7, $\ker f$ is a normal subgroup of G and hence $G/\ker f$ is a group. We write $K = \ker f$ and define a mapping $\phi : G/K \to H$ by the rule $\phi(xK) = f(x)$. Because two distinct elements in G can determine the same coset, we must show that the mapping ϕ is well defined. Let $x, y \in G$ such that $xK = yK$. Then $x^{-1}yK = K$, and so $x^{-1}y \in K$. Thus $f(x)^{-1}f(y) = f(x^{-1}y) = e'$, proving $f(x) = f(y)$. This proves that ϕ is well defined. To show that ϕ is injective, suppose $\phi(xK) = \phi(yK)$ for some $x, y \in G$. Then $f(x) = f(y)$, so $f(x^{-1}y) = (f(x))^{-1}f(y) = e'$, which implies that $x^{-1}y \in K$, so $xK = yK$. Hence ϕ is injective. For all $xK, yK \in G/K$, $\phi((xK)(yK)) = \phi(xyK) = f(xy) = f(x)f(y) = \phi(xK)\phi(yK)$. Hence ϕ is an injective homomorphism. Clearly Im $\phi = $ Im f. Hence $G/K \simeq \text{Im } f$.   ∎

Given groups G and H, if there exists a surjective homomorphism $f : G \to H$, we say that H is a *homomorphic image* of G. Theorem 0.5.10 shows that every homomorphic image of G is isomorphic to a quotient group of G. In Theorem 0.5.8, on the other hand, we saw that every quotient group of G is a homomorphic image of G.

As an example of Theorem 0.5.10, consider the mapping $f : \mathbb{Z} \to \mathbb{Z}_n$ given by $f(i) = res_n(i)$. It is easily shown that f is a surjective homomorphism and $\ker f = n\mathbb{Z}$. Hence, by Theorem 0.5.10, $\mathbb{Z}_n \simeq \mathbb{Z}/(n\mathbb{Z})$. [It can be directly checked that the mapping $\phi : \mathbb{Z}_n \to \mathbb{Z}/(n\mathbb{Z})$ given by $\phi(i) = n\mathbb{Z} + i$ is an isomorphism.]

EXERCISES 0.5

1. Let H, N be subgroups of G, and suppose N is normal. Show that HN is a subgroup of G. Further show that $N \lhd HN$ and $H \cap N \lhd H$.

2. If H and N are normal subgroups of G, show that HN is a normal subgroup.

3. Let G be the group of all invertible $n \times n$ matrices with real entries. Let $N = \{A \in G \mid \det A = 1\}$. Show that $N \lhd G$.

4. Let G be the group of all invertible $n \times n$ matrices with real entries. Let $N = \{A \in G \mid \det A > 0\}$. Show that $N \lhd G$. Find $[G : N]$.

5. Show that every subgroup of the quaternion group Q is normal.

6. Show that for every positive integer m, the mapping $f : \mathbb{Z} \to \mathbb{Z}$ given by $f(x) = mx$ is a monomorphism.

7. Let G be an abelian group. Show that for every positive integer m, the mapping $f : G \to G$ given by $f(x) = x^m$ is a homomorphism.

8. Let G be an abelian group of order n. Show that for every positive integer m relatively prime to n, the mapping $f : G \to G$ given by $f(x) = x^m$ is an isomorphism.

9. Let G be a group. Show that for every $a \in G$, the mapping $f : G \to G$ given by $f(x) = axa^{-1}$ is an isomorphism.

10. Let $f : G \to H$ be a homomorphism. Show that for every $a \in G$, $o(f(a)) \mid o(a)$. Further show that if f is injective, then $o(f(a)) = o(a)$.

11. Show that every group of order 6 is either cyclic or isomorphic to S_3.

12. Show that every nonabelian group of order 10 is generated by two elements, say a and b, such that $o(a) = 5$, $o(b) = 2$, and $ab = ba^{-1}$.

0.6 APPLICATIONS OF GROUPS IN NUMBER THEORY

In this section we consider applications of groups to prove some classical results in number theory. The first application is a group-theoretic proof of the well-known *Euler–Fermat theorem*. Recall that Euler's phi-function $\varphi(n)$ denotes the number of positive integers not exceeding n and relatively prime to n.

THEOREM 0.6.1 Let n be a positive integer, and let a be any integer relatively prime to n. Then $a^{\varphi(n)} \equiv 1 \pmod{n}$.

In particular, if p is prime and p does not divide a, then $a^{p-1} \equiv 1 \pmod{p}$.

Proof: Let $a' = res_n(a)$. Because a is relatively prime to n, it follows that a' is also relatively prime to n. Hence $a' \in \mathbb{Z}_n^*$ and $a \equiv a' \pmod{n}$. Now, by Theorem 0.2.8, $\mathbb{Z}_n^*$ is a multiplicative group of order $\varphi(n)$. Hence, by Theorem 0.4.10, $a'^{\varphi(n)} = 1$ in the group $\mathbb{Z}_n$; that is, $a'^{\varphi(n)} \equiv 1 \pmod{n}$ in $\mathbb{Z}$. Hence $a^{\varphi(n)} \equiv a'^{\varphi(n)} \equiv 1 \pmod{n}$.

The second part follows from the fact that $\varphi(p) = p - 1$.   ∎

THEOREM 0.6.2 (Wilson's theorem) Let p be a prime. Then $(p-1)! \equiv -1 \pmod{p}$.

Proof: By Theorem 0.2.8, $\mathbb{Z}_p^* = \{1, 2, \ldots, p-1\}$ is a group under multiplication modulo p. Let $a \in \mathbb{Z}_p^*$ and suppose $a^2 = 1$ (in the group $\mathbb{Z}_p^*$). Then $a^2 \equiv 1 \pmod{p}$ in $\mathbb{Z}$, so p divides $a^2 - 1$. Since p is prime, it follows that either $a = 1$ or $a = p - 1$. Hence, for every $a \in \mathbb{Z}_p^*$ other than 1 and $p - 1$, $a^{-1} \neq a$. So, if we pair each such a with its inverse, the product $(p-1)!$ (in the group $\mathbb{Z}_p^*$) can be expressed as

$$(p-1)! = 1 \cdot 2 \cdots (p-1) = 1 \cdot (2 \cdot 2^{-1}) \cdots (a \cdot a^{-1}) \cdots (p-1) = p - 1$$

Hence $(p-1)! \equiv -1 \pmod{p}$.   ∎

THEOREM 0.6.3 Let m, n be relatively prime positive integers. Then $\varphi(mn) = \varphi(m)\varphi(n)$.

Proof: Let G, H be cyclic groups of orders m, n, respectively. Then, by Theorem 0.3.10, their direct product $G \times H$ is a cyclic group of order mn. Further, (a, b) is a generator of $G \times H$ if and only if a is a generator of G and b is a generator of H. By Theorem 0.3.7, the groups G, H, and $G \times H$ have $\varphi(m)$, $\varphi(n)$ and $\varphi(mn)$ generators, respectively. Hence it follows that $\varphi(mn) = \varphi(m)\,\varphi(n)$.   ∎

The foregoing result can be used to compute $\varphi(n)$ for any positive integer n by decomposing n into prime factors. If p is prime and r is a positive integer, then there are exactly p^{r-1} positive integers not exceeding p^r and *not* relatively prime to p^r—namely, integers of the form qp, where $q = 1, 2, \ldots, p^{r-1}$. Hence $\varphi(p^r) = p^r - p^{r-1}$. Given a positive integer n, suppose $n = p_1^{r_1} \cdots p_k^{r_k}$ (where $p_1, \ldots, p_k$ are distinct primes). Then, by Theorem 0.6.3,

$$\begin{aligned}
\varphi(n) &= \varphi\left(p_1^{r_1}\right) \cdots \varphi\left(p_k^{r_k}\right) \\
&= \left(p_1^{r_1} - p_1^{r_1-1}\right) \cdots \left(p_k^{r_k} - p_k^{r_k-1}\right) \\
&= n\left(1 - \frac{1}{p_1}\right) \cdots \left(1 - \frac{1}{p_k}\right)
\end{aligned}$$

THEOREM 0.6.4 *(Chinese remainder theorem)* Let $n_1, \ldots, n_r$ be pairwise relatively prime positive integers, and $N = n_1 \cdots n_r$. Then, given any integers $k_1, \ldots, k_r$, there is a unique integer m, $0 \le m \le N - 1$, such that

$$m \equiv k_i \pmod{n_i} \text{ for each } i = 1, \ldots, r$$

Moreover, if x is any integer such that $x \equiv k_i \pmod{n_i}$ for each $i = 1, \ldots, r$, then $x \equiv m \pmod{N}$.

Proof: Let $G_1, \ldots, G_r$ be cyclic groups of orders $n_1, \ldots, n_r$, respectively. Then, by generalization of Theorem 0.3.10, their direct product $G = G_1 \times \cdots \times G_r$ is a cyclic group of order $N = n_1 \cdots n_r$. Let $a_1, \ldots, a_r$ be generators of groups $G_1, \ldots, G_r$, respectively. Then $a = (a_1, \ldots, a_r)$ is a generator of G, and $G = \{e, a, \ldots, a^{N-1}\}$. Given any integers $k_1, \ldots, k_r$, $(a_1^{k_1}, \ldots, a_r^{k_r}) \in G$. Hence there is a unique integer m, $0 \le m \le N - 1$, such that $(a_1^{k_1}, \ldots, a_r^{k_r}) = a^m = (a_1^m, \ldots, a_r^m)$. So for each $i = 1, \ldots, r$, we have $a_i^{k_i} = a_i^m$, which implies $n_i \mid (m - k_i)$. Hence $m \equiv k_i \pmod{n_i}$ for each $i = 1, \ldots, r$.

Let x be any integer such that $x \equiv k_i \pmod{n_i}$ for each $i = 1, \ldots, r$. We write $m' = res_N(x)$. Then $0 \le m' \le N - 1$, and $x \equiv m' \pmod{N}$. So $N \mid (x - m')$. Hence, for each $i = 1, \ldots, r$, $n_i \mid (x - m')$, so $m' \equiv x \pmod{n_i}$, which implies $m' \equiv k_i \pmod{n_i}$. By the uniqueness of m, it follows that $m' = m$ and hence $x \equiv m \pmod{N}$. ∎

Theorem 0.6.4 ensures the existence of an integer m such that the congruences $m \equiv k_i \pmod{n_i}$, $i = 1, \ldots, r$, hold. But the proof of the theorem does not provide a practical method for computing such a number m. We give an algorithm for doing so.

We write $N_i = N/n_i$, $i = 1, \ldots, r$. Since $n_1, \ldots, n_r$ are pairwise relatively prime, it follows that n_i, N_i are relatively prime. Hence there exist integers s_i, t_i such that $s_i n_i + t_i N_i = 1$; then $t_i N_i \equiv 1 \pmod{n_i}$. We find these integers s_i, t_i by using the iterative division algorithm explained earlier (see Example 0.2.10). Now we set

$$x = \sum_{i=1}^{r} k_i t_i N_i \quad \text{and} \quad m = res_N(x)$$

Because $n_i \mid N_j$ for all $j \ne i$, it follows that $m \equiv k_i t_i N_i \equiv k_i \pmod{n_i}$ for each $i = 1, \ldots, r$.

The following example illustrates this procedure.

Example 0.6.5 Find the smallest positive integer m such that

$$m \equiv 4 \pmod{5}, \quad m \equiv 3 \pmod{6}, \quad m \equiv 5 \pmod{7}$$

Solution. With the notation used above, we have here $k_1 = 4, k_2 = 3, k_3 = 5$; $n_1 = 5, n_2 = 6, n_3 = 7$; $N = 210, N_1 = 42, N_2 = 35, N_3 = 30$. By the iterative division algorithm, $42 = 8(5) + 2$ and $5 = 2(2) + 1$. Hence $1 =$

$5 - 2(2) = 5 - 2[42 - 8(5)] = 17(5) - 2(42)$.   Similarly,   we   obtain $1 = 6(6) - 1(35)$ and $1 = 13(7) - 3(30)$. So we take

$$x = \sum_{i=1}^{3} k_i t_i N_i = 4(-2)(42) + 3(-1)(35) + 5(-3)(30) = -891$$

Hence $m = res_N(x) = 159$.

0.7 RINGS AND FIELDS

In this section we introduce another algebraic system of fundamental importance in abstract algebra.

Definition 0.7.1 A *ring* is an algebraic system $(R, +, \cdot)$ such that

(a) $(R, +)$ is an additive abelian group.
(b) $(R, \cdot)$ is a multiplicative semigroup.
(c)   For all $a, b, c \in R$,
    (1) $a(b + c) = ab + ac$ (left distributive law).
    (2) $(a + b)c = ac + bc$ (right distributive law).

Further, a ring R is said to be commutative if $ab = ba$ for all $a, b \in R$.

We refer to a ring $(R, +, \cdot)$ as ring R. The notational conventions for a group or semigroup explained in Section 0.2 are also followed here. The identity of the group $(R, +)$ is written 0 and is called the *zero* element of the ring R. The inverse of an element a in $(R, +)$ is written $-a$, and $b + (-a)$ is written $b - a$. The identity of the semigroup $(R, \cdot)$, if it exists, is called the *identity* or *unity* of the ring R and written e or 1. If an element $a \in R$ has an inverse in $(R, \cdot)$, it is called a *unit*, and its inverse is written a^{-1}. We write $a \cdot b$ as ab.

EXAMPLES

1. The most familiar examples of a ring are $\mathbb{Z}$, $\mathbb{Q}$, $\mathbb{R}$, and $\mathbb{C}$, under the usual addition and multiplication.
2. For every positive integer n, $\mathbb{Z}_n$ is a ring under addition and multiplication modulo n. We have shown already (Theorem 0.2.7) that $(\mathbb{Z}_n, +)$ is an abelian group and $(\mathbb{Z}_n, \cdot)$ is a semigroup with identity. The distributive property is easily shown to hold.
3. The set $F^{n \times n}$ of all $n \times n$ matrices over F, where $F = \mathbb{Q}$, $\mathbb{R}$, or $\mathbb{C}$, is a ring under addition and multiplication of matrices. If $n > 1$, the ring is noncommutative.

The following theorem states some elementary properties of a ring. The proof is left as an exercise.

THEOREM 0.7.2 Let R be a ring. Then for all $a, b, c \in R$,

(a) $a0 = 0 = 0a$.
(b) $a(-b) = -(ab) = (-a)b$; hence $(-a)(-b) = ab$.
(c) $a(b - c) = ab - ac$; $(a - b)c = ac - bc$.

Definition 0.7.3 A ring R is called a *field* if the nonzero elements in R form a group under multiplication.

In other words, every nonzero element in a field has a multiplicative inverse. For example, the rings $\mathbb{Q}$, $\mathbb{R}$, and $\mathbb{C}$ are fields. For every prime p, the ring $\mathbb{Z}_p$ is a field (Theorem 0.2.8).

Definition 0.7.4 A ring R is called an *integral domain* if for any $a, b \in R$, $ab = 0 \Rightarrow a = 0$ or $b = 0$.

Every field is an integral domain (but not conversely). Let F be a field. Let $a, b \in F$, and suppose $ab = 0$ and $a \neq 0$. Then $b = a^{-1}(ab) = a^{-1}0 = 0$. Hence F is an integral domain. The ring $\mathbb{Z}$ is an integral domain but not a field.

Let R be a ring. The powers and multiples of an element $a \in R$ are defined as explained in Section 0.3. In addition to formulas (1) and (2) stated there, we have, as a consequence of the distributive property, the following result: For all $a, b \in R$ and all integers m, n,

$$(ma)(nb) = (mn)(ab)$$

Definition 0.7.5 Let R be a ring. The *characteristic* of R, written $\text{char } R$, is the smallest positive integer m such that $ma = 0$ for all $a \in R$. If no such integer exists, R is said to have characteristic 0.

For example, the ring $\mathbb{Z}$ has characteristic zero. The ring $\mathbb{Z}_n$ has characteristic n.

THEOREM 0.7.6 Let F be a field. Then $\text{char } F$ is either 0 or a prime number.

Proof: Let $\text{char } F = m \neq 0$, and suppose m is not prime. Then $m = pq$, where p, q are positive integers less than n. Let e be the identity of F. Then $(pe)(qe) = (pq)e = 0$. Hence $pe = 0$ or $qe = 0$. Therefore, for all $a \in F$, $pa = (pe)a = 0$ or $qa = (qe)a = 0$, which contradicts the fact that m is the least positive integer such that $ma = 0$ for all a. Hence m must be prime. ∎

A homomorphism of rings is defined in the same manner as a homomorphism of groups. The general idea of a homomorphism for any algebraic system is that it is mapping that preserves the operations in that system.

Definition 0.7.7 Let R and S be rings. A mapping $f : R \rightarrow S$ is called a *homomorphism* if for all $a, b \in R$,

$$f(a + b) = f(a) + f(b) \quad \text{and} \quad f(ab) = f(a)\,f(b)$$

A bijective homomorphism is called an *isomorphism*.

   Given any rings R and S, the mapping $z : R \rightarrow S$ given by $f(a) = 0$ for all $a \in R$ is a homomorphism. It is called a *zero homomorphism*.

   If there exists an isomorphism from a ring R to a ring S, we say that R is isomorphic to S and write $R \simeq S$. As in the case of groups, isomorphism of rings is an equivalence relation. If there exists a surjective homomorphism from R to S, then S is called a *homomorphic image* of R.

0.7.1 Subrings, Ideals, and Quotient Rings

Definition 0.7.8 Let $(R, +, \cdot)$ be a ring. A nonempty subset S of R is called a *subring* of R if S is itself a ring under the operations $+$ and $\cdot$ restricted to S.

   A *subfield* of a field is defined similarly.

   Every ring R has two trivial subrings—namely, $\{0\}$ [also written as (0) or simply 0] and R itself.

   The following theorem gives a necessary and sufficient condition for a subset of a ring to be a subring.

THEOREM 0.7.9 A nonempty subset S of a ring R is a subring if and only if for all $a, b \in S$,

$$a - b \in S \quad \text{and} \quad ab \in S$$

Proof: Suppose the conditions hold. By Theorem 0.4.3, the first condition implies that $(S, +)$ is a subgroup of the group $(R, +)$. The second condition implies that $(S, \cdot)$ is a semigroup. Since the two distributive laws hold in R, they also hold in S. Hence $(S, +, \cdot)$ is a subring of $(R, +, \cdot)$. The converse follows from the definition of a subring.   ∎

   It is worth noting that a subring S of a ring R may have an identity that is different from the identity of R, or S may have an identity even though R does not have one. For example, let $R = \mathbb{R}^{2 \times 2}$ and S be the set of all matrices in R of the form $\begin{bmatrix} a & 0 \\ 0 & 0 \end{bmatrix}$. Then S is obviously a subring with identity $\begin{bmatrix} 1 & 0 \\ 0 & 0 \end{bmatrix}$, but the identity of R is $\begin{bmatrix} 1 & 0 \\ 0 & 1 \end{bmatrix}$. (See Exercise 3 for a less trivial and more interesting example.)

Definition 0.7.10 A nonempty subset I of a ring R is called an *ideal* in R if for all $a, b \in I$ and $r \in R$,

$$a - b \in I, \quad ar \in I, \quad ra \in I$$

It is clear that every ideal in R is a subring of R, but the converse need not be true. For example, the set $S = \left\{ \begin{bmatrix} a & b \\ 0 & 0 \end{bmatrix} \mid a, b \in \mathbb{R} \right\}$ is a subring of the ring $\mathbb{R}^{2 \times 2}$ but not an ideal.

The ideals in a ring play the same role as normal subgroups of a group. We saw in Theorem 0.5.4 that a normal subgroup N of a group G generates a quotient group G/N. In the same manner, an ideal in a ring generates a quotient ring, as explained below.

Let I be an ideal in a ring R. Since I is a subgroup of the additive group $(R, +)$, we define for every $a \in R$ the coset $a + I$ as

$$a + I = \{a + x \mid x \in I\}$$

By Theorem 0.4.8, these cosets form a partition of R. Two elements a, b lie in the same coset if and only if $a - b \in I$. The set of all cosets of I is written R/I. We define addition and multiplication in R/I as follows: For all $a, b \in R$,

$$(a + I) + (b + I) = (a + b) + I$$
$$(a + I) \cdot (b + I) = (ab) + I$$

It can be easily shown that these operations are well defined. One can now verify that $(R/I, +, \cdot)$ satisfies all the conditions for being a ring. It is called the *quotient ring* of R modulo I. The zero element of the ring is the coset $0 + I = I$. If R has an identity e, then $e + I$ is the identity of the ring R/I. We record this fact in the following theorem.

THEOREM 0.7.11 Let I be an ideal in a ring R. Then R/I is a ring.

We often use a simpler notation for the elements of R/I and write $\bar{a}$ to denote the coset $a + I$. Then the rules for the operations in R/I assume the form

$$\bar{a} + \bar{b} = \overline{a + b}, \quad \bar{a} \cdot \bar{b} = \overline{ab}$$

Note that if $a \in I$, then $\bar{a} = a + I = I = \bar{0}$.

Let R be a commutative ring, and let $a \in R$. It is easily shown that the set $aR = \{ar \mid r \in R\}$ is an ideal in R. Further, if the ring R has an identity e, then $a = ae \in aR$. But if R does not have an identity, then a is not necessarily an element of R. Suppose R has an identity. Then aR is the smallest ideal containing a, by which we mean that if I is any ideal containing a, then $aR \subset I$. In this case, we say that aR is the ideal generated by a. But in general, when R is an arbitrary ring, the set aR is not necessarily an ideal. However, there does exist a smallest ideal containing a. The intersection of all ideals containing a is indeed the smallest ideal containing a.

Definition 0.7.12 Let R be a ring, and let $a \in R$. The smallest ideal in R containing the element a is called the ideal *generated by a* and denoted by (a). An ideal I in R is called a *principal ideal* if $I = (a)$ for some $a \in R$.

We have seen that if R is a commutative ring with identity, then the ideal generated by a is $(a) = aR$. It is easily shown that if R is a commutative ring without identity, then

$$(a) = \{ar + na \mid r \in R,\ n \in \mathbb{Z}\} = aR + \mathbb{Z}a$$

A ring R is called a *principal ideal ring* if every ideal in R is a principal ideal. The simplest example of such a ring is $\mathbb{Z}$, which we show in the following theorem.

THEOREM 0.7.13 Every ideal in the ring $\mathbb{Z}$ is a principal ideal.

Proof: Let I be an ideal in $\mathbb{Z}$. If $I = \{0\}$, then $I = (0)$. Suppose $I \neq (0)$. Then there exists a smallest positive integer a in I. We claim that $I = a\mathbb{Z}$. Let $b \in I$. By Euclidean algorithm $b = aq + r$, where $0 \leq r < a$. Now $a, b \in I$ and therefore $b - aq \in I$. Hence $r \in I$. But $r < a$ and a is the smallest positive integer in I; hence $r = 0$. Thus $b = aq \in a\mathbb{Z}$ and $I \subset a\mathbb{Z}$. Since $\mathbb{Z}$ is a commutative ring with identity, $a\mathbb{Z}$ is the smallest ideal containing a; hence $a\mathbb{Z} \subset I$. This proves $I = a\mathbb{Z} = (a)$. ∎

EXERCISES 0.7

1. A ring R is called a *Boolean ring* if $a^2 = a$ for all $a \in R$. Show that a Boolean ring R is commutative and $a + a = 0$ for all $a \in R$.

2. Let $R = \mathcal{P}(X)$ be the power set of X. Define addition and multiplication in R as follows.

$$a + b = (a \cap b') \cup (a' \cap b)$$
$$a \cdot b = a \cap b$$

   Show that $(R, +, \cdot)$ is a Boolean ring.

3. Prove that $S = \{0, 2, 4, 6, 8\}$ is a subring of $\mathbb{Z}_{10}$, and show that S has an identity that is different from the identity of $\mathbb{Z}_{10}$.

4. *(Generalization of Exercise 3.)* Show that for every positive integer n, the ring $\mathbb{Z}_{4n+2}$ has a nontrivial subring whose identity is different from the identity of $\mathbb{Z}_{4n+2}$.

5. Show that the set $U(R)$ of all units (invertible elements) in a ring R is a group under multiplication. Find the group of units of the following rings: (a) $\mathbb{Z}_{12}$, (b) $\mathbb{Z}_{25}$, and (c) $\mathbb{Z}_p$ (p prime).

6. An element a in a ring R is called *nilpotent* if $a^n = 0$ for some positive integer n. Let R be a commutative ring. Let $a, b \in R$ be nilpotent elements. Show that

$a + b$ is also nilpotent. Give an example to show that if R is not commutative, then $a + b$ need not be nilpotent.

7. Show that the set of nilpotent elements in a commutative ring is an ideal.

8. Show that a finite nonzero commutative integral domain is a field.

9. Let R be a commutative ring with identity. Suppose (0) and R are the only ideals in R. Prove that R is a field.

10. Find all ideals in $\mathbb{Z}$ and $\mathbb{Z}_{15}$.

11. Let $C\,[0, 1]$ be the set of all real-valued continuous functions defined on the closed interval $[0, 1]$, where $f + g$ and fg are defined as usual by $(f + g)(x) = f(x) + g(x)$ and $(fg)(x) = f(x)\,g(x)$.
    (a) Show that $C\,[0, 1]$ is not an integral domain.
    (b) Let M denote the set of all functions $f \in C[0, 1]$ that vanish on some fixed point $a \in [0, 1]$. Show that M is an ideal in $C[0, 1]$.
    (c) Show that the quotient ring $C[0, 1]/M$ is a field.

12. Show that the only ring homomorphisms from $\mathbb{Z}$ to itself are the zero and identity homomorphisms.

13. Show that there is no homomorphism $f : \mathbb{Z}_{10} \rightharpoonup \mathbb{Z}_{15}$ such that $f(1) = 1$.

14. Show that there exists a ring homomorphism $f : \mathbb{Z}_m \rightharpoonup \mathbb{Z}_n$ such that $f(1) = 1$ if n divides m.

15. Show that every nonzero homomorphism from a field to a ring is injective.

16. Show that every nonzero homomorphism from a finite field to itself is an isomorphism.

17. Show that if $\gcd(m, n) = 1$, then $\mathbb{Z}_{mn} \simeq \mathbb{Z}_m \times \mathbb{Z}_n$.

0.8 FINITE FIELDS

In this section we consider fields that have a finite number of elements. So far, the only finite fields we have come across are those of the form $\mathbb{Z}_p$, where p is any prime. We now intend to find all finite fields. To begin, we state, without proof, the following basic result about the order of a finite field. (The proof may be found in any standard text on abstract algebra.)

THEOREM 0.8.1 The number of elements in any finite field is p^r, where p is some prime and r is a positive integer.

A finite field is also known as a *Galois field*. It can be shown that any two finite fields of the same order are isomorphic. We denote a finite field of order $q(= p^r)$ by $\mathbb{F}_q$ or $GF(q)$.

We will show that for any prime p and any positive integer r, we can construct a field of order p^r. To obtain this construction, we make use of rings of polynomials, which we now describe.

We assume the reader is familiar with polynomials and operations on them as taught in high school algebra. In ordinary algebra, we use polynomials in which the coefficients are real numbers, most commonly integers. We generalize this concept of polynomials and allow the coefficients to be elements of any ring.

Let R be a ring. A *polynomial* $p(x)$ over R is an expression of the form

$$p(x) = a_0 + a_1 x + a_2 x^2 + \cdots + a_m x^m \tag{1}$$

where $a_0, a_1, a_2, \ldots, a_m \in R$. The symbol x is called the *indeterminate* in the polynomial. (The indeterminate is usually written x, but one may use any other symbol for it.) The order in which the terms in a polynomial are written is immaterial. We usually write the terms either in ascending order of powers of x or in descending order. Thus the polynomial (1) may be written as

$$p(x) = a_m x^m + a_{m-1} x^{m-1} + \cdots + a_1 x + a_0$$

As in ordinary algebra, if a coefficient a_i is 1 for any $i > 0$, we don't write it. If $a_i = 0$, we omit the term $a_i x^i$. For example, the polynomial $1 + 1x + 0x^2 + 1x^3 + 0x^4$ is written $1 + x + x^3$ or $x^3 + x + 1$. If every coefficient in (1) is zero, we call $p(x)$ the *zero polynomial* and write $p(x) = 0$.

If the coefficient a_m in (1) is nonzero, then m is called the *degree* of the polynomial $p(x)$, written deg $p(x)$, and a_m is called the *leading coefficient*. The degree of the zero polynomial is undefined. A polynomial of degree 0 is simply $p(x) = a$, where a is any nonzero element in R. The zero polynomial and any polynomial of degree 0 are called *constant polynomials*.

We write $R[x]$ to denote the set of all polynomials over the ring R. Two polynomials in $R[x]$ are said to be *equal* if they consist of the same terms. Addition and multiplication of two polynomials are defined as in ordinary algebra. Let $f(x), g(x) \in R[x]$, and suppose

$$f(x) = a_0 + a_1 x + a_2 x^2 + \cdots + a_m x^m$$
$$g(x) = b_0 + b_1 x + b_2 x^2 + \cdots + b_n x^n$$

Then

$$f(x) + g(x) = (a_0 + b_0) + (a_1 + b_1)x + (a_2 + b_2)x^2 + \cdots$$

and

$$f(x) \cdot g(x) = a_0 b_0 + (a_0 b_1 + a_1 b_0)x + \cdots + a_m b_n x^{m+n} \tag{2}$$

The coefficient of x^i in the product $f(x) \cdot g(x)$ is $\sum a_j b_k$, the summation being over all j, k such that $0 \le j \le m, 0 \le k \le n$, and $j + k = i$.

It is easily verified that, with these operations, $(R[x], +, \cdot)$ satisfies all the postulates of a ring. The ring $R[x]$ is called the *ring of polynomials* over R in indeterminate x.

The rule for the product of two polynomials shows that if R is a commutative ring, then the ring $R[x]$ is also commutative. If R has an identity e, then the constant polynomial $f(x) = e$ is the identity of $R[x]$. If R is an integral domain, then $R[x]$

is also an integral domain. Moreover, in this case, if $f(x)$ and $g(x)$ are polynomials of degree m and n, respectively, then $f(x)g(x)$ has degree $m + n$.

Let F be a field. Then it follows from the foregoing observations that the ring $F[x]$ is a commutative integral domain with identity. It is shown in high school algebra that, by using the so-called method of long division, we can divide one polynomial by another and get a quotient and a remainder. The same method works for polynomials in $F[x]$. In other words, the Euclidean algorithm holds in $F[x]$. We record this fact in the following theorem.

THEOREM 0.8.2 Let F be a field. Then $F[x]$ is a commutative integral domain with identity. Moreover, given polynomials $f(x)$, $g(x) \in F[x]$, there exist polynomials $q(x)$ and $r(x)$ such that

$$f(x) = q(x)g(x) + r(x)$$

and either $r(x) = 0$ or $\deg r(x) < \deg g(x)$.

Given a nonzero polynomial $f(x) \in F[x]$, we say that $f(x)$ is *reducible over* F if there exist polynomials $g(x)$, $h(x) \in F[x]$ of degree greater than 0, such that $f(x) = g(x)h(x)$. If no such polynomials exist, $f(x)$ is said to be *irreducible*. In other words, $f(x)$ is irreducible if $f(x) = g(x)h(x)$ implies that $f(x)$ or $g(x)$ is a constant polynomial.

It should be noted that the same polynomial may be reducible over one field but irreducible over another. For example, the polynomial $x^2 + 1$ is reducible over the field $\mathbb{C}$ for $x^2 + 1 = (x + i)(x - i)$. But it is irreducible over the field $\mathbb{R}$.

Given a nonzero polynomial $f(x) \in F[x]$, an element $\alpha \in F$ is called a *root* of $f(x)$ in F if $f(\alpha) = 0$. If α is a root of $f(x)$, then, by using the Euclidean algorithm, we find that $f(x)$ can be factored as $f(x) = q(x)(x - \alpha)$. Hence it follows that if $f(x)$ is of degree n, then $f(x)$ cannot have more than n roots.

Suppose $f(x)$ is a polynomial of degree 2 or 3. If it is reducible, one of the factors must be of degree 1. Hence $f(x)$ is reducible over F if and only if it has a root in F.

The greatest common divisor (gcd) of two polynomials is defined in the same manner as for integers. Given $f(x)$, $g(x) \in F[x]$, we can find their gcd by using the Euclidean algorithm. If $\gcd(f(x), g(x))$ is a constant, then there exist polynomials $a(x)$, $b(x) \in F[x]$ such that $a(x)f(x) + b(x)g(x) = 1$.

After this preliminary discussion of polynomial rings, we now turn to our main objective of constructing finite fields. We first show that given any field F, we can construct a larger field K that contains F as its subfield. This process is referred to as *field extension*.

We saw in Theorem 0.7.13 that the ring $\mathbb{Z}$ is a principal ideal ring. The proof of that theorem shows that this property is a consequence of the division algorithm in $\mathbb{Z}$ and the well-ordering principle for positive integers. Hence, by using the same arguments, we can similarly prove that every ideal in the ring $F[x]$ is a principal ideal. If I is an ideal in the ring $F[x]$, then $I = (p(x))$, where $p(x)$ is a polynomial of the smallest degree in I.

Let F be a field. Let $p(x) \in F[x]$, and let $I = (p(x))$ be the ideal generated by $p(x)$; that is,

$$I = p(x)F[x] = \{p(x)g(x) \mid g(x) \in F[x]\}$$

Then, by Theorem 0.7.11, we get the quotient ring $F[x]/I$. We show in the following theorem that the quotient ring $F[x]/I$ is a field if the polynomial $p(x)$ is irreducible.

In the sequel, we denote the coset $f(x) + I$ by $\overline{f(x)}$; that is,

$$\overline{f(x)} = f(x) + (p(x)) = \{f(x) + p(x)g(x) \mid g(x) \in F[x]\}$$

So $F[x]/I = \{\overline{f(x)} \mid f(x) \in F[x]\}$. For any $f(x) \in I$, $\overline{f(x)} = I = \overline{0}$. In particular, $\overline{p(x)} = \overline{0}$.

THEOREM 0.8.3 Let F be a field. Let $p(x) \in F[x]$ be an irreducible polynomial over F. Then the quotient ring $K = F[x]/(p(x))$ is a field. Moreover, the field K has a subfield isomorphic with the field F.

Proof: Let $f(x) \in F[x]$, and suppose $\overline{f(x)} \neq \overline{0}$. Then $p(x)$ is not a factor of $f(x)$. Since $p(x)$ is irreducible, $\gcd(f(x), p(x))$ is a constant. Hence there exist polynomials $a(x), b(x) \in F[x]$ such that

$$a(x)f(x) + b(x)p(x) = 1$$

Writing the coset determined by each side of this equation, we have

$$\overline{a(x)}\,\overline{f(x)} + \overline{b(x)}\,\overline{p(x)} = \overline{1}$$

But $\overline{p(x)} = \overline{0}$, so $\overline{a(x)}\,\overline{f(x)} = \overline{1}$. Thus every nonzero element in K is invertible, which proves that the ring K is a field.

We can treat every $a \in F$ as a constant polynomial in $F[x]$. We write $\bar{F} = \{\bar{a} \in K \mid a \in F\}$ and define the mapping $\phi : F \to K$ by $\phi(a) = \bar{a}$. Then ϕ is easily seen to be an injective homomorphism. Hence $F \simeq \text{Im}\,\phi = \bar{F}$, and $\bar{F}$ is a subfield of K. This completes the proof of the theorem. ∎

Because of the isomorphism $F \simeq \bar{F}$ shown above, we can identify each element a in F with the corresponding element $\bar{a}$ in K. Then the field F itself becomes a subfield of K. This process is similar to the way in which we identify each real number a with the complex number $a + 0i$ and thereby make the field $\mathbb{R}$ a subfield of the field $\mathbb{C}$. Thus we have shown that, given any field F, we can construct a larger field K that contains F as a subfield.

We now give a representation of the field K obtained above that makes it convenient to write its elements and add or multiply two elements. Let $p(x) \in F[x]$

be an irreducible polynomial of degree m, and suppose

$$p(x) = c_0 + c_1 x + \cdots + c_m x^m$$

Given $f(x) \in F[x]$, by Euclidean algorithm, there exist $q(x), r(x) \in F(x)$ such that $f(x) = q(x)p(x) + r(x)$, where $r(x) = 0$ or $\deg r(x) < m$. So $r(x)$ is a polynomial of the form

$$r(x) = a_0 + a_1 x + \cdots + a_{m-1} x^{m-1}$$

where $a_0, a_1, \ldots, a_{m-1} \in F$. Hence, on using $\overline{p(x)} = \overline{0}$, we have

$$\overline{f(x)} = \overline{q(x)} \cdot \overline{p(x)} + \overline{r(x)}$$
$$= \overline{a_0} + \overline{a_1} \cdot \overline{x} + \cdots + \overline{a_{m-1}} \cdot \overline{x}^{m-1}$$

Let us write $t = \overline{x}$. Then $\overline{f(x)} = \overline{a_0} + \overline{a_1} t + \cdots + \overline{a_{m-1}} t^{m-1}$. Thus we have

$$K = \{\overline{f(x)} \mid f(x) \in F[x]\}$$
$$= \{\overline{a_0} + \overline{a_1} t + \cdots + \overline{a_{m-1}} t^{m-1} \mid a_0, a_1, \ldots, a_{m-1} \in F\}$$

Further, since $\overline{p(x)} = \overline{0}$, we have $\overline{c_0} + \overline{c_1} \cdot t + \cdots + \overline{c_{m-1}} \cdot t^{m-1} = \overline{0}$.

   As we have explained, we can identify $\overline{a}$ with a for each $a \in F$. Thus we get the following result:

THEOREM 0.8.4 Let F be a field. Let $p(x) = c_0 + c_1 x + \cdots + c_m x^m \in F[x]$ be an irreducible polynomial of degree m. Then the field $K = F[x]/(p(x))$ can be represented as

$$K = \{a_0 + a_1 t + \cdots + a_{m-1} t^{m-1} \mid a_0, a_1, \ldots, a_{m-1} \in F\}$$

where t satisfies the relation

$$c_0 + c_1 t + \cdots + c_m t^m = 0$$

   With this representation of the field K, the addition of any two elements in K is straightforward. To multiply two elements, we proceed as follows: Let $\alpha, \beta \in K$, and let

$$\alpha = a_0 + a_1 t + \cdots + a_{m-1} t^{m-1}$$
$$\beta = b_0 + b_1 t + \cdots + b_{m-1} t^{m-1}$$

Then, on using the distributive property, we have

$$\alpha\beta = a_0 b_0 + (a_0 b_1 + a_1 b_0) + \cdots + a_{m-1} b_{m-1} t^{2m-2}$$

Using the Euclidean algorithm, we can express the right-hand side as

$$q(t)(c_0 + c_1 t + \cdots + c_m t^m) + d_0 + d_1 t + \cdots + d_{m-1} t^{m-1}$$

So $\alpha\beta = d_0 + d_1 t + \cdots + d_{m-1} t^{m-1}$.

   Theorem 0.8.4 provides the basis for the construction of finite fields. Before doing that, however, we give another interesting application of the theorem.

THEOREM 0.8.5 The quotient ring $K = \mathbb{R}[x]/(x^2 + 1)$ is a field isomorphic with the field $\mathbb{C}$.

Proof: The polynomial $p(x) = x^2 + 1$ is obviously irreducible over $\mathbb{R}$. Hence K is a field. Moreover $\deg p(x) = 2$. Hence, by Theorem 0.8.4, the field K can be represented as

$$K = \{a_0 + a_1 t \mid a_0, a_1 \in \mathbb{R}\}$$

where t satisfies the relation $t^2 + 1 = 0$. We define the mapping $f : K \to \mathbb{C}$ by

$$f(a + bt) = a + ib$$

It is easily checked that f is an isomorphism. ∎

To construct a finite field of order p^r, we take the field F in Theorem 0.8.4 to be the field $\mathbb{Z}_p = \{0, 1, \ldots, p - 1\}$ and $p(x) \in \mathbb{Z}_p[x]$ to be an irreducible polynomial of degree r. (It can be proved that for any prime p and any positive integer r, there exists an irreducible polynomial of degree r over $\mathbb{Z}_p$.) Without loss of generality we can assume the leading coefficient in $p(x)$ to be 1. (A polynomial in which the leading coefficient is 1 is called a *monic polynomial*.)

Let $p(x) = c_0 + c_1 x + \cdots + x^r$. Then the field $K = \mathbb{Z}_p[x]/(p(x))$ can be represented as

$$K = \{a_0 + a_1 t + \cdots + a_{r-1} t^{r-1} \mid a_0, a_1, \ldots, a_{r-1} \in \mathbb{Z}_p\}$$

where t satisfies the relation $c_0 + c_1 t + \cdots + t^r = 0$. Since $\mathbb{Z}_p$ has p elements, it follows that the field K has p^r elements. We record this result in the following theorem.

THEOREM 0.8.6 Let p be a prime and r a positive integer. Then there exists a finite field $\mathbb{F}_{p^r}$ of order p^r given by

$$\mathbb{F}_{p^r} = \{a_0 + a_1 t + \cdots + a_{r-1} t^{r-1} \mid a_0, a_1, \ldots, a_{r-1} \in \mathbb{Z}_p\}$$

and t satisfies the relation $p(t) = 0$, where $p(x) \in \mathbb{Z}_p[x]$ is a monic irreducible polynomial of degree r.

It is easy to check whether a polynomial $p(x)$ of degree 2 or 3 is irreducible over $\mathbb{Z}_p$ because $p(x)$ is reducible if and only if it has a root in $\mathbb{Z}_p$. So we have to check for each $a \in \mathbb{Z}_p$ whether $p(a) = 0$. For example, consider the polynomial $x^2 + 1$. It is reducible over $\mathbb{Z}_2$ since $1^2 + 1 = 0$ (in $\mathbb{Z}_2$). Indeed, we verify that $x^2 + 1 = (x + 1)(x + 1)$ in $\mathbb{Z}_2[x]$. But $x^2 + 1$ is irreducible in $\mathbb{Z}_3[x]$ because none of the elements 0, 1, 2 is a root. We can similarly show that $x^2 + x + 1$ is irreducible in $\mathbb{Z}_2[x]$ but reducible in $\mathbb{Z}_3[x]$.

Example 0.8.7 Find a field of nine elements.

Solution. Since $9 = 3^2$, we need an irreducible polynomial of degree 2 in $\mathbb{Z}_3[x]$. We have seen that $x^2 + 1$ satisfies the requirement. Hence $\mathbb{F}_9$ is given by

$$\mathbb{F}_9 = \{a + bt \mid a, b \in \mathbb{Z}_3\}$$

where t satisfies the relation $t^2 + 1 = 0$. Assigning all possible values to a, b, we have

$$\mathbb{F}_9 = \{0, 1, 2, t, 1 + t, 2 + t, 2t, 1 + 2t, 2 + 2t\}$$

To multiply two elements in $\mathbb{F}_9$, we make use of the relation $t^2 + 1 = 0$. For example, $(1 + t)(1 + 2t) = 1 + t + 2t + 2t^2 = 1 + 2t^2 = 1 + 2(-1) = 1 - 2 = 2$.

We now prove an important property of the multiplicative group of a finite field. By definition, the nonzero elements of a field F form a group under multiplication. The next theorem shows that if F is a finite field, then this group is cyclic. To prove it, we need the following result in group theory. Let a, b be elements of orders m, n, respectively, in an abelian group G. Then there exists an element in G whose order is $\text{lcm}(m, n)$.

We write $d = \gcd(m, n)$ and $q = mn/d = \text{lcm}(m, n)$. Let $c = a^d b$. Then $c^q = a^{dq} b^q = (a^m)^n (b^n)^{m/d} = e$ and hence $o(c) \mid q$. We write $r = o(c)$. Then $c^r = e$ and hence $a^{dr} = b^{-r}$. So $o(a^{dr}) = o(b^{-r}) = o(b^r) = t$ (say). Now $o(a^d) = m/d$ and $o(b) = n$. Hence t divides m/d and n. But m/d and n are relatively prime. Hence $t = 1$, so $(a^d)^r = e = b^r$. Hence $\frac{m}{d} \mid r$ and $n \mid r$. Therefore $q \mid r$. This proves that $o(c) = r$.

THEOREM 0.8.8 Let F be a finite field, and let F^* denote the set of nonzero elements in F. Then $(F^*, \cdot)$ is a cyclic group.

Proof: Let $|F| = n$. Then F^* is a finite abelian group of $n - 1$ elements. Let q be the lcm of the orders of all elements of F^*. Then $a^q = 1$ for all $a \in F^*$. Hence every element in F^* is a root of the polynomial $x^q - 1$. But a polynomial of degree q can have at most q roots. Hence $n - 1 \leq q$. By the result proved above, there exists an element c in F^* of order q. Hence q divides $n - 1$, and this proves $q = n - 1$. F^* is a cyclic group generated by c. ∎

Definition 0.8.9 Let F be a finite field. An element $a \in F$ is called a *primitive element* if a is a generator of the cyclic group $(F^*, \cdot)$.

An irreducible polynomial $p(x) \in \mathbb{Z}_p[x]$ is called *primitive* if $\overline{x} = x + (p(x))$ is a primitive element in the field $\mathbb{Z}_p[x]/(p(x))$.

For example, the primitive elements in the field $\mathbb{Z}_{11}$ are 2, 6, 7, and 8 (see Example 0.3.9). In the group $\mathbb{F}_9$, described in Example 0.8.7, $t^2 = 2$ and $t^4 = 1$;

hence $o(t) = 4$. Thus t is not a primitive element in the field. Hence $x^2 + 1$ is not a primitive polynomial in $\mathbb{Z}_3[x]$. Further, $(t + 1)^2 = 2t$, $(t + 1)^4 = t^2 = 2$, and $(t + 1)^8 = 1$; hence $o(t + 1) = 8$. Thus $t + 1$ is a primitive element. By Theorem 0.3.7, the other primitive elements in the field are $(t + 1)^3$, $(t + 1)^5$, and $(t + 1)^7$—that is, $2t + 1$, $2t + 2$, and $t + 2$.

EXERCISES 0.8

1. Find all ideals in the polynomial ring $F[x]$ over a field F.

2. Prove that $\mathbb{Z}[x]/(x^2 + 1) \simeq \mathbb{Z}[i]$, where $\mathbb{Z}[i] = \{a + bi \mid a, b \in Z\}$ (ring of *Gaussian integers*).

3. Show that $x^3 + x + 1$ is irreducible over $\mathbb{Z}_2$. Is it also primitive? Use this polynomial to construct a field F of eight elements. Write the addition and multiplication tables for F, and find all primitive elements in F.

4. Find all irreducible polynomials of degree 3 over $\mathbb{Z}_2$.

5. Show that every irreducible polynomial of degree 2, 3, 5, or 7 over $\mathbb{Z}_2$ is primitive.

6. Find all irreducible polynomials of degrees 2 or 3 over $\mathbb{Z}_3$. Which of these are primitive?

7. Construct a field of nine elements by using a primitive polynomial.

8. Construct a field of 27 elements. Find all primitive elements in the field.

CHAPTER

1

BOOLEAN ALGEBRAS AND SWITCHING CIRCUITS

Boolean algebra had its origin in the ideas presented by the philosopher George Boole in his book *Investigation of the Laws of Thought* (1854) on treating logic with algebraic methods. In recent years, however, it has found more interesting applications in the theory of switching circuits used in electrical engineering and computer science. The simplest mathematical model for a Boolean algebra is furnished by the algebra of sets.

1.1 BOOLEAN ALGEBRAS

Definition 1.1.1 A *Boolean algebra* is an algebraic system $(B, \vee, \wedge, ')$, where B is a nonempty set, $\vee$ and $\wedge$ are binary operations on B, and $'$ is a unary operation on B, with the following properties:

(B1) $x \vee y = y \vee x$ and $x \wedge y = y \wedge x$ for all $x, y \in B$.

(B2) $x \wedge (y \vee z) = (x \wedge y) \vee (x \wedge z)$ and $x \vee (y \wedge z) = (x \vee y) \wedge (x \vee z)$ for all $x, y, z \in B$.

(B3) There exist elements $0, 1 \in B, 0 \neq 1$, such that $0 \vee x = x$ and $1 \wedge x = x$ for all $x \in B$.

(B4) $x \wedge x' = 0$ and $x \vee x' = 1$ for all $x \in B$.

The binary operations $\vee$ and $\wedge$ are called *join* and *meet*, respectively. (The symbols $\vee$ and $\wedge$ may be read "cup" and "cap.") Postulates (B1) and (B2) in the definition say that join and meet are both commutative operations and each is distributive over the other. Postulate (B3) says that both the operations have identities, 0 being the identity for $\vee$ and 1 the identity for $\wedge$; 0 and 1 are called *zero* and *unity,* respectively. (Note that the symbols 0 and 1 here do not stand for the integers 0 and 1.)

The unary operation $'$ is called *complementation.* Given an element x in B, if there exists an element $y \in B$ such that $x \vee y = 1$ and $x \wedge y = 0$, then y is called a *complement* of x. Thus postulate (B4) says that each element x in B has a complement—namely x'. We shall presently show that x' is in fact the unique complement of x.

The following theorem furnishes the most important example of a Boolean algebra.

THEOREM 1.1.2 Let $B = \mathcal{P}(X)$ be the set of all subsets of a nonempty set X. For all $a, b \in B$, define

$$a \vee b = a \cup b, \quad a \wedge b = a \cap b, \quad a' = X - a = \{x \in X \mid x \notin a\}$$

Then $(B, \vee, \wedge)$ is a Boolean algebra.

Proof: We know that union and intersection of sets are commutative operations, and each is distributive over the other. Hence postulates (B1) and (B2) in Definition 1.1.1 are satisfied here. For any subset a of X, $a \cup \emptyset = a$ and $a \cap X = a$. Moreover $a \cup a' = X$ and $a \cap a' = \emptyset$. Hence postulates (B3) and (B4) are satisfied on taking $0 = \emptyset$ and $1 = X$. This proves that B is a Boolean algebra with $\emptyset$ and X as the zero and unity elements. ∎

This Boolean algebra is called the *power set algebra* of X. It is in fact the model on which the general definition of a Boolean algebra is based. The symbols $\vee, \wedge$ and the terms *join, meet* are echos of the corresponding symbols $\cup, \cap$ and the terms *union, intersection.* Moreover, it can be shown that any Boolean algebra is isomorphic with a subset of the power set algebra $\mathcal{P}(X)$ for some X.

Consider now the special case where X is a singleton. Then $\mathcal{P}(X) = \{\emptyset, X\} = \{0, 1\}$. We shall write $\mathbb{B}$ for this Boolean algebra consisting of only two elements 0, 1. The operations $\vee, \wedge,$ and $'$ in $\mathbb{B}$ are as shown here:

$$0 \vee 0 = 0, \quad 0 \vee 1 = 1 = 1 \vee 0, \quad 1 \vee 1 = 1$$
$$0 \wedge 0 = 0, \quad 0 \wedge 1 = 0 = 1 \wedge 0, \quad 1 \wedge 1 = 1$$
$$0' = 1, \quad 1' = 0$$

Since every Boolean algebra must contain the elements 0 and 1, $\mathbb{B}$ is the smallest Boolean algebra. Somewhat surprisingly, this simple algebra of only two elements has important applications in the field of switching circuits.

EXERCISES 1.1

1. Let a, b be elements in a Boolean algebra. Show that $a = a \wedge b \Leftrightarrow a \vee b = b$.

2. Let B be a Boolean algebra. For $a, b \in B$, let $a P b$ mean that $a \vee b = b$. Show that P is a partial order in the set B.

3. Let n be a product of r distinct primes. Let B be the set of all positive divisors of n. For all $a, b \in B$, define

$$a \wedge b = \gcd(a, b), \quad a \vee b = \operatorname{lcm}(a, b)$$

Prove that $(B, \vee, \wedge)$ is a Boolean algebra. What is the number of elements in B?

4. Let R be a Boolean ring with identity. (A ring R is called *Boolean* if $a^2 = a$ for all $a \in R$.) For all $a, b \in R$, define

$$a \vee b = a + b + ab, \quad a \wedge b = ab$$

Prove that $(R, \vee, \wedge)$ is a Boolean algebra.

5. Let B be a Boolean algebra. For all $a, b \in B$, define

$$a + b = (a \wedge b') \vee (a' \wedge b), \quad a \cdot b = a \wedge b$$

Prove that $(B, +, \cdot)$ is a Boolean ring.

1.2 SWITCHES AND LOGIC GATES

1.2.1 Switches

A *switch* is a device for turning electric current on and off. The ordinary (mechanical) switch consists of two terminals connected by means of conducting wires to a power source (battery or generator). When the switch is in the *off* state, there is a gap between the terminals and the current cannot flow through it. When the switch is turned *on,* the gap is closed by a conducting rod, which enables the current to flow. Therefore the two states—namely, off and on—are referred to as *open* and *closed* states, respectively. Since a switch (unless it is defective) can stay in only two equilibrium states, off and on, it is called a *bistable* device.

The diagram below is a schematic representation of a circuit that contains a single switch (in open and closed states) connected to a battery.

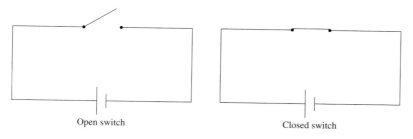

Open switch   Closed switch

We use the same terms, *open* and *closed,* for the state of a circuit that contains one or many switches. When the current can flow through the circuit or a part of it, it is said to be closed. When no current can flow, it said to be open. In the circuit shown above, the circuit is open when the switch is open, and closed when the switch is closed.

There are two ways in which two switches can be combined in a circuit, in *parallel* and in *series,* as shown below.

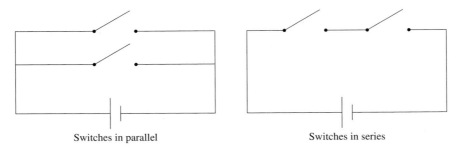

Switches in parallel   Switches in series

Let us call the two switches *A* and *B*. The following table lists the states of the two circuits for all possible states of the two switches.

Switch *A*	Switch *B*	Parallel Circuit	Series Circuit
open	open	open	open
open	closed	closed	open
closed	open	closed	open
closed	closed	closed	closed

Compare this table with the table below, which gives the join and meet of two elements in the Boolean algebra $\mathbb{B}$.

a	*b*	$a \vee b$	$a \wedge b$
0	0	0	0
0	1	1	0
1	0	1	0
1	1	1	1

If we identify the open and closed states with 0 and 1, respectively, then the two tables are identical. The parallel and series arrangements correspond with the join and meet operations. Furthermore, we can define the open and closed states as complements of each other. Then the correspondence is complete. This is the basis for the application of Boolean algebra to switching circuits.

We can express the above correspondence more mathematically as follows: Let *S* be the set consisting of the two states open and closed; that is,

$$S = \{\text{open, closed}\}$$

Let us define two binary operations, *par* and *ser* (short for *parallel* and *series*) on
S, as follows:

$$\text{open } par \text{ open} = \text{open, open } par \text{ closed} = \text{closed}$$
$$\text{closed } par \text{ open} = \text{closed, closed } par \text{ closed} = \text{closed}$$
$$\text{open } ser \text{ open} = \text{open, open } ser \text{ closed} = \text{open}$$
$$\text{closed } ser \text{ open} = \text{open, closed } ser \text{ closed} = \text{closed}$$

Further, let us define (open)$'$ = closed and (closed)$'$ = open. Then the set S, with
the operations par, ser, and $'$, is called the *switching algebra*. The switching algebra
S is isomorphic with the Boolean algebra $\mathbb{B}$.

The application of Boolean algebra to switching circuits is based on the identification of the switching algebra S with the Boolean algebra $\mathbb{B}$. We represent the
open and closed states of a switch or a circuit by the symbols 0 and 1, respectively.
A switch is then a variable x that assumes the values 0, 1 $\in \mathbb{B}$. A circuit that contains
two switches x, y is a function f of the variables x, y. If the two switches are in
parallel, then $f = x \vee y$. If they are in series, then $f = x \wedge y$.

Before considering circuits with more than two switches, let us see how the
complement of a given switch is obtained. By a complement of a switch x we mean
a switch x' whose state is always opposite to that of x; that is; x' is closed when x
is open, and open when x is closed. Such a switch is constructed by using a device
called a *relay*.

The main component in a relay, shown schematically in the diagram below,
is an elecromagnet (EM), which is essentially an iron bar with conducting wire
coiled around it. When the current flows through the wire, the EM behaves like a
magnet; otherwise, it has no magnetic property. One end of the wire is connected
to a terminal of the switch x. The switch y, which is placed against one end of the
iron bar, is closed ordinarily. When the switch x is open, no current flows through
the EM, and so y is closed. But when the switch x is turned on, the EM is activated
and its magnetic attraction pulls the lever of the switch y so that y becomes open.
Thus we see that the state of the switch y is always opposite to that of x. So y is
the complement of x; that is, $y = x'$.

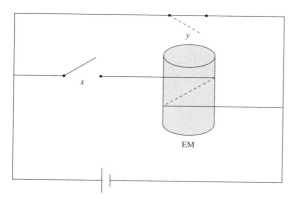

A relay provides not only the complement of a switch x but also a copy of x. Consider the switch z in the diagram below. Like the switch x, z is ordinarily in the open state. When x is closed, under the action of the EM, z is also closed. Thus the switches x and z are always in the same state. Mathematically, $z = x$.

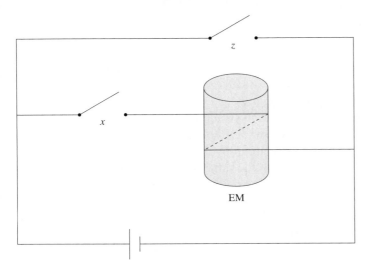

By using relays, we can obtain any number of *slave* switches, all controlled by a *master* switch x, each being a complement or a copy of x.

1.2.2 Logic Gates

Switches are mechanical devices for controlling the flow of current in a circuit. We now describe electronic devices called *gates,* which are the main part of a computer chip. Generally speaking, a *gate* is an electronic device that has one or more inputs and one output. An *input* or *output* here means electric current at either of two specific voltages. The actual values of these two voltages are immaterial for our purpose. We represent them symbolically as 0 and 1, so each input or output is 0 or 1.

The three basic gates are called *NOT gate* (or *inverter*), *AND gate,* and *OR gate*. Here are their schematic representations:

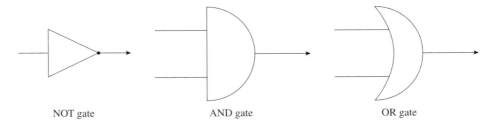

NOT gate   AND gate   OR gate

The *NOT gate* is a gate with one input and one output. When the input is 0, the output is 1; when the input is 1, the output is 0. The *AND gate* has two inputs and one output. The output is 1 if and only if both the inputs are 1; otherwise, it is 0. The *OR gate* also has two inputs and one output. The output is 0 if and only if both the inputs are 0; otherwise, it is 1. The tables below show the relation between the inputs and outputs for the three gates.

NOT Gate		AND Gate		OR Gate	
Input	**Output**	**Inputs**	**Output**	**Inputs**	**Output**
x	x'	$x\ y$	$x \wedge y$	$x\ y$	$x \vee y$
0	1	0 0	0	0 0	0
1	0	0 1	0	0 1	1
		1 0	0	1 0	1
		1 1	1	1 1	1

If we denote the inputs by x, y and the output by f, we have the following rules for the output as a function of the input variables in the three cases:

NOT gate: input x, output $f = x'$

AND gate: inputs x, y, output $f = x \wedge y$

OR gate: inputs x, y, output $f = x \vee y$

Before proceeding further with switching circuits and logic gates, we present some more properties of a Boolean algebra.

1.3 LAWS OF BOOLEAN ALGEBRA

We notice that the definition of a Boolean algebra is symmetric with respect to the operations $\vee$ and $\wedge$. If we interchange $\vee$, $\wedge$ and the corresponding identities 0, 1, the definition remains unaltered. This fact has the following important consequence, referred to as the *duality principle*. Let T be a theorem that holds in general for any Boolean algebra. Let $\bar{T}$ be the statement obtained on interchanging $\vee$ with $\wedge$ and 0 with 1 in the statement of the theorem T. Then $\bar{T}$ is also a theorem that holds for every Boolean algebra. It is called the *dual* of T. Thus for every theorem we prove, we get its dual as a bonus. If $\bar{T}$ is identical with T, then T is said to be *self-dual*.

We take advantage of the duality principle in proving the following theorem. Every statement in this theorem [except (b5), which is self-dual] consists of two parts, each being the dual of the other. By the duality principle, it suffices to prove any one of the two.

THEOREM 1.3.1 Let B be a Boolean algebra, and let x, y, $z \in B$. Then

(b1) $x \vee x = x = x \wedge x$
(b2) $0 \wedge x = 0$, $1 \vee x = 1$
(b3) $x \wedge (x \vee y) = x = x \vee (x \wedge y)$
(b4) $x \vee (y \vee z) = (x \vee y) \vee z$, $x \wedge (y \wedge z) = (x \wedge y) \wedge z$
(b5) $(x')' = x$
(b6) $0' = 1$, $1' = 0$
(b7) $(x \vee y)' = x' \wedge y'$, $(x \wedge y)' = x' \vee y'$

Proof:

(b1)

$$\begin{aligned}
x \vee x &= 1 \wedge (x \vee x) \quad \text{(by B3 in Definition 1.1.1)} \\
&= (x \vee x') \wedge (x \vee x) \quad \text{(by B4)} \\
&= x \vee (x' \wedge x) \quad \text{(by B2)} \\
&= x \vee 0 \quad \text{(by B4)} \\
&= x \quad \text{(by B3)}
\end{aligned}$$

(b2)

$$\begin{aligned}
0 \wedge x &= 0 \vee (0 \wedge x) \quad \text{(by B3)} \\
&= (x' \wedge x) \vee (0 \wedge x) \quad \text{(by B4)} \\
&= (x' \vee 0) \wedge x \quad \text{(by B2)} \\
&= x' \wedge x \quad \text{(by B3)} \\
&= 0 \quad \text{(by B4)}
\end{aligned}$$

(b3)

$$\begin{aligned}
x \wedge (x \vee y) &= (x \vee 0) \wedge (x \vee y) \quad \text{(by B3)} \\
&= x \vee (0 \wedge y) \quad \text{(by B2)} \\
&= x \vee 0 \quad \text{(by b2)} \\
&= x \quad \text{(by B3)}
\end{aligned}$$

(b4)   Write $a = (x \vee y) \vee z$ and $b = x \vee (y \vee z)$. Then

$$\begin{aligned}
x \wedge a &= x \wedge ((x \vee y) \vee z) \\
&= (x \wedge (x \vee y)) \vee (x \wedge z) \quad \text{(by B2)} \\
&= x \vee (x \wedge z) \quad \text{(by b3)} \\
&= x \quad \text{(by b3)}
\end{aligned}$$

Similarly $y \wedge a = y$ and $z \wedge a = z \wedge (z \vee (x \wedge y)) = z$. Hence

$$a \wedge b = a \wedge (x \vee (y \vee z))$$
$$= (a \wedge x) \vee (a \wedge (y \vee z)) \quad \text{(by B2)}$$
$$= (a \wedge x) \vee ((a \wedge y) \vee (a \wedge z)) \quad \text{(by B2)}$$
$$= x \vee (y \vee z)$$
$$= b$$

We can similarly prove $b \wedge a = a$. Hence $a = b \wedge a = a \wedge b = b$.

(b5) We first show that x' is the unique complement of x. Suppose y is a complement of x. Then $x \vee y = 1$ and $x \wedge y = 0$. Hence

$$y = 1 \wedge y \quad \text{(by B3)}$$
$$= (x \vee x') \wedge y \quad \text{(by B4)}$$
$$= (x \wedge y) \vee (x' \wedge y) \quad \text{(by B2)}$$
$$= 0 \vee (x' \wedge y)$$
$$= (x' \wedge x) \vee (x' \wedge y) \quad \text{(by B4)}$$
$$= x' \wedge (x \vee y) \quad \text{(by B2)}$$
$$= x' \wedge 1$$
$$= x' \quad \text{(by B3)}$$

So x' is the unique complement of x. This holds for any element x in B. Therefore $(x')'$ is the unique complement of x'. By B4, x is a complement of x', hence $x = (x')'$.

(b6) By B3, $0 \vee 1 = 1, 0 \wedge 1 = 0$, and hence 0 and 1 are complements of each other. Therefore, by the uniqueness of a complement, $0' = 1$ and $1' = 0$.

(b7) We show that $x' \wedge y'$ is the complement of $x \vee y$. Now

$$(x \vee y) \vee (x' \wedge y') = ((x \vee y) \vee x') \wedge ((x \vee y) \vee y') \quad \text{(by B2)}$$
$$= ((x \vee x') \vee y) \wedge (x \vee (y \vee y')) \quad \text{(by b4)}$$
$$= (1 \vee y) \wedge (x \vee 1) \quad \text{(by B4)}$$
$$= 1 \wedge 1 \quad \text{(by b2)}$$
$$= 1 \quad \text{(by B3)}$$

and

$$(x \vee y) \wedge (x' \wedge y') = (x \wedge (x' \wedge y')) \vee (y \wedge (x' \wedge y')) \quad \text{(by B2)}$$
$$= ((x \wedge x') \wedge y') \vee (x' \wedge (y \wedge y')) \quad \text{(by b4)}$$
$$= (0 \wedge y') \vee (x' \wedge 0) \quad \text{(by B4)}$$
$$= 0 \vee 0 \quad \text{(by b2)}$$
$$= 0 \quad \text{(by B3)}$$

So $x' \wedge y'$ is the complement of $x \vee y$. Hence $(x \vee y)' = x' \wedge y'$.
This completes the proof of the theorem. ∎

We shall refer to the four postulates (B1)–(B4) in Definition 1.1.1 and the seven statements (b1)–(b7) in Theorem 1.3.1 as the *laws of a Boolean algebra*.

1.3.1 New Notation

The notation that we have used so far is the one that is normally used by algebraists. But computer scientists and electrical engineers use a different notation, in which the join and meet of x, y are written as $x + y$ and xy and called *sum* and *product*, respectively. Since we intend to study the application of Boolean algebra to switching circuits and computer circuits, we shall now switch to the latter notation. For ready reference, we present the laws of Boolean algebra in the new notation.

(**B1**) $x + y = y + x$; $\quad xy = yx$ (commutative laws)
(**B2**) $x(y + z) = xy + xz$; $\quad x + yz = (x + y)(x + z)$ (distributive laws)
(**B3**) $0 + x = x = 1x$ (identity elements)
(**B4**) $x + x' = 1$; $\quad xx' = 0$ (complements)
(**b1**) $x + x = x = xx$ (idempotency)
(**b2**) $0x = 0$; $\quad 1 + x = 1$ (dominant elements)
(**b3**) $x(x + y) = x = x + xy$ (absorption)
(**b4**) $(x + y) + z = x + (y + z)$; $\quad (xy)z = x(yz)$ (associative laws)
(**b5**) $(x')' = x$ (involution)
(**b6**) $0' = 1$; $\quad 1' = 0$
(**b7**) $(x + y)' = x'y'$; $\quad (xy)' = x' + y'$ (DeMorgan's laws)

Note that we assume the usual arithmetic convention about the precedence of multiplication over addition. Thus, in (B2), we have written simply $xy + xz$ and $x + yz$ instead of $(xy) + (xz)$ and $x + (yz)$.

Some of these laws of Boolean algebra are amenable to generalization. To begin with, the associative laws (b4) have an obvious generalization as follows: Given any n elements $x_1, \ldots, x_n$ in B, the sum $x_1 + \cdots + x_n$ and the product $x_1 \cdots x_n$ have unambiguous meanings without the parentheses. Further, it follows from the commutative laws (B1) that the order in which the elements $x_1, \ldots, x_n$ are written in the sum or the product is immaterial.

The distributive laws (B2) can be generalized as follows: The proof requires straightforward induction and is left as an exercise.

$$x(y_1 + \cdots + y_n) = xy_1 + \cdots + xy_n$$
$$x + y_1 \cdots y_n = (x + y_1) \cdots (x + y_n)$$

Likewise, by commutative laws, we have

$$(x_1 + \cdots + x_n)y = x_1 y + \cdots + x_n y$$
$$x_1 \cdots x_n + y = (x_1 + y) \cdots (x_n + y)$$

More generally, combining these results, we have

$$(x_1 + \cdots + x_m)(y_1 + \cdots + y_n) = \sum_{i=1}^{m} \sum_{j=1}^{n} x_i y_j$$

$$x_1 \cdots x_m + y_1 \cdots y_n = \prod_{i=1}^{m} \prod_{j=1}^{n} (x_i + y_j)$$

DeMorgan's laws (b7) are generalized as follows.

$$(x_1 + \cdots + x_n)' = x_1' \cdots x_n'$$
$$(x_1 \cdots x_n)' = x_1' + \cdots + x_n'$$

EXERCISES 1.3

Express each Boolean expression as a product of the sums of the variables.

1. $x_2 + x_1 x_3 x_4$

2. $x_1 x_2 + x_1' x_2'$

3. $x_1 x_2 + x_1' x_3$

4. $x_1 x_2' x_3 + x_1' x_2 x_3$

5. $x_1 x_2 + x_2 x_3 + x_1 x_3$

1.4 BOOLEAN POLYNOMIALS AND BOOLEAN FUNCTIONS

1.4.1 Boolean Polynomials

Speaking informally, a Boolean polynomial in n indeterminates $x_1, \ldots, x_n$ is any expression formed by using the symbols $x_1, \ldots, x_n$ with $+, \cdot, '$ (and parentheses wherever necessary to prevent ambiguity). Here $\cdot$ stands for multiplication (that is, meet), but usually it is not written. For example, $x(y + z') + xy'z$ is a Boolean polynomial in x, y, z. The formal definition follows.

Definition 1.4.1 A *Boolean polynomial* in n indeterminates $x_1, \ldots, x_n$ is defined recursively by the following rules.

(a) x_i is a Boolean polynomial for each $i = 1, \ldots, n$.

(b) If p and q are Boolean polynomials, then $p + q$ and pq are Boolean polynomials.

(c)   If p is a Boolean polynomial, then p' is a Boolean polynomial.
(d)   0 and 1 are Boolean polynomials.

Two Boolean polynomials are said to be equal if one of them can be transformed into the other by using the laws of Boolean algebra, (B1)–(B4) and (b1)–(b7), and treating the indeterminates $x_1, \ldots, x_n$ as if they were elements of a Boolean algebra. For example, $x(yz)' = xy' + xz'$.

In applications of Boolean algebra to switching circuits, it is often required to simplify a given polynomial. If p is a Boolean polynomial in indeterminates $x_1, \ldots, x_n$, we refer to the symbols $x_1, \ldots, x_n$ and their primes $x_1', \ldots, x_n'$ as *literals*. In counting the number of literals in a polynomial, we count each literal as many times as it occurs. By *simplifying* a polynomial p, we mean finding a polynomial q that is equal to p and has fewer literals than p. In addition to the distributive and absorption properties, (B2) and (b3), the following identities are helpful in simplifying a Boolean polynomial:

THEOREM 1.4.2   Let x, y, z be elements in a Boolean algebra. Then

(a) $xy + xy' = x = (x + y)(x + y')$
(b) $x + x'y = x + y, \quad x(x' + y) = xy$
(c) $xy + x'z + yz = xy + x'z, \quad (x + y)(x' + z)(y + z) = (x + y)(x' + z)$

Proof:

(a) $xy + xy' = x(y + y') = x1 = x.$
   $(x + y)(x + y') = x + yy' = x + 0 = x.$
(b) $x + x'y = (x + x')(x + y) = 1(x + y) = x + y.$
   $x(x' + y) = xx' + xy = 0 + xy = xy.$
(c) $xy + x'z + yz = xy + x'z + (x + x')yz = xy + xyz + x'z + x'zy$
   $= xy + x'z.$

The second part follows by duality.   ∎

Example 1.4.3   Simplify the polynomial

$$p = yz + xy'zw + zw' + xz' + x'yz' + y'z'w'$$

Solution.   Using the distributive property (B2), we get

$$p = (y + xy'w + w')z + (x + x'y + y'w')z'$$

Now, using Theorem 1.4.2, we have

$$y + xy'w + w' = y + xw + w' = y + x + w'$$

and

$$x + x'y + y'w' = x + y + y'w' = x + y + w'$$

Hence

$$p = (x + y + w')(z + z') = x + y + w'$$

1.4.2 Canonical Forms of a Boolean Polynomial

We now introduce two canonical forms of a Boolean polynomial, called disjunctive normal form and conjunctive normal form.

A *minterm* in n indeterminates $x_1, \ldots, x_n$ is defined to be a Boolean polynomial m of the form

$$m = \xi_1 \cdots \xi_n \quad \text{where } \xi_i = x_i \text{ or } x_i' \qquad \text{for each } i = 1, \ldots, n$$

Because each factor in m can be chosen in two ways, there are exactly 2^n distinct minterms in $x_1, \ldots, x_n$.

Definition 1.4.4 A Boolean polynomial p in n indeterminates $x_1, \ldots, x_n$ is said to be in *disjunctive normal form* if $p = 0$ or if p is of the form

$$p = m_1 + \cdots + m_k$$

where $m_1, \ldots, m_k$ are distinct minterms in $x_1, \ldots, x_n$ ($1 \leq k \leq 2^n$).

The *conjunctive normal form* is defined dually.

A *maxterm* in n indeterminates $x_1, \ldots, x_n$ is a Boolean polynomial M of the form

$$M = \xi_1 + \cdots + \xi_n \quad \text{where } \xi_i = x_i \text{ or } x_i' \qquad \text{for each } i = 1, \ldots, n$$

As before, there are 2^n distinct maxterms in $x_1, \ldots, x_n$.

A Boolean polynomial p in n indeterminates $x_1, \ldots, x_n$ is said to be in *conjunctive normal form* if $p = 1$, or if p is of the form

$$p = M_1 \cdots M_k$$

where $M_1, \ldots, M_k$ are distinct maxterms in $x_1, \ldots, x_n$ ($1 \leq k \leq 2^n$).

THEOREM 1.4.5 Every Boolean polynomial can be expressed in disjunctive normal form as well as in conjunctive normal form.

Proof: Let p be a Boolean polynomial in n indeterminates $x_1, \ldots, x_n$. If $p = 0$, then p is in disjunctive normal form, by definition. If $p = 1$, we have

$$p = 1 \cdots 1 = (x_1 + x_1') \cdots (x_n + x_n')$$

By expanding the product on the right-hand side by using the distributive law, we obtain p as the sum of all the 2^n minterms. Now suppose p is neither 0 nor 1.

Then p can be expressed in disjunctive normal form by using the following steps in order.

Step 1. If $'$ occurs anywhere other than directly on the indeterminates $x_1, \ldots,$ x_n, we remove it by using DeMorgan's laws.

Step 2. By using the distributive law, we express p as a sum of products.

Step 3. If any product in this sum contains neither x_i nor x_i' (for some $i = 1, \ldots, n$), we multiply it by $(x_i + x_i')$, which is equal to 1, and then apply the distributive law. By repeated application of this step, we obtain p as a sum of minterms.

Step 4. If any minterm occurs more than once, we use the idempotent property (b1) to delete the repetitions.

This proves that every Boolean polynomial can be expressed in disjunctive normal form. By following a dual procedure, we can express a Boolean polynomial in conjunctive normal form. In particular, if $p = 0$, then p is equal to the product of all the 2^n maxterms.   ∎

The following example illustrates the procedure outlined in the proof of the theorem.

Example 1.4.6 Express the Boolean polynomial

$$p = x(yz')'$$

in **(a)** disjunctive normal form and **(b)** conjunctive normal form.

Solution. **(a)** Using DeMorgan's law (b7) and then the distributive property (B2), we have

$$p = x(y' + z) = xy' + xz$$

which expresses p as a sum of products. The first term xy' in this sum does not contain z or z', and the second term xz is free of y, y'. Hence we multiply the first by $z + z'$ and the second by $y + y'$, which gives us

$$p = xy'(z + z') + x(y + y')z$$
$$= xy'z + xy'z' + xyz + xy'z$$

The term $xy'z$ occurs twice. Deleting one of these [using the idempotent property (b1)], we get

$$p = xy'z + xy'z' + xyz$$

This is the disjunctive normal form of p.

(b) The conjunctive normal form of p is obtained as follows.

$$
\begin{aligned}
p &= x(y' + z) \\
&= (x + yy' + zz')(xx' + y' + z) \\
&= (x + y + z)(x + y + z')(x + y' + z)(x + y' + z')(x + y' + z)(x' + y' + z) \\
&= (x + y + z)(x + y + z')(x + y' + z)(x + y' + z')(x' + y' + z)
\end{aligned}
$$

1.4.3 Boolean Functions

A Boolean polynomial in n indeterminates determines a function of n variables on an arbitrary Boolean algebra as follows: Given a Boolean polynomial p in indeterminates $x_1, \ldots, x_n$ and a Boolean algebra B, we define

$$
p_B : B^n \to B
$$
$$
p_B(a_1, \ldots, a_n) = p(a_1, \ldots, a_n)
$$

where $p(a_1, \ldots, a_n)$ denotes the result of replacing $x_1, \ldots, x_n$ with $a_1, \ldots, a_n$, respectively, in the expression for the polynomial p. Then p_B is a function of n variables on B and is called a *Boolean function*. Thus a Boolean function is simply a Boolean polynomial where we treat the indeterminates as variables that take values in B.

Let f be a Boolean function of n variables $x_1, \ldots, x_n$ on a Boolean algebra B (that is, $f = p_B$ for some Boolean polynomial p). If we assign to each independent variable x_i the value 0 or 1, it is clear that the function f also takes the value 0 or 1; that is, $f(a_1, \ldots, a_n) \in \{0, 1\}$ for all $a_1, \ldots, a_n \in \{0, 1\}$. We shall refer to an n-tuple $(a_1, \ldots, a_n)$, where $a_1, \ldots, a_n \in \{0, 1\}$, as a *0-1 assignment* for the independent variables. Clearly, there are 2^n such assignments.

It is easily seen that for any given 0-1 assignment $(a_1, \ldots, a_n)$, there is exactly one minterm that takes the value 1—namely, the minterm

$$
m = \xi_1 \cdots \xi_n \quad \text{where } \xi_i = \begin{cases} x_i & \text{if } a_i = 1 \\ x_i' & \text{if } a_i = 0 \end{cases}
$$

All other minterms take the value 0. Likewise, there is exactly one maxterm that takes the value 0 for a given 0-1 assignment $(a_1, \ldots, a_n)$—namely,

$$
M = \xi_1 + \cdots + \xi_n \quad \text{where } \xi_i = \begin{cases} x_i & \text{if } a_i = 0 \\ x_i' & \text{if } a_i = 1 \end{cases}
$$

It is clear that the maxterm that corresponds to any given 0-1 assignment is the complement of the corresponding minterm.

Example 1.4.7 The table shows the eight ($=2^3$) 0-1 assignments for three independent variables x, y, z and the corresponding minterms and maxterms.

x	y	z	Minterm	Maxterm
1	1	1	xyz	$x' + y' + z'$
1	1	0	xyz'	$x' + y' + z$
1	0	1	$xy'z$	$x' + y + z'$
1	0	0	$xy'z'$	$x' + y + z$
0	1	1	$x'yz$	$x + y' + z'$
0	1	0	$x'yz'$	$x + y' + z$
0	0	1	$x'y'z$	$x + y + z'$
0	0	0	$x'y'z'$	$x + y + z$

THEOREM 1.4.8 Let f be a Boolean function of n independent variables. Then f can be expressed uniquely in disjunctive normal form (except for the order in which the minterms are written). Moreover,

$$f = m_1 + \cdots + m_k$$

where $m_1, \ldots, m_k$ are precisely the minterms which correspond to the 0-1 assignments where f takes the value 1.

Dually, f can be expressed uniquely in conjunctive normal form:

$$f = M_1 \cdots M_k$$

where $M_1, \ldots, M_k$ are the maxterms corresponding to the 0-1 assignments where f takes the value 0.

Proof: We have already shown in Theorem 1.4.5 that every Boolean polynomial (Boolean function) can be expressed in disjunctive normal form. To prove the uniqueness, let $f = m_1 + \cdots + m_k$, where $m_1, \ldots, m_k$ are minterms. Consider any 0-1 assignment $(a_1, \ldots, a_n)$ for the independent variables. If the function f takes the value 1 at this assignment, then one (and only one) of the minterms $m_1, \ldots, m_k$ must take the value 1 at this assignment. On the other hand, if f takes the value 0, then $m_1, \ldots, m_k$ must be all 0. Hence $m_1, \ldots, m_k$ are precisely the minterms that correspond to the 0-1 assignments where f takes the value 1. This proves that the disjunctive normal form of f is unique (except for the order in which the minterms are written). ∎

COROLLARY 1.4.9 A Boolean function is uniquely determined by the values taken by it at all the 0-1 assignments of the independent variables. Consequently, there are exactly 2^{2^n} distinct Boolean functions of n variables.

Proof: The first statement is obvious from Theorem 1.4.8. The second statement follows from the fact that there are exactly 2^n 0-1 assignments, and for each such assignment a function f can take the value 0 or 1. ∎

Note that these properties of a Boolean function are independent of the Boolean algebra B on which the function is defined. Therefore, without loss of generality, we may assume B to be $\mathbb{B} = \{0, 1\}$.

Earlier in this section (Theorem 1.4.5), we obtained a method for expressing a Boolean polynomial in disjunctive (conjunctive) normal form. Theorem 1.4.8 provides another method for doing that. Given a Boolean polynomial p in indeterminates $x_1, \ldots, x_n$, we evaluate the Boolean function $f = p(x_1, \ldots, x_n)$ at each 0-1 assignment of the variables $x_1, \ldots, x_n$. The disjunctive normal form of p (or f) is the sum of the minterms that correspond to the 0-1 assignments where f takes the value 1. Likewise, its conjunctive normal form is the product of the maxterms that correspond to the 0-1 assignments where f takes the value 0.

Example 1.4.10 Use Theorem 1.4.8 to obtain the disjunctive and conjunctive normal forms of the Boolean function

$$f = x(yz')'$$

Solution. The table below gives the value of the function $f(x, y, z) = x(yz')'$ at each 0-1 assignment of x, y, z.

x	y	z	$x(yz')'$
1	1	1	$1(11')' = (10)' = 0' = 1$
1	1	0	$1(10')' = (11)' = 1' = 0$
1	0	1	$1(01')' = (00)' = 0' = 1$
1	0	0	$1(00')' = (01)' = 0' = 1$
0	1	1	$0(11')' = 0$
0	1	0	$0(10')' = 0$
0	0	1	$0(01')' = 0$
0	0	0	$0(00')' = 0$

Thus f takes the value 1 at the assignments $(1, 1, 1)$, $(1, 0, 1)$, and $(1, 0, 0)$. The minterms corresponding to these 0-1 assignments are xyz, $xy'z$, and $xy'z'$. Hence, by Theorem 1.4.8, the disjunctive normal form of f is

$$f = xyz + xy'z + xy'z'$$

Likewise, taking the product of the maxterms corresponding to the 0-1 assignments where f has the value 0, we obtain the conjunctive normal form as

$$f = (x' + y' + z)(x + y' + z')(x + y' + z)(x + y + z')(x + y + z)$$

EXERCISES 1.4

Simplify the Boolean polynomials in Exercises 1–11.

1. $x_1x_3 + x_1x_4 + x_2x_3 + x_2x_4$
2. $(x' + y')(x' + y)(x + y)$
3. $yz + xy'zw + zw' + xz' + x'yz' + y'z'w'$
4. $xyz' + xy'z + x'y'z + x'y'z'f$
5. $xyz + xyz' + xy'z + x'yz$
6. $(x'y + x + z)(x + z)(x'y + y + z)(x + y + z)$
7. $(xu + xv + yu + yv)(x + y + v)(y + z + u)$
8. $(x_1 + x_2)(x_1' + x_3)(x_2' + x_3)$
9. $(x_1 + x_2 + x_3)(x_1 + x_2')x_3'$
10. $x_1 + x_2(x_3x_4 + x_3'x_4')$
11. $x_1'x_2' + x_1x_2 + x_1x_2'$
12. Given the Boolean equation $x_1 + x_2 + x_3 = x_1x_2x_3$, prove that $x_1 = x_2 = x_3$.
13. Prove that the Boolean equation $x_1x_2 + x_3x_2 = 1$ implies $x_1 + x_3 = 1$.
14. Let $p = x'y'z' + x'y'z + xyz'$. Find p'. Also, obtain the disjunctive and conjunctive normal forms of p'.

Express the Boolean polynomials in Exercises 15–23 in (a) disjunctive normal form and (b) conjunctive normal form.

15. $(xy' + xz)' + x'$
16. $xyz + (x + y)(x + z)$
17. $(x'y + xyz' + xy'z + x'y'z'w + w')'$
18. $(x + y')(y + z')(z + x')(x'y')$
19. $(x + y)(x + y')(x' + z)$
20. $x + y$ (in indeterminates x, y, and z)
21. $x'y'z' + x'y'z + xyz'$
22. $(x + y + z')(x' + y' + z')(x' + y' + z)$
23. $xyz + xy'z' + x'yz' + x'y'z$

1.5 SWITCHING CIRCUITS AND GATE NETWORKS

1.5.1 Switching Circuits

We saw in Section 1.2 that two switches can be combined in a circuit either in parallel or in series. Such parallel or series arrangements can be extended to more than two switches. For the sake of convenience, from now on we will show a switch in a circuit diagram by simply leaving a gap in a line and inserting the variable that represents the switch in the gap. We will also omit the battery and show only the part of the circuit that contains the switches.

The following diagram shows three switches x, y, z in parallel and four switches a, b, c, d in series.

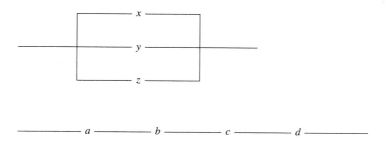

These two circuits are represented by the Boolean functions $x + y + z$ and $abcd$, respectively. The circuits may themselves be combined in parallel or in series, as shown below.

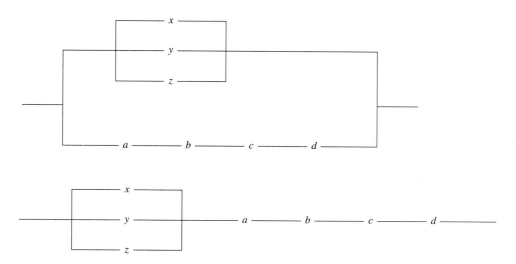

The Boolean functions that represent these circuits are $x + y + z + abcd$ and $(x + y + z)abcd$, respectively. This process can be repeated. In general, starting from simple parallel and series circuits of several switches, and successive combinations of these in parallel and series arrangements, we can obtain a complex switching circuit. Any circuit obtained in this manner is called a *series-parallel* circuit.

It is clear from the examples that any series-parallel circuit is represented by a Boolean function. We refer to this function as the *switching function* of the circuit.

Conversely, given any Boolean function f, we can draw a series-parallel circuit represented by f, provided f does not contain complements of sums or products of independent variables—that is, ' occurs on the variables alone. In case f contains a complement of a sum or a product, we first use DeMorgan's laws to remove the complementation.

It was explained in Section 1.4 that by using relays we can obtain several copies of a switch x as well as of its complement x'. This makes it possible for a circuit to have the same switch x at several places and also to have x and x' at different places. All switches marked x in the circuit are linked through relays so that they are always in the same state. Likewise, switches marked x and x' are always in opposite states.

Example 1.5.1 Write the switching function of the following circuit:

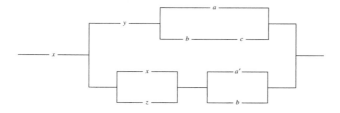

Solution.

$$f = x\{y(a + bc) + (x + z)(a' + b)\}$$

Example 1.5.2 Draw the circuit for the Boolean function

$$f = (x + a'b)' + x(y + bc)$$

Solution. We first remove the complementation over the bracket. By DeMorgan's laws,

$$(x + a'b)' = x'(a'b)' = x'(a + b')$$

Hence

$$f = x'(a + b') + x(y + bc)$$

The circuit represented by the Boolean function f is shown below.

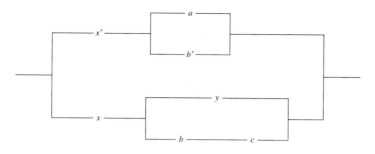

1.5.2 Gate Networks

In Section 1.2 we defined the AND gate and the OR gate with two inputs. In general, these gates can have an arbitrary number of inputs. The output of an AND gate is 1 if each input is 1; otherwise, it is 0. The output of an OR gate is 0 if each input is 0; otherwise, it is 1. In other words, the output of an AND gate with inputs $x_1, \ldots, x_n$ is given by the function $f = x_1 \cdots x_n$, and the output of an OR gate with inputs $x_1, \ldots, x_n$ is given by the function $f = x_1 + \cdots + x_n$.

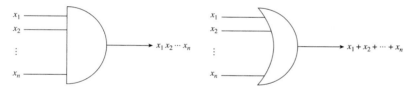

By the use of several leads, the same input can be fed into several gates. Moreover, the output from one gate can be used as input for another gate. In this manner, by interconnecting several gates, we can construct a complex network of gates that has several inputs and a single output, the output being a Boolean function of the input variables. (We can also have a network with more than one ouput.) For example, the following diagram shows a network consisting of three gates. The output of the AND gate, given by xy, and the output of the NOT gate, z', are the inputs for the OR gate, whose output is therefore given by the function $f = xy + z'$.

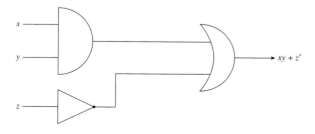

Any gate network with a single output is represented by a Boolean function that gives its output as a function of its inputs. This function is called the *output function* (or the *switching function*) of the network.

Conversely, any Boolean function f can be realized as a gate network whose output is given by the function f. (Here it is not necessary that f is free of complements of sums and products.)

Mathematically, there is no difference between switching circuits and gate networks. A Boolean function f of independent variables $x_1, \ldots, x_n$ can be represented by a switching circuit or a gate network. In the former case, the variables $x_1, \ldots, x_n$ are shown as switches and f gives the state of the circuit. In the latter, the variables are the inputs and f gives the output. Therefore the term *switching circuit* is often used to mean a gate network as well as a circuit consisting of electromechanical switches.

When the NOT gate is combined with one of the other two gates, there is a brief notation for its representation in the diagram. Instead of the usual symbol for the NOT gate, we simply draw a black disc • immediately next to the gate with which the NOT gate is combined. For example,

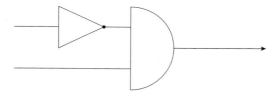

is shown as

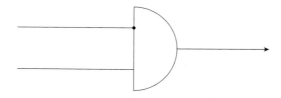

The *NAND gate* and the *NOR gate* are obtained by putting a NOT gate immediately after an AND gate and an OR gate, respectively, as shown in the diagram below.

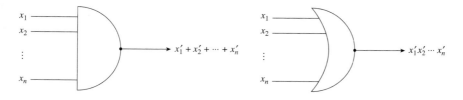

The output of the NAND gate with input variables $x_1, \ldots, x_n$ is given by the function

$$f = (x_1 \cdots x_n)' = x_1' + \cdots + x_n'$$

The output of the NOR gate with input variables $x_1, \ldots, x_n$ is given by the function

$$f = (x_1 + \cdots + x_n)' = x_1' \cdots x_n'$$

Example 1.5.3 Draw the gate network for Boolean function

$$f = (x + a'b)' + x(y + bc)$$

Solution.

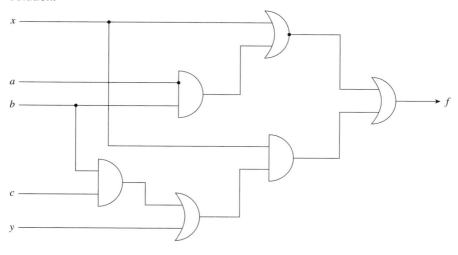

1.6 SIMPLIFICATION OF CIRCUITS

We shall use the term *contact* to denote a switch or a relay (used to obtain a duplicate copy or a complement of a switch) in a circuit. Thus a contact corresponds to a literal in the switching function. The cost of constructing a switching circuit depends on the number of contacts in it. Therefore, when a circuit is constructed for a given purpose, it is desirable to have the fewest possible contacts in it.

Two circuits are said to be *equivalent* if they function identically, in the sense that they are in identical states for every possible combination of the states of the given switches. By *simplifying* a circuit, we mean obtaining an equivalent circuit that has fewer contacts than the given circuit.

It is clear that two circuits are equivalent if and only if their corresponding switching functions are equal. Therefore, to simplify a given circuit, we first write its switching function and then try to simplify it and express it in a form with the fewest possible literals. Then we draw the circuit that realizes this simpler switching function.

Similar remarks apply to gate networks.

Example 1.6.1 Simplify the switching circuit

Solution. The switching function of the given circuit is

$$f = x + x'y + y'z$$

which, using Theorem 1.4.2, can be simplified to

$$f = x + y + y'z = x + y + z$$

Hence the required circuit is

Example 1.6.2 Simplify the gate network

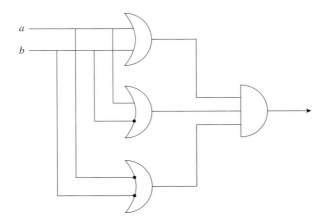

Solution. The output function of the given network is

$$f = (a + b)(a + b')(a' + b')$$

Using Theorem 1.4.2, we obtain

$$f = a(a' + b') = ab'$$

which is realized by this gate:

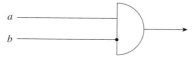

EXERCISES 1.6

Write the Boolean functions of the circuits in Exercises 1–3.

1.

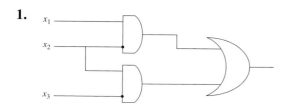

2.

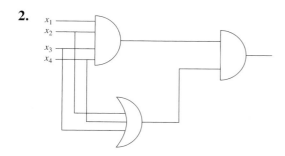

3.

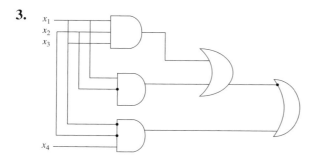

In Exercises 4–6, draw the series-parallel circuits and the corresponding gate representations of the given Boolean functions.

4. $p = x_1 x_2 + x_2 x_3 + x_3 x_1$

5. $p = x_1' x_2 x_3 + x_1 x_2' x_3 + x_1 x_2 x_3$

6. $p = x_1 x_2 x_3 + x_1' x_2 x_3' + x_1' x_2' x_3' + x_1 x_2' x_3'$

7. Draw the series-parallel circuits (as simple as possible) for each of the functions f, g, and h given in the table.

x	y	z	f	g	h
1	1	1	1	1	0
1	1	0	0	0	1
1	0	1	0	1	1
1	0	0	1	1	0
0	1	1	2	0	?
0	1	0	1	0	0
0	0	1	0	1	?
0	0	0	0	1	1

8. Draw the corresponding gate representations for the circuits in Exercise 7.

9. Draw the series-parallel circuits (as simple as possible) for the functions k, p, and q.

x	y	z	k	p	q
1	1	1	1	1	0
1	1	0	0	1	1
1	0	1	0	0	1
1	0	0	0	1	0
0	1	1	?	0	1
0	1	0	?	1	0
0	0	1	0	0	0
0	0	0	?	0	1

10. Draw the corresponding gate representations for the circuits in Exercise 9.

Simplify the circuits in Exercises 11–15.

11.

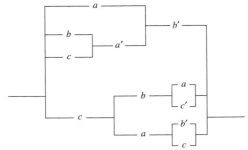

12.

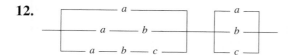

13.

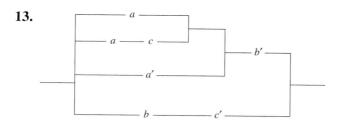

14.

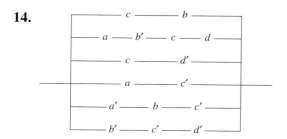

15.

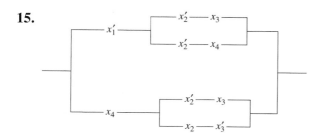

1.7 DESIGNING CIRCUITS

1.7.1 Switching Function of Circuits

We now consider the problem of designing a circuit that performs some given function. To obtain the switching function of the circuit, we need to know the state of the circuit for every possible combination of the states of the switches in the circuit. Knowing these, we prepare a table with the values of the switching function f that correspond to the 2^n 0-1 assignments of the switch variables $x_1, \ldots, x_n$. This table is referred to as the table of *closure properties* (or *functional values*) of the circuit. From this table, by using Theorem 1.4.8, we can write the function f in disjunctive (or conjunctive) normal form. We simplify this function to the extent possible and then draw the switching circuit that realizes the function.

The procedure is the same if we want to design a gate network. Here we deal with the input variables $x_1, \ldots, x_n$ and the output function f.

Example 1.7.1 Design a circuit, with the fewest possible contacts, that has the closure properties given by the values of f in the table.

x	y	z	f
1	1	1	0
1	1	0	1
1	0	1	1
1	0	0	0
0	1	1	0
0	1	0	0
0	0	1	1
0	0	0	1

Solution. Using Theorem 1.4.8 and taking the sum of the minterms corresponding to the rows where f is 1, we obtain the disjunctive normal form of f as

$$f = xyz' + xy'z + x'y'z + x'y'z'$$

Using Theorem 1.4.2, we simplify f as follows.

$$f = xyz' + xy'z + x'y'$$
$$= xyz' + y'(xz + x')$$
$$= xyz' + y'(z + x')$$

The required circuit, with six contacts, is shown below.

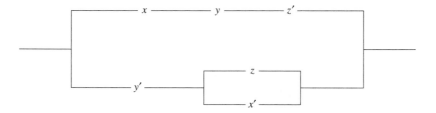

Example 1.7.2 Design a circuit for a staircase light that can be turned on and off independently from each of two switches located at the bottom and the top of the staircase.

Solution. Suppose the light is turned on when both the switches x and y are in closed state. Then, if either of the switches is turned off (the other being on), the light is turned off. When both the switches are turned off, the light is turned on again. The values of the switching function f are given in the table.

x	y	f
1	1	1
1	0	0
0	1	0
0	0	1

Hence, by Theorem 1.4.8, the function f is given by the disjunctive normal form

$$f = xy + x'y'$$

Here is the required circuit:

We note that this circuit has four contacts. But, interestingly, it can be constructed more simply by using two 3-pin switches:

Example 1.7.3 An electronic voting system is required for a committee consisting of three members. Each member is provided with a switch that is turned on or off at the proper time according to whether the member votes for or against the motion. Design a circuit such that an electric bulb lights up whenever two or all three members vote in favor of a motion.

Solution. From the conditions stated in the problem, it follows that the switching function f has the value 1 whenever any two or all three of the variables

x, y, z are 1. Otherwise, f has the value 0. The function values are given in the table.

x	y	z	f
1	1	1	1
1	1	0	1
1	0	1	1
1	0	0	0
0	1	1	1
0	1	0	0
0	0	1	0
0	0	0	0

Hence, by Theorem 1.4.8, on taking the sum of the minterms corresponding to the assignments where f is 1, we get the disjunctive normal form of f as

$$f = xyz + xyz' + xy'z + x'yz$$

We now simplify f. By using idempotency, we have

$$f = xyz + xyz + xyz + xyz' + xy'z + x'yz$$
$$= (xyz + xyz') + (xyz + xy'z) + (xyz + x'yz)$$

So, by Theorem 1.4.2,

$$f = xy + xz + yz = x(y + z) + yz$$

The required circuit has five contacts:

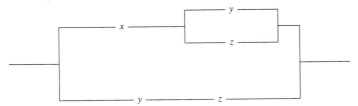

The function f can also be realized by the gate network shown below.

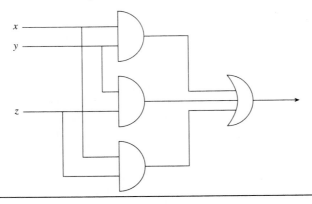

1.7.2 Exclusive-OR Gate, Half-Adder, and Adder

The *exclusive-OR gate,* briefly called the XOR gate, is a gate with two inputs such that the output is equal to 0 when the inputs are equal, and equal to 1 when they are not equal. It is symbolically represented by the diagram.

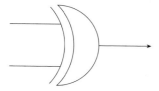

The XOR gate can be constructed by using the three basic gates as follows. The table gives the functional values of the output f of an XOR gate with inputs x, y:

x	y	f
1	1	0
1	0	1
0	1	1
0	0	0

Hence, by Theorem 1.4.8, f is given by the disjunctive normal form

$$f = xy' + x'y$$

This function is realized by the network

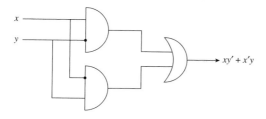

Alternatively, the function f can be expressed in the conjunctive normal form

$$f = (x' + y')(x + y)$$

So the XOR gate can also be constructed as

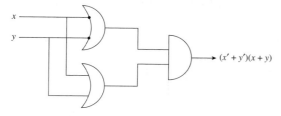

Moreover, since $x' + y' = (xy)'$, we have

$$f = (xy)'(x + y)$$

This gives yet another design for the XOR gate:

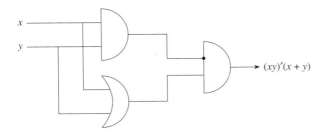

These are equivalent ways of constructing the XOR gate.

The output function of an XOR gate suggests the definition of a new binary operation in a Boolean algebra B. For any $x, y \in B$, we define $xy' + x'y$ to be the *XOR sum* of x and y and denote it by $x \oplus y$. Thus

$$x \oplus y = xy' + x'y = (x' + y')(x + y)$$

One of the essential components of the central processing unit of a computer is a circuit for adding two numbers. We give a brief description of its design. Let us examine the usual method of adding two numbers manually. Consider the sum

$$\begin{array}{r} 835 \\ 748 \\ \hline 1583 \end{array}$$

We first add the two rightmost digits 5 and 8, which gives 13. The first digit 3 is called the *local sum* and is written below the digits being added. The second digit 1 is called the *carry* and is carried over to the next pair of digits to be added. In the next step, we add 1, 3, and 4, which gives 8. Here there is no carry. In the last step, we add 8 and 7, which gives 15. Thus, at each step we add a pair of digits in the given numbers together with the carry (if any) from the previous step.

The computer follows the same iterative procedure, except that the numbers are in binary form. We refer to the digits 0 and 1 as *bits* (short for binary digits). An m-bit number $a_{m-1} \ldots a_1 a_0$, when expressed in decimal form, is equal to $a_0 + a_1 2 + \cdots + a_{m-1} 2^{m-1}$. By using m bits, we can express 2^m decimal numbers 0 to $2^m - 1$ in binary form. The table below lists the binary representations of the numbers 0 to 7.

Decimal	Binary
0	000
1	001
2	010
3	011
4	100
5	101
6	110
7	111

We first design a circuit, called a *half-adder,* to add two 1-bit numbers a, b. Four sums are involved here—namely,

$$0 + 0 = 0, \quad 0 + 1 = 1, \quad 1 + 0 = 1, \quad 1 + 1 = 10$$

The following table lists the functional values of the local sum s and the carry c in the four cases.

a	b	s	c
0	0	0	0
0	1	1	0
1	0	1	0
1	1	0	1

Hence, by Theorem 1.4.8, the functions s and c are given by the disjunctive normal forms

$$s = a'b + ab'$$
$$c = ab$$

Now $a'b + ab' = a \oplus b$ is the output of the XOR gate with inputs a, b, and ab is the output of the AND gate. The circuit for adding two 1-bit numbers is shown here.

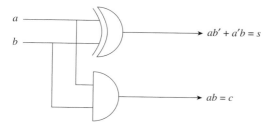

The design for a half-adder can be simplified by making use of the last of the three equivalent designs obtained earlier for an XOR gate. That design has within itself an AND gate that can give the output ab required for the carry c. Therefore, the half-adder can be built like this:

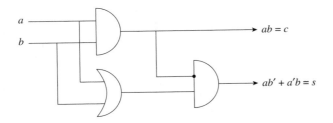

Thus we see that the XOR gate itself can function as a half-adder. We need only an extra lead for an output. We use the following symbolic representation for a half-adder.

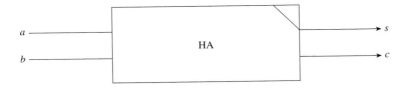

We now consider the circuit for adding three 1-bit numbers, referred to as an *adder* (or full-adder). We saw that in the iterative procedure for adding two numbers, at each step we have to add the digits a, b together with the carry c from the previous step. The following table lists the functional values of the local sum s and the new carry $\bar{c}$ when three 1-bit numbers a, b, c are added.

a	b	c	s	$\bar{c}$
0	0	0	0	0
0	0	1	1	0
0	1	0	1	0
0	1	1	0	1
1	0	0	1	0
1	0	1	0	1
1	1	0	0	1
1	1	1	1	1

Hence, by Theorem 1.4.8, the functions s and $\bar{c}$ are given by the disjunctive normal forms

$$s = a'b'c + a'bc' + ab'c' + abc$$
$$\bar{c} = a'bc + ab'c + abc' + abc$$

We simplify these functions in such a manner that we can make use of half-adders to obtain the design of a full-adder. Now

$$a'b + ab' = a \oplus b = (a + b)(a' + b')$$

and

$$a'b' + ab = (a + b)' + (a' + b')' = ((a + b)(a' + b'))' = (a \oplus b)'$$

Hence

$$s = (a'b' + ab)c + (a'b + ab')c'$$
$$= (a \oplus b)'c + (a \oplus b)c'$$
$$= (a \oplus b) \oplus c$$

Further

$$\bar{c} = (a'b + ab')c + ab(c + c')$$
$$= (a \oplus b)c + ab$$

We can obtain the outputs s and $\bar{c}$ by using two half-adders and an OR gate as follows. The first half-adder, with inputs a, b, gives outputs $(a \oplus b)$ and ab. The second half-adder, with inputs $(a \oplus b)$ and c, gives outputs $s = (a \oplus b) \oplus c$ and $(a \oplus b)c$. Finally, an OR gate, with inputs $(a \oplus b)c$ and ab, gives the output $\bar{c} = (a \oplus b)c + ab$. The diagram below shows the circuit for an adder.

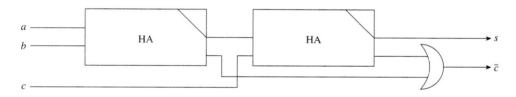

Replacing the symbol for the half-adder with its actual circuit in the above diagram, we get the following diagram for the full-adder:

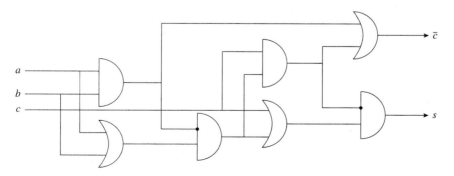

We refer to c and $\bar{c}$ as carry-in and carry-out, respectively, and denote them by c_{in} and c_{out}. We represent the adder symbolically as shown below.

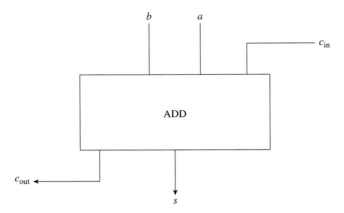

To add two m-bit numbers $a_{m-1} \ldots a_1 a_0, b_{m-1} \ldots b_1 b_0$, we use this array of m adders:

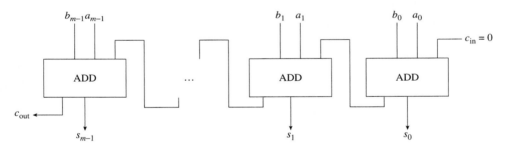

1.8 BRIDGE CIRCUITS

We have so far considered series-parallel circuits. We now give a brief account of nonseries-parallel circuits. The simplest example of such a circuit is the *bridge* circuit with five switches:

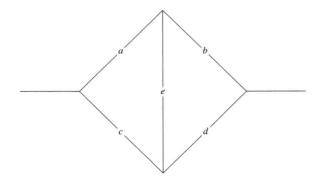

This circuit can also be drawn as

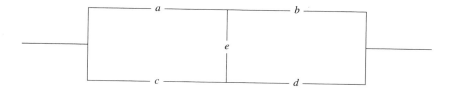

We can obtain a series-parallel circuit equivalent to this bridge circuit as follows. We first obtain the switching function f of the circuit by writing its table of closure properties, and then using Theorem 1.4.8. Since the circuit has five switches, this requires a table with $2^5 = 32$ rows. However, there is an easier approach.

Suppose the switching function f is expressed as a sum of products. Consider all possible paths that the electric current can take in flowing through the circuit. Clearly there are just four paths—namely, –a–b–, –c–d–, –a–e–d–, –c–e–b–. The current can flow along the path –a–b– if and only if switches a and b are both closed, irrespective of the states of the other switches. Hence, if f contains a term ab, then $f = 1$ when $a = b = 1$, irrespective of the values of the other variables. The same argument applies to the other paths. Hence we have

$$f = ab + cd + aed + ceb \tag{1}$$

We can now draw the series-parallel circuit represented by this switching function f. But let us first simplify f to reduce the number of literals, as follows. By absorption, $ab = ab + e$. Hence

$$f = ab + cd + aed + ceb + abe$$
$$= ab + (c + ae)(d + be)$$

Here is the series-parallel circuit representing this function:

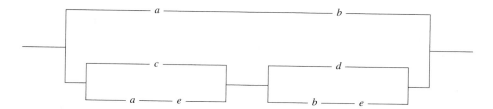

This circuit is equivalent to the bridge circuit. Since it has eight contacts, it follows that the bridge circuit is more economical than the equivalent series-parallel circuit.

More generally, any circuit obtained by replacing one or more individual switches in the 5-switch bridge circuit described above by series-parallel circuits is also called a *bridge circuit*. Given any such circuit, its switching function can be easily obtained by making appropriate substitutions in the function f in (1). Conversely, given an arbitrary Boolean function f, there is no method for determining whether f can be realized by a bridge circuit. If we can express f in a form that, with appropriate substitutions, is identical with (1), then f can be realized by a bridge circuit.

Example 1.8.1 Write the switching function of the bridge circuit

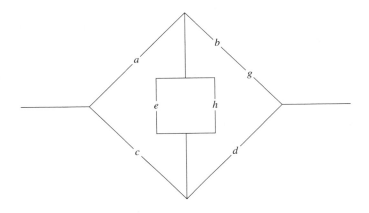

Solution. This circuit is the result of replacing switches b and e in the bridge circuit described above by switches b, g in series and switches e, h in parallel, respectively. Hence its switching function is obtained by replacing b and e in (1) with bg and $e + h$, respectively. So

$$f = abg + cd + a(e + h)d + c(e + h)bg$$

Example 1.8.2 Simplify the following function as a sum of products, and draw a bridge circuit to realize it:

$$f = (xu + xv + yu + yv)(x + z + v)(y + z + u)$$

Solution. Using the distributive laws, we have

$$f = (x + y)(u + v)(z + (x + v)(y + u))$$
$$= (x + y)(u + v)z + (x + y)(u + v)(x + v)(y + u)$$

Now

$$(x + y)(u + v)(x + v)(y + u) = (x + y)(x + v)(u + v)(u + y)$$
$$= (x + yv)(u + yv)$$
$$= xu + yv$$

Hence, using absorption, we have

$$f = (xv + yu)z + xu + yv$$
$$= xu + yv + xzv + yzu$$

which is of the same form as the function in (1). Hence it is realized by this bridge circuit:

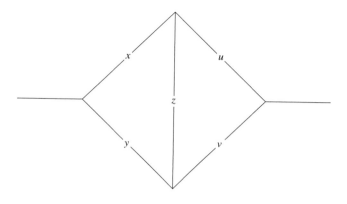

EXERCISES 1.8

1. Draw a circuit for a light bulb that can be turned on and off from each of three switches independently.

2. Draw a circuit for a machine in a factory that runs by three generators so that the machine keeps running if at least two generators work. Find the Boolean function $p(a_1, a_2, a_3)$ that gives the state of the machine, where a_i represents the state of the ith generator. Assume that $a_i = 1, 0$ according to whether the ith generator is working or not working, and $p(a_1, a_2, a_3) = 1, 0$ according to whether the machine has been operating or has stopped.

Draw bridge circuits for the Boolean functions in Exercises 3 and 4.

3. $f = xw' + y'uv + (xz + y')(zw' + uv)$

4. $f = (x'u + x'v's + yu + yv's)(x' + z + w' + v's)(y + z + w' + u)$

5. By considering the ways in which the flow of current can be blocked in the bridge circuit, show that its switching function can be expressed in the product-of-sums form

$$f = (a + b)(c + d)(a + e + d)(b + e + c)$$

Simplify f and express it as a sum of products. Hence show that the bridge circuit is equivalent to the series-parallel circuit shown on the right.

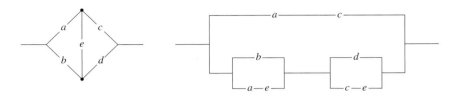

6. Define an exclusive-OR gate and show how it can be constructed from the basic three gates. Show that the XOR gate can be used as a half-adder.

7. Draw a full-adder circuit for adding three 1-digit binary numbers.

CHAPTER
2

BALANCED INCOMPLETE BLOCK DESIGNS

2.1 BASIC DEFINITIONS AND RESULTS

Suppose we have v different brands of chemical fertilizers that we wish to test on b different crops. If we test each fertilizer on each crop, we need vb plots of land. Suppose we do not have enough land or other resources to carry out the complete experiment. We decide to test each fertilizer on only some of the crops. To avoid bias for any fertilizer or any crop, it is desirable that each fertilizer be tested on the same number r of crops, and each crop should be subjected to the same number k of fertilizers. Further, to ensure that we can make a fair comparison between any two fertilizers, it is necessary that any two fertilizers be tested together on the same number λ of crops.

Problems of this kind, which are dealt with in the statistical design of experiments, lead to the concept of a balanced incomplete block design, defined below.

Definition 2.1.1 Let V be a set of v elements, $v > 2$. A set D consisting of b subsets of V (hereafter referred to as *blocks* of D) is called a *balanced incomplete block design* (in short *BIBD*) on V with parameters (v, b, r, k, λ) if the following conditions hold:

(BD1) Each element of V occurs in r blocks of D.
(BD2) Each block of D has k elements, where $1 < k < v$.
(BD3) Every two elements of V occur together in λ blocks of D.

Moreover, if $v = b$, then D is called a *symmetric BIBD*.

A BIBD with parameters (v, b, r, k, λ) is briefly referred to as a (v, b, r, k, λ)-design.

Example 2.1.2 Let $V = \{a_1, a_2, a_3\}$ and $D = \{B_1, B_2, B_3\}$, where

$$B_1 = \{a_2, a_3\}, \qquad B_2 = \{a_1, a_3\}, \qquad B_3 = \{a_1, a_2\}$$

Show that D is a symmetric BIBD on V, and find its parameters.

Solution. By direct checking, we find that each element of V occurs in two blocks of D, and every two elements of V occur together in exactly one block. Further, each block has two elements. This proves that D is a BIBD with parameters $(3, 3, 2, 2, 1)$. Moreover, since $v = 3 = b$, D is a symmetric BIBD.

The following theorem generalizes Example 2.1.2.

THEOREM 2.1.3 Let V be a set of v elements, and let $D = V^{(k)}$ be the set of all subsets of V having k elements, $1 < k < v$. Then D is a balanced incomplete block design on V with parameters (v, b, r, k, λ) where

$$b = \binom{v}{k}, \qquad r = \binom{v-1}{k-1}, \qquad \lambda = \binom{v-2}{k-2}$$

In particular, if $k = v - 1$, then D is a symmetric BIBD with parameters $(v, v, v - 1, v - 1, v - 2)$.

Proof: The number of ways in which k elements can be selected from a set of v elements is $\binom{v}{k}$. Hence D consists of $\binom{v}{k}$ blocks. Given an element a in V, let B be any block containing a. There are $k - 1$ elements in B besides a, and these can be selected from the remaining $v - 1$ elements in V in $\binom{v-1}{k-1}$ ways. Hence a occurs in $\binom{v-1}{k-1}$ blocks of D. Finally, given two elements a, b in V, let B be any block that contains both a and b. Now there are $k - 2$ elements in B besides a, b, and these can be selected from the remaining $v - 2$ elements in V in $\binom{v-2}{k-2}$ ways. Hence a, b occur together in $\binom{v-2}{k-2}$ blocks of D. This proves that D is a balanced incomplete block design on V with parameters (v, b, r, k, λ), where

$$b = \binom{v}{k}, \quad r = \binom{v-1}{k-1}, \quad \lambda = \binom{v-2}{k-2}.$$

If $k = v - 1$, then $b = \binom{v}{v-1} = v$, $r = \binom{v-1}{v-2} = v - 1$, and $\lambda = \binom{v-2}{v-3} = v - 2$. Hence D is a symmetric BIBD with parameters $(v, v, v - 1, v - 1, v - 2)$.

∎

Example 2.1.4 An advertising agency wishes to test five brands of a beauty lotion on 10 subjects. Each of the subjects is given three brands to try. Find a scheme for allocating the brands to subjects such that each brand is tried by the same number of subjects and every two brands are tried together by the same number of subjects.

How many subjects try any given brand? How many subjects try any given pair of brands?

Solution. The problem requires finding a BIBD on the given set V of five brands of beauty lotion, with the parameters $v = 5$, $b = 10$, $k = 3$. Now $b = 10 = \binom{5}{3} = \binom{v}{k}$. By Theorem 2.1.3, the set of all subsets of V of order 3 is a BIBD with parameters $(5, 10, 6, 3, 3)$. So each brand is tried by six subjects, and every two brands are tried together by three subjects. Numbering the brands as 1, 2, 3, 4, 5 and similarly numbering the subjects, we have the following table for the allocation of brands to subjects:

Subject	Brands
1	1, 2, 3
2	1, 2, 4
3	1, 2, 5
4	1, 3, 4
5	1, 3, 5
6	1, 4, 5
7	2, 3, 4
8	2, 3, 5
9	2, 4, 5
10	3, 4, 5

The parameters of a BIBD cannot have arbitrary values. The following theorem shows that these parameters must satisfy certain relations.

THEOREM 2.1.5 If there exists a BIBD with parameters (v, b, r, k, λ), then the following relations must hold:

(a) $vr = bk$

(b) $r(k - 1) = \lambda(v - 1)$

(c) $b > r > \lambda$

[Subsequently, we shall also prove (Theorem 2.2.4) that $v \leq b$ and hence $r \geq k$.]

Proof: Let $V = \{a_1, \ldots, a_v\}$, and suppose $D = \{B_1, \ldots, B_b\}$ is a balanced incomplete block design on V with parameters (v, b, r, k, λ).

(a) We count in two ways the number N of ordered pairs (i, j) such that $a_i \in B_j$. Since each a_i, $i = 1, \ldots, v$, occurs in r blocks of D, we have $N = vr$. On the other hand, each block B_j, $j = 1, \ldots, b$, contains k elements of V and hence $N = bk$. This proves $vr = bk$.

(b) We count in two ways the number M of ordered pairs (i, j) such that $a_1, a_i \in B_j$ ($i \neq 1$). For each $i = 2, \ldots, v$, the elements a_1 and a_i occur together in λ blocks of D; hence $M = (v - 1)\lambda$. On the other hand, a_1 occurs in r blocks, and in each such block there are $k - 1$ elements besides a_1. Therefore $M = r(k - 1)$. This proves $r(k - 1) = \lambda(v - 1)$.

(c) Finally, since $k < v$, it follows from parts (a) and (b) that $b > r > \lambda$. ∎

The following corollary is an immediate consequence of the relation $vr = bk$ proved in Theorem 2.1.5.

COROLLARY 2.1.6 A BIBD with parameters (v, b, r, k, λ) is symmetric if and only if $r = k$.

Because of this result, we need to specify only three parameters (v, k, λ) for a symmetric BIBD. A symmetric BIBD with parameters (v, k, λ) is briefly referred to as a symmetric (v, k, λ)-design. By Theorem 2.1.5, a symmetric (v, k, λ)-design must satisfy the condition $k(k - 1) = \lambda(v - 1)$.

It should be noted that the conditions in Theorem 2.1.5 are necessary but not sufficient for the existence of a BIBD with given parameters. For example, it is known that no BIBD has the parameters $v = b = 43$, $r = k = 7$, $\lambda = 1$, even though these satisfy the conditions of Theorem 2.1.5.

Example 2.1.7 Show that it is not possible to have a BIBD with parameters (v, b, r, k, λ) if $v = 10$, $b = 12$, $k = 5$.

Solution. From the equality $vr = bk$ it follows that $r = 6$. But then the equality $r(k - 1) = \lambda(v - 1)$ required by Theorem 2.1.5 yields $\lambda = \frac{8}{3}$, which is not possible, since λ has to be an integer.

In the following theorem we show that we can obtain another BIBD from a given BIBD by taking the complements of all the blocks.

THEOREM 2.1.8 Let $D = \{B_1, \ldots, B_b\}$ be a BIBD on the set V with parameters (v, b, r, k, λ), where $1 < k < v - 1$. Let

$$B_i' = V - B_i = \{x \in V \mid x \notin B_i\}, \qquad i = 1, \ldots, b$$

Then $D' = \{B_1', \ldots, B_b'\}$ is a BIBD on V with parameters $(v, b, b - r, v - k, b - 2r + \lambda)$.

(The design D' is called the *complement* of D.)

Proof: It is clear that every block B_i' of D' has $v - k$ elements. An element a in V occurs in B_i' if and only if $a \notin B_i$. Since a occurs in exactly r blocks of D, it follows that a occurs in $b - r$ blocks of D'. Now consider any two elements a, b in V. Let P, Q be the sets of blocks of D' that contain a, b, respectively, so $P \cap Q$ is the set of blocks of D' in which a and b occur together. If a, b are both present in a block B_i of D, then a, b are both absent in B_i'. Hence the number of blocks of D' in which a, b are both absent is λ. Therefore $|P \cup Q| = b - \lambda$. Now $|P \cup Q| = |P| + |Q| - |P \cap Q|$, so

$$|P \cap Q| = |P| + |Q| - |P \cup Q| = (b - r) + (b - r) - (b - \lambda) = b - 2r + \lambda$$

Thus every two elements in V occur together in $b - 2r + \lambda$ blocks of D'. This proves that D' is a balanced incomplete block design on V with parameters $(v, b, b - r, v - k, b - 2r + \lambda)$. ∎

The following corollary is an immediate consequence of Theorem 2.1.8.

COROLLARY 2.1.9 If D is a symmetric BIBD with parameters (v, k, λ), then its complement D' is a symmetric BIBD with parameters $(v, v - k, v - 2k + \lambda)$.

Example 2.1.10 Let $V = \{1, 2, 3, 4, 5, 6, 7\}$, and let $D = \{B_1, B_2, B_3, B_4, B_5, B_6, B_7\}$, where

$$B_1 = \{1, 2, 4\}, \quad B_2 = \{2, 3, 5\}, \quad B_3 = \{3, 4, 6\}, \quad B_4 = \{4, 5, 7\},$$
$$B_5 = \{5, 6, 1\}, \quad B_6 = \{6, 7, 2\}, \quad B_7 = \{7, 1, 3\}$$

Show that D is a symmetric BIBD, and find its complement D'. What are the parameters of D'?

Solution. By direct checking we verify that each element of V occurs in three blocks, and every two elements occur together in one block. Hence D is a symmetric $(7, 3, 1)$-design. Taking the complements of the blocks of D, we have the blocks of D':

$$B_1' = \{3, 5, 6, 7\}, \quad B_2' = \{4, 6, 7, 1\}, \quad B_3' = \{5, 7, 1, 2\}, \quad B_4' = \{6, 1, 2, 3\},$$
$$B_5' = \{7, 2, 3, 4\}, \quad B_6' = \{1, 3, 4, 5\}, \quad B_7' = \{2, 4, 5, 6\}$$

D' is a symmetric design with parameters $(7, 4, 2)$. This is easily verified by direct checking.

Given a BIBD D on a set V with parameters (v, b, r, k, λ), we can obtain several BIBDs with the same parameters by permuting the elements of V. All these BIBDs are said to be *isomorphic* to one another. For example, on interchanging the elements 1 and 2 in the design D of Example 2.1.10, we get the BIBD with blocks

$$\{1, 2, 4\}, \{1, 3, 5\}, \{3, 4, 6\}, \{4, 5, 7\}, \{5, 6, 2\}, \{6, 7, 1\}, \{7, 2, 3\}$$

More generally, we define isomorphism of BIBDs as follows:

Definition 2.1.11 Let $D = \{B_1, \ldots, B_b\}$, $D' = \{B_1', \ldots, B_b'\}$ be BIBDs on the sets V, V', respectively, with parameters (v, b, r, k, λ). The designs D, D' are said to be *isomorphic* if there exists a bijective mapping $f : V \to V'$ such that for each $i = 1, \ldots, b$, there exists j such that

$$B_j' = f(B_i) = \{f(x) \mid x \in B_i\}$$

2.2 INCIDENCE MATRIX OF A BIBD

We now introduce a matrix representation of a BIBD that is helpful in proving several results. We give below a general definition of an incidence matrix that can in particular be applied to a BIBD. A matrix in which every entry is 0 or 1 is called a *0-1 matrix*.

Definition 2.2.1 Let $X = \{x_1, \ldots, x_m\}$ be a set of m elements, and let $S_1, \ldots, S_n$ be a list of subsets of X. The *incidence matrix* of the list $S_1, \ldots, S_n$ is defined to be the $m \times n$ 0-1 matrix $A = (a_{ij})$ given by

$$a_{ij} = \begin{cases} 1 & \text{if } x_i \in S_j \\ 0 & \text{if } x_i \notin S_j \end{cases}$$

In particular, if $D = \{B_1, \ldots, B_b\}$ is a BIBD on a set V, then the incidence matrix of the list $B_1, \ldots, B_b$ of subsets of V is called the *incidence matrix* of D.

The following theorem is an immediate consequence of the definitions of a BIBD and its incidence matrix.

THEOREM 2.2.2 Let $D = \{B_1, \ldots, B_b\}$ be a BIBD on V with parameters (v, b, r, k, λ), and let A be the incidence matrix of D. Then A is a $v \times b$ 0-1 matrix with the following properties:

(a) $\displaystyle\sum_{j=1}^{b} a_{ij} = r \quad (i = 1, \ldots, v)$

(b) $\displaystyle\sum_{i=1}^{v} a_{ij} = k \quad (j = 1, \ldots, b)$

(c) $\displaystyle\sum_{j=1}^{b} a_{pj} a_{qj} = \lambda \quad (p, q = 1, \ldots, v; \ p \neq q)$

Proof:

(a) Since each element in V occurs in r blocks of D, there are exactly r 1's in each row of A. So the sum of each row in A is r.

(b) Since each block of D contains k elements, there are exactly k 1's in each column of A. So the sum of each column in A is k.

(c) Since every two elements in V occur together in λ blocks of D, for any two rows in A there are exactly λ columns where there is a 1 in both the rows. Hence the inner product of any two rows of A is λ. $\blacksquare$

The following theorem gives an important property of the incidence matrix of a BIBD. As usual, A^T denotes the transpose of A.

THEOREM 2.2.3 Let A be the incidence matrix of a BIBD. Then AA^T is a non-singular matrix.

Proof: Let D be a BIBD with parameters (v, b, r, k, λ). Then its incidence matrix A is a $v \times b$ matrix, so AA^T is a $v \times v$ matrix. The (p, q)th entry in AA^T is equal to $\sum_{j=1}^{b} a_{pj} a_{qj}$. Since A is a 0-1 matrix, $(a_{pj})^2 = a_{pj}$. Hence, by Theorem 2.2.2(a),

$$\sum_{j=1}^{b} (a_{pj})^2 = \sum_{j=1}^{b} a_{pj} = r \quad \text{for all } p = 1, \ldots, v$$

Therefore, from Theorem 2.2.2(c), it follows that for all $p, q = 1, \ldots, v$,

$$\sum_{j=1}^{b} a_{pj} a_{qj} = \begin{cases} r & \text{if } p = q \\ \lambda & \text{if } p \neq q \end{cases}$$

Hence

$$AA^T = \begin{bmatrix} r & \lambda & \cdots & \lambda \\ \lambda & r & \cdots & \lambda \\ \vdots & \vdots & \ddots & \vdots \\ \lambda & \lambda & \cdots & r \end{bmatrix}$$

We evaluate the determinant of AA^T as follows. Adding to the first row all other rows, and then subtracting the first column from each of the other columns, we get

$$\det(AA^T) = \begin{vmatrix} r + (v-1)\lambda & 0 & \cdots & 0 \\ \lambda & r - \lambda & \cdots & 0 \\ \vdots & \vdots & \ddots & \vdots \\ \lambda & 0 & \cdots & r - \lambda \end{vmatrix}$$

$$= (r + (v-1)\lambda)(r - \lambda)^{v-1}$$

The last step follows from the property that the determinant of a triangular matrix is equal to the product of its diagonal entries.

By Theorem 2.1.5, $r > \lambda$, so $\det(AA^T) > 0$. Hence AA^T is a nonsingular $v \times v$ matrix. ∎

The foregoing theorem yields another constraint on the parameters of a BIBD, as anticipated earlier (see Theorem 2.1.5).

THEOREM 2.2.4 (Fisher's inequality) If there exists a BIBD with parameters (v, b, r, k, λ), then $v \le b$ and hence $r \ge k$.

Proof: Let D be a BIBD with parameters (v, b, r, k, λ). Then its incidence matrix A is a $v \times b$ matrix; hence *rank* $A \le b$. By Theorem 2.2.3, AA^T is a nonsingular $v \times v$ matrix; hence $rank(AA^T) = v$.

It is a well-known property of the rank of a matrix that $rank(AB) \le \min(rank\ A, rank\ B)$. Therefore

$$v = rank(AA^T) \le rank\ A \le b$$

Hence $v \le b$. By Theorem 2.1.5, $vr = bk$ and hence $r \ge k$. ∎

We now consider the special case of a symmetric BIBD.

THEOREM 2.2.5 Let D be a symmetric BIBD with parameters (v, k, λ), and let A be the incidence matrix of D. Then A is nonsingular and $A^T A = AA^T$.

Proof: Since D is symmetric, A and A^T are both $v \times v$ matrices. So it follows by Theorem 2.2.3 that $\det A \det A^T = \det(AA^T) > 0$. Hence $\det A \ne 0$ and so A is nonsingular. Hence A^{-1} exists. So $A^T A$ may be expressed as

$$A^T A = A^{-1}(AA^T)A$$

Now, as shown in the proof of Theorem 2.2.3,

$$AA^T = \begin{bmatrix} k & \lambda & \cdots & \lambda \\ \lambda & k & \cdots & \lambda \\ \vdots & \vdots & \ddots & \vdots \\ \lambda & \lambda & \cdots & k \end{bmatrix}$$

$$= (k - \lambda)I + \lambda J$$

where I is the $v \times v$ identity matrix and J is the $v \times v$ matrix in which every entry is 1. By Theorem 2.2.2, the sum of each row as well as of each column in A is k. So it follows that $JA = kJ = AJ$ and hence $A^{-1}JA = J$. Therefore

$$A^T A = A^{-1}(AA^T)A$$
$$= A^{-1}[(k - \lambda)I + \lambda J]A$$
$$= (k - \lambda)I + \lambda J$$
$$= AA^T$$ ∎

THEOREM 2.2.6 Let D be a symmetric BIBD with parameters (v, k, λ). Then every two blocks of D have exactly λ elements in common.

Proof: By Theorem 2.2.5, $A^T A = AA^T$. If we equate the (p, q)th entries in the two sides of this equality, it follows that for all $p, q = 1, \ldots, v$ and $p \neq q$,

$$\sum_{j=1}^{v} a_{jp} a_{jq} = \sum_{j=1}^{v} a_{pj} a_{qj}$$
$$= \lambda \quad \text{(by Theorem 2.2.2)}$$

Hence the inner product of any two columns of A is equal to λ, which implies that any two blocks of D have λ elements in common. ∎

The following theorem enables us to obtain another symmetric BIBD from a given symmetric BIBD.

THEOREM 2.2.7 Let $D = \{B_1, \ldots, B_v\}$ be a symmetric BIBD on the set $V = \{a_1, \ldots, a_v\}$ with parameters (v, k, λ). Let $B_1^*, \ldots, B_v^*$ be subsets of V given by

$$a_i \in B_j^* \Leftrightarrow a_j \in B_i \qquad i, j = 1, \ldots, v$$

Then $D^* = \{B_1^*, \ldots, B_v^*\}$ is a symmetric BIBD with parameters (v, k, λ).

(The symmetric design D^* is called the *dual* of D.)

Proof: Let A be the incidence matrix of D. By Theorem 2.2.2, the sum of each row and of each column in A is k. Hence A^T has the same property. Moreover, by

Theorem 2.2.6, the inner product of any two columns of A is equal to λ. Hence, from the equality $A^T A = A A^T$ obtained in Theorem 2.2.5, it follows that the inner product of any two rows of A^T is equal to λ.

It is clear that the incidence matrix of the list $B_1^*, \ldots, B_v^*$ is A^T. Hence D^* is a symmetric design with parameters (v, k, λ).   ∎

Example 2.2.8 Find the dual of the symmetric BIBD of Example 2.1.10.

Solution. Examining the blocks of D, we find that 1 belongs to the blocks B_1, B_5, B_7. Hence $B_1^* = \{1, 5, 7\}$. Using the same procedure for the other elements in V, we obtain $D^* = \{B_1^*, B_2^*, B_3^*, B_4^*, B_5^*, B_6^*, B_7^*\}$, where

$$B_1^* = \{1, 5, 7\}, \quad B_2^* = \{1, 2, 6\}, \quad B_3^* = \{3, 7, 2\}, \quad B_4^* = \{4, 1, 3\},$$
$$B_5^* = \{5, 2, 4\}, \quad B_6^* = \{6, 3, 5\}, \quad B_7^* = \{7, 4, 6\}$$

The following theorem shows that a symmetric BIBD with $\lambda > 1$ determines a BIBD on each of its blocks.

THEOREM 2.2.9 Let $D = \{B_1, \ldots, B_v\}$ be a symmetric design with parameters (v, k, λ), where $\lambda > 1$. Let D_1 consist of the sets $B_1 \cap B_2, \ldots, B_1 \cap B_v$. Then D_1 is a BIBD on the set B_1 with parameters $(k, v - 1, k - 1, \lambda, \lambda - 1)$.

Thus every symmetric (v, k, λ)-design with $\lambda > 1$ determines a BIBD with parameters $(k, v - 1, k - 1, \lambda, \lambda - 1)$.

Proof: The blocks of D_1 are all subsets of B_1. By Theorem 2.2.6, every two blocks of D have λ common elements, and hence every block of D_1 has λ elements. Let x be any element in B_1. Now x occurs in $k - 1$ blocks of D other than B_1; hence x occurs in $k - 1$ blocks of D_1. Further, any two elements $x, y \in B_1$ occur together in $\lambda - 1$ blocks of D other than B_1; hence x, y occur together in $\lambda - 1$ blocks of D_1. This proves that D_1 is a BIBD on the set B_1 with parameters $(k, v - 1, k - 1, \lambda, \lambda - 1)$.   ∎

The design D_1 is called the BIBD on the set B_1 *derived* from D.

EXERCISES 2.2

1. Show that conditions (BD2) and (BD3) in Definition 2.1.1 of a BIBD imply condition (BD1); that is, if every block of D has k elements and every two elements of V occur in λ blocks, then every element of V occurs in the same number of blocks.

2. Show that the following are equivalent conditions for a BIBD with parameters (v, b, r, k, λ) to be symmetric: (a) $v = b$, (b) $r = k$, (c) $\lambda(v - 1) = k(k - 1)$, and (d) every two blocks have λ common elements.

3. Let D be a BIBD on V with parameters (v, b, r, k, λ), and let x, y be distinct elements in V. Find the number of blocks that contain (a) x or y and (b) x or y but not both.

4. Given any pair of blocks of a symmetric (v, k, λ)-design, find the number of elements that do not occur in either block.

5. Show that there is no BIBD with parameters $(12, 18, 6, 4, 1)$.

6. Show that there is no (v, b, r, k, λ)-BIBD with $v = 10, k = 7, \lambda = 1$.

7. Show that there is no BIBD consisting of seven blocks of five elements each.

8. Suppose D_1, D_2 are BIBDs on the same set with parameters $(v, b_1, r_1, k, \lambda_1)$, $(v, b_2, r_2, k, \lambda_2)$, respectively. Let D consist of blocks of D_1, D_2 taken together. Show that D is a BIBD with parameters $(v, b_1 + b_2, r_1 + r_2, k, \lambda_1 + \lambda_2)$.

9. Construct a symmetric $(7, 4, 2)$-design and obtain the derived BIBD.

10. Construct a BIBD with parameters $(7, 14, 6, 3, 2)$.

11. Show that the following sets constitute a BIBD, and find its parameters:

$$\{1, 2, 3\}, \ \{1, 4, 7\}, \ \{1, 5, 9\}, \ \{1, 6, 8\}, \ \{4, 5, 6\}, \ \{2, 5, 8\},$$
$$\{2, 6, 7\}, \ \{2, 4, 9\}, \ \{7, 8, 9\}, \ \{3, 6, 9\}, \ \{3, 4, 8\}, \ \{3, 5, 7\}$$

12. A BIBD on a set V is said to be *resolvable* if its blocks can be grouped into several sets each of which is a partition of the set V. Show that if a BIBD with parameters (v, b, r, k, λ) is resolvable, then k divides v.

13. Show that the BIBD in Exercise 11 is resolvable.

14. Construct a BIBD with parameters $(9, 12, 8, 6, 5)$.

15. Construct a BIBD with parameters $(9, 24, 8, 3, 2)$.

16. If D is a symmetric BIBD with parameters (v, k, λ) and v is even, show that $k - \lambda$ is a perfect square.

17. Show that there is no symmetric BIBD with parameters $(15, 8, 3)$.

18. Show that there is no symmetric BIBD with parameters $(46, 10, 2)$.

19. Show that there is no symmetric BIBD with parameters $(34, 12, 4)$.

20. An advertising agency wishes to test six brands of toothpaste on six subjects, with each subject receiving five tubes in five different colors. Design a scheme for distributing brands among subjects so that each brand is used by the same number of subjects, each pair of brands is used by the same number of subjects, and each brand is given each color the same number of times.

21. A number of volleyball teams of five students each are formed from a group of 15 students in such a way that every student plays on the same number of teams, and any pair of students play together on exactly two teams. How many teams are there?

22. Is it possible to design an experiment to test 22 brands of breakfast cereal by distributing them to several households in such a way that each household tests seven brands over a period of seven weeks, and each pair of brands is tested by exactly two households? If so, how many households are needed for the experiment?

23. A number of subcommittees of three persons each are formed from a board of six persons in such a way that each person is a member of five subcommittees, and any two persons are together in the same number of subcommittees. How many subcommittees are there? In how many subcommittees are any two given persons together?

24. A schoolmistress takes nine girls out for a daily walk, with the girls arranged in rows of three each. Design an arrangement of rows for four consecutive days so that any two girls walk together in the same row exactly once.

25. (*"Kirkman schoolgirl problem"*) A schoolmistress takes 15 girls out for a daily walk, with the girls arranged in rows of three each. Design an arrangement of rows for seven consecutive days so that any two girls walk together in the same row exactly once.

2.3 CONSTRUCTION OF BIBDs FROM DIFFERENCE SETS

No general method is known for constructing BIBDs; nor are general sufficient conditions known for the existence of a BIBD with given parameters. Only some ad hoc methods for constructing BIBDs in special cases are available that make use of other algebraic systems like groups, rings, and fields. In this section we consider the construction of a BIBD from a difference set in a group.

Definition 2.3.1 Let G be an additive abelian group of order v. A subset S of G of order k is called a *difference set* in G with parameters (v, k, λ) if for each nonzero $a \in G$, there are exactly λ ordered pairs (x, y) of elements in S such that $a = x - y$.

Example 2.3.2 Let $G = \mathbb{Z}_7$ be the additive group of integers modulo 7. Let $S = \{1, 2, 4\}$. Show that S is a difference set in G, and find its parameters.

Solution. Writing the nonzero differences of all ordered pairs of elements in S, we have

$$1 - 2 = 6, \quad 2 - 1 = 1,$$
$$1 - 4 = 4, \quad 4 - 1 = 3,$$
$$2 - 4 = 5, \quad 4 - 2 = 2$$

Thus we see that each nonzero element in G occurs exactly once as a difference of two elements of S. Hence S is a difference set in G with parameters $(7, 3, 1)$.

The following theorem gives a necessary condition on the parameters of a difference set.

THEOREM 2.3.3 Let S be a difference set in a group G with parameters (v, k, λ). Then

$$k(k - 1) = \lambda(v - 1)$$

Proof: Because S has k elements, the number of ordered pairs $(x, y) \in S \times S$ such that $x \ne y$ is equal to $k(k - 1)$. On the other hand, G has $v - 1$ nonzero elements, and for each nonzero element $a \in G$, there are λ ordered pairs $(x, y) \in S \times S$ such that $a = x - y$. Hence $k(k - 1) = \lambda(v - 1)$. ∎

Given a subset S of G and an element g in G, we write

$$g + S = \{g + s \mid s \in S\}$$

It is clear that if S is a (v, k, λ)-difference set, then for every $g \in G$, $g + S$ is also a (v, k, λ)-difference set in G. We have

$$a = x - y \Leftrightarrow a = (g + x) - (g + y)$$

Thus S determines v difference sets $g + S$, $g \in G$. The following theorem shows that these difference sets form a symmetric BIBD on the set G.

THEOREM 2.3.4 Let $S = \{s_1, \ldots, s_k\}$ be a difference set in group $G = \{g_1, \ldots, g_v\}$ with parameters (v, k, λ). For each $i = 1, \ldots, v$, let

$$B_i = g_i + S = \{g_i + s \mid s \in S\}$$

Then $D = \{B_1, \ldots, B_v\}$ is a symmetric BIBD on the set G with parameters (v, k, λ).

Proof: It is clear that each block of D has k elements. If $g_i + x = g_i + y$, then $x = y$ and hence $g_i + s_1, \ldots, g_i + s_k$ are k distinct elements in B_i. Let a be any element in G. If $a \in B_i$, then $a = g_i + s$ for some $s \in S$. Now $a = (a - s_j) + s_j \in (a - s_j) + S$ for each $j = 1, \ldots, k$. Since $a - s_1, \ldots, a - s_k$ are k distinct elements in G, it follows that each element in G occurs in k blocks of D.

Let a and b be any two distinct elements in G. Then $a - b \ne 0$, and hence there are exactly λ ordered pairs $(x_i, y_i) \in S \times S$, $i = 1, \ldots, \lambda$, such that $a - b = x_i - y_i$. So $a - x_i = b - y_i = t_i$ (say) and hence $a = t_i + x_i$, $b = t_i + y_i$. Thus $a, b \in t_i + S$ for each $i = 1, \ldots, \lambda$. Conversely, if a, b occur in the same block $g + S$, then $a = g + x$, $b = g + y$ for some $x, y \in S$ and hence $a - b = x - y$. So $(x, y) = (x_i, y_i)$ for some $i = 1, \ldots, \lambda$ and $g = t_i$. Thus a, b occur together in exactly λ blocks of D. This proves that D is a symmetric BIBD with parameters (v, k, λ). ∎

We saw in Corollary 2.1.9 that the complement D' of a symmetric (v, k, λ)-design D is a symmetric $(v, v - k, v - 2k + \lambda)$-design. Further, by Theorem 2.2.9, a symmetric (v, k, λ)-BIBD D with $\lambda > 1$ determines a BIBD D_1 with parameters $(k, v - 1, k - 1, \lambda, \lambda - 1)$. Thus a (v, k, λ)-difference set generates two symmetric BIBDs with parameters (v, k, λ), $(v, v - k, v - 2k + \lambda)$, and two more BIBDs with parameters $(k, v - 1, k - 1, \lambda, \lambda - 1)$, $(v - k, v - 1, v - k - 1, v - 2k + \lambda, v - 2k + \lambda - 1)$, provided $\lambda > 1$ and $v - 2k + \lambda > 1$. The last two BIBDs themselves generate their complements.

Example 2.3.5 Find the BIBDs determined by the difference set of Example 2.3.2.

Solution. We saw in Example 2.3.2 that $S = \{1, 2, 4\}$ is a $(7, 3, 1)$-difference set in the additive group $\mathbb{Z}_7 = \{0, 1, 2, 3, 4, 5, 6\}$. By Theorem 2.3.4, S determines the symmetric BIBD D with the following blocks:

$0 + S = \{1, 2, 4\}, \quad 1 + S = \{2, 3, 5\}, \, 2 + S = \{3, 4, 6\}, \quad 3 + S = \{4, 5, 0\},$
$4 + S = \{5, 6, 1\}, \quad 5 + S = \{6, 0, 2\}, 6 + S = \{0, 1, 3\}$

So

$$D = \{\{1, 2, 4\}, \{2, 3, 5\}, \{3, 4, 6\}, \{0, 4, 5\}, \{1, 5, 6\}, \{0, 2, 6\}, \{0, 1, 3\}\}$$

is a symmetric BIBD on the set $\mathbb{Z}_7$ with parameters $(7, 3, 1)$, and its complement D' is the symmetric $(7, 4, 2)$-design

$$D' = \{\{0, 3, 5, 6\}, \{0, 1, 4, 6\}, \{0, 1, 2, 5\}, \{1, 2, 3, 6\}, \{0, 2, 3, 4\},$$
$$\{1, 3, 4, 5\}, \{2, 4, 5, 6\}\}$$

Taking the intersections of $\{0, 3, 5, 6\}$ with the other blocks of D', we get the derived BIBD

$$D_1' = \{\{0, 6\}, \{0, 5\}, \{3, 6\}, \{0, 3\}, \{3, 5\}, \{5, 6\}\}$$

which has parameters $(4, 6, 3, 2, 1)$.

2.4 CONSTRUCTION OF BIBDs USING QUADRATIC RESIDUES

No general method is known for constructing difference sets. Quadratic residues, which we define below, provide difference sets in some special cases.

By the Euclidean division algorithm, given integers m, n, with $m > 0$, there exist integers q, r, such that

$$n = qm + r \quad \text{and} \quad 0 \leq r \leq m - 1$$

The integer r is called the *remainder* on dividing n by m, or *residue modulo m* of n, and is written $res_m(n)$.

Definition 2.4.1 Let p be a prime number greater than 2. A positive integer a is called a *quadratic residue modulo p* if there exists an integer n such that $a = res_p(n^2)$.

Let Q_p denote the set of all quadratic residues modulo p.

THEOREM 2.4.2 Let p be a prime number greater than 2. Then there are $(p-1)/2$ quadratic residues modulo p, and

$$Q_p = \left\{ res_p(n^2) \mid 1 \leq n \leq \frac{p-1}{2} \right\}$$

Proof: Given any integer n, let $\bar{n} = res_p(n)$. Then $n = qp + \bar{n}$ for some q, so $n^2 = q^2 p^2 + 2\bar{n}qp + \bar{n}^2$ and hence $res_p(n^2) = res_p(\bar{n}^2)$. Therefore, for obtaining the quadratic residues modulo p, we need to consider the numbers $res_p(n^2)$ for $n = 1, \ldots, p-1$ only. Moreover these $p-1$ numbers are not all distinct. Suppose $res_p(x^2) = res_p(y^2)$ for some integers $x, y \in \{1, \ldots, p-1\}$. Then $res_p(x^2 - y^2) = 0$ and hence p divides $(x^2 - y^2) = (x-y)(x+y)$. Since p is prime, it follows that either $x - y = 0$ or $x + y = p$. Thus either $y = x$ or $y = p - x$. Therefore, for each $a \in Q_p$, there are exactly two integers $n = x, p - x \in \{1, \ldots, p-1\}$ such that $a = res_p(n^2)$. It follows that there are exactly $(p-1)/2$ quadratic residues modulo p and

$$Q_p = \left\{ res_p(n^2) \mid 1 \leq n \leq \frac{p-1}{2} \right\}$$ ∎

Example 2.4.3 Find the set of all quadratic residues modulo 7.

Solution. $Q_7 = \{res_7(n^2) \mid n = 1, 2, 3\}$. Now $res_7(1^2) = 1$, $res_7(2^2) = 4$, $res_7(3^2) = 2$; hence

$$Q_7 = \{1, 2, 4\}$$

We know that for any positive integer n, the set $\mathbb{Z}_n = \{0, 1, \ldots, n-1\}$ is a ring under addition and multiplication modulo n. Moreover, for any prime p, the ring $\mathbb{Z}_p$ is a field. Hence $\mathbb{Z}_p^* = \{1, \ldots, p-1\}$ is a group under multiplication modulo p. The relation $a = bc$ in the group $\mathbb{Z}_p^*$ is equivalent to $a = res_p(bc)$ in $\mathbb{Z}$. So a is a quadratic residue modulo p if and only if $a \in \mathbb{Z}_p^*$ such that $a = x^2$ for some $x \in \mathbb{Z}_p^*$.

Thus Q_p is a subset of $\mathbb{Z}_p^*$ given by

$$Q_p = \{a \in \mathbb{Z}_p^* \mid a = x^2 \text{ for some } x \in \mathbb{Z}_p^*\}$$
$$= \left\{x^2 \in \mathbb{Z}_p^* \mid x = 1, \ldots, \frac{p-1}{2}\right\}$$

THEOREM 2.4.4 Let p be a prime number greater than 2. Then Q_p is a subgroup of $\mathbb{Z}_p^*$ of index 2.

Proof: Q_p is obviously nonempty. Let $a, b \in Q_p$. Then $a = x^2$, $b = y^2$ for some $x, y \in \mathbb{Z}_p^*$, so $a^{-1}b = (x^2)^{-1}y^2 = (x^{-1}y)^2 \in Q_p$. Hence Q_p is a subgroup of $\mathbb{Z}_p^*$. Moreover Q_p has $(p-1)/2$ elements, and $\mathbb{Z}_p^*$ is of order $p-1$. Hence Q_p is a subgroup of index 2. ∎

The fact that Q_p is a subgroup of index 2 in $\mathbb{Z}_p^*$ leads to an interesting property of quadratic residues. We prove this result in the following theorem for a subgroup of index 2 in general.

THEOREM 2.4.5 Let N be a subgroup of index 2 in a group G. Let $x, y \in G$. Then $xy \in N$ if and only if either x and y are both in N or neither of them is in N.

Proof: N, being a subgroup of index 2, is a normal subgroup of G, and the quotient group G/N is of order 2. Hence $(xN)^2 = N$ for all $x \in G$. Let $b \in G$ such that $b \notin N$. Then $G/N = \{N, bN\}$. If $x, y \in N$, then clearly $xy \in N$. If $x, y \notin N$, then $xN = bN = yN$. Therefore $xyN = (xN)(yN) = (xN)^2 = N$ and hence $xy \in N$.

Conversely, suppose $xy \in N$. Then $xyN = N$. Therefore $xN = (xN)\,N = (xN)(xyN) = (xN)(xN)(yN) = N(yN) = yN$. So either $xN = yN = N$ or $xN = yN = bN$. Hence either x and y are both in N or neither is in N. This proves the theorem. ∎

It follows from Theorem 2.4.5 that for any x, y in $\mathbb{Z}_p^*$, $xy \in Q_p$ if and only if x and y are both in Q_p or neither of them is in Q_p. Let us briefly refer to an element in $\mathbb{Z}_p^*$ as a residue or nonresidue according to whether or not it is a quadratic residue modulo p. Then we can paraphrase the result just proved as follows:

COROLLARY 2.4.6 The product of two residues is a residue. The product of two nonresidues is a residue. The product of a residue and a nonresidue is a nonresidue.

Now we prove the main theorem that enables us to obtain difference sets from quadratic residues.

THEOREM 2.4.7 Let p be a prime such that $p \equiv 3 \pmod 4$, $p > 3$. Then Q_p is a difference set in the additive group $\mathbb{Z}_p$ with parameters $\left(p, \dfrac{p-1}{2}, \dfrac{p-3}{4}\right)$.

Proof: We first show that $-1 \notin Q_p$. (Note that $-1 = p - 1$ in $\mathbb{Z}_p$.) Suppose, on the contrary, that $-1 = x^2$ for some $x \in \mathbb{Z}_p^*$. Since $p \equiv 3 \pmod 4$, we have $(p - 1)/2$ is an odd integer. Hence, working in the field $\mathbb{Z}_p$, we have

$$-1 = (-1)^{(p-1)/2} = (x^2)^{(p-1)/2} = x^{p-1}$$

But $\mathbb{Z}_p^*$ is a group of order $p - 1$ and hence $x^{p-1} = 1$, a contradiction. Thus -1 cannot be a quadratic residue modulo p.

Let $a \in \mathbb{Z}_p^*$. If $a \in Q_p$, then it follows from Corollary 2.4.6 that $-a = (-1)a \notin Q_p$. If, on the contrary, $a \notin Q_p$, then $-a = (-1)a \in Q_p$. Thus for any a in $\mathbb{Z}_p^*$, either $a \in Q_p$ or $-a \in Q_p$.

We now show that Q_p is a difference set in the additive group $\mathbb{Z}_p$. For any nonzero a in $\mathbb{Z}_p$, let $S(a)$ denote the set of ordered pairs (x, y) such that $x, y \in Q_p$, $x - y = a$. We claim that there is a one-to-one correspondence between $S(1)$ and $S(a)$.

Consider first the case where $a \in Q_p$. Let $(x, y) \in S(1)$. Then $1 = x - y$ and hence $a = a(x - y) = ax - ay$. Now $a, x, y \in Q_p$ and hence $ax, ay \in Q_p$. So $(ax, ay) \in S(a)$. Conversely, let $(x', y') \in S(a)$. Then $a = x' - y'$, and hence $1 = a^{-1}a = a^{-1}(x' - y') = a^{-1}x' - a^{-1}y'$. Since Q_p is a subgroup of $\mathbb{Z}_p^*$, $a^{-1} \in Q_p$, so $(a^{-1}x', a^{-1}y') \in S(1)$. Thus we have a one-to-one correspondence between $S(1)$ and $S(a)$, given by $(x, y) \mapsto (ax, ay)$.

Now suppose $a \notin Q_p$. Then $-a \in Q_p$. If $(x, y) \in S(1)$, then $a = a(x - y) = ax - ay = (-a)y - (-a)x$. Hence $(-ay, -ax) \in S(a)$. Conversely, if $(x'y') \in S(a)$, then $1 = a^{-1}a = a^{-1}(x' - y') = (-a^{-1})y' - (-a^{-1})x'$. Since $a \notin Q_p$, $a^{-1} \notin Q_p$ and hence $-a^{-1} \in Q_p$. Thus $(-a^{-1}y', -a^{-1}x') \in S(1)$. So there is a one-to-one correspondence between $S(1)$ and $S(a)$, given by $(x, y) \mapsto (-ay, -ax)$.

We have thus proved that for every nonzero a in $\mathbb{Z}_p$, there is a one-to-one correspondence between $S(1)$ and $S(a)$. We write $\lambda = |S(1)|$. Then $|S(a)| = \lambda$ for every nonzero $a \in \mathbb{Z}_p$. Hence, for every nonzero a in $\mathbb{Z}_p$, there are exactly λ ordered pairs (x, y) of elements in $\mathbb{Z}_p$ such that $a = x - y$. This proves that Q_p is a difference set in the additive group $\mathbb{Z}_p$.

Now $|\mathbb{Z}_p| = p$ and $|Q_p| = \dfrac{p-1}{2}$. Hence, by Theorem 2.3.3, $\lambda(p - 1) = \left(\dfrac{p-1}{2}\right)\left(\dfrac{p-1}{2} - 1\right)$, which yields $\lambda = \dfrac{p-3}{4}$. Thus Q_p is a difference set with parameters $\left(p, \dfrac{p-1}{2}, \dfrac{p-3}{4}\right)$. ∎

By Theorem 2.3.4, a (v, k, λ)-difference set determines a symmetric BIBD with parameters (v, k, λ). Hence we have the following corollary from the foregoing theorem.

COROLLARY 2.4.8 For every positive integer t such that $4t + 3$ is a prime, there exists a symmetric BIBD with parameters $(4t + 3, 2t + 1, t)$.

Example 2.4.9 Find the set of quadratic residues modulo 11, and construct the symmetric BIBD determined by it.

Solution. By Theorem 2.4.2, the set of quadratic residues modulo 11 is

$$Q_{11} = \{1^2, 2^2, 3^2, 4^2, 5^2\} \quad (\bmod\, 11)$$
$$= \{1, 4, 9, 5, 3\}$$

Since $11 \equiv 3 (\bmod\, 4)$, Q_{11} is a difference set with parameters $(11, 5, 2)$ in the additive group $\mathbb{Z}_{11}$. The blocks of the symmetric design D on the set $\mathbb{Z}_{11}$ determined by the difference set $S = Q_{11}$ are $i + S$, $i = 0, \ldots, 10$. So

$$D = \left\{ \begin{array}{l} \{1, 3, 4, 5, 9\}, \{2, 4, 5, 6, 10\}, \{3, 5, 6, 7, 0\}, \ \{4, 6, 7, 8, 1\}, \\ \{5, 7, 8, 9, 2\}, \{6, 8, 9, 10, 3\}, \{7, 9, 10, 0, 4\}, \{8, 10, 0, 1, 5\}, \\ \{9, 0, 1, 2, 6\}, \{10, 1, 2, 3, 7\}, \{0, 2, 3, 4, 8\} \end{array} \right\}$$

D is a symmetric BIBD with parameters $(11, 5, 2)$.

Example 2.4.10 Construct a BIBD on the set $\{1, 2, 3, 4, 5, 6\}$ with parameters $(6, 10, 5, 3, 2)$.

Solution. In Example 2.4.9, we obtained a symmetric BIBD D with parameters $(11, 5, 2)$. By Theorem 2.1.8, its complement D' is a symmetric BIBD with parameters $(11, 6, 3)$. Further, by Theorem 2.2.9, the BIBD derived from D' has parameters $(6, 10, 5, 3, 2)$.

Now the complement of D is

$$D' = \left\{ \begin{array}{l} \{0, 2, 6, 7, 8, 10\}, \{0, 1, 3, 7, 8, 9\}, \{1, 2, 4, 8, 9, 10\}, \{0, 2, 3, 5, 9, 10\}, \\ \{0, 1, 3, 4, 6, 10\}, \{0, 1, 2, 4, 5, 7\}, \{1, 2, 3, 5, 6, 8\}, \ \{2, 3, 4, 6, 7, 9\}, \\ \{3, 4, 5, 7, 8, 10\}, \{0, 4, 5, 6, 8, 9\}, \{1, 5, 6, 7, 9, 10\} \end{array} \right\}$$

Taking the intersections of the block $\{1, 2, 3, 5, 6, 8\}$ with the other blocks in D', we obtain the derived BIBD

$$D_1' = \left\{ \begin{array}{l} \{2, 6, 8\}, \{1, 3, 8\}, \{1, 2, 8\}, \{2, 3, 5\}, \{1, 3, 6\}, \\ \{1, 2, 5\}, \{2, 3, 6\}, \{3, 5, 8\}, \{5, 6, 8\}, \{1, 5, 6\} \end{array} \right\}$$

This a BIBD on the set $\{1, 2, 3, 5, 6, 8\}$ with parameters $(6, 10, 5, 3, 2)$. To obtain a BIBD with the same parameters on the set $\{1, 2, 3, 4, 5, 6\}$, we simply have to replace 8 with 4. So we obtain

$$D_1' = \left\{ \begin{array}{l} \{2, 6, 4\}, \{1, 3, 4\}, \{1, 2, 4\}, \{2, 3, 5\}, \{1, 3, 6\}, \\ \{1, 2, 5\}, \{2, 3, 6\}, \{3, 5, 4\}, \{5, 6, 4\}, \{1, 5, 6\} \end{array} \right\}$$

which is the required BIBD on the set $\{1, 2, 3, 4, 5, 6\}$.

2.5 DIFFERENCE SET FAMILIES

The concept of a difference set in an additive group can be extended to a difference set family.

Definition 2.5.1 Let G be an additive group. A set $\{S_1, \ldots, S_t\}$ consisting of t subsets of G, each having k elements, is called a *t-fold (v, k, λ)-difference set family* in G if for each nonzero element $a \in G$, there are exactly λ triples (x, y, j) such that $x, y \in S_j$ and $a = x - y$.

For $t = 1$, the difference set family is just a difference set. If $t > 1$, we may assume that none of the sets in a difference set family is itself a difference set. Suppose that S_1 is a (v, k, λ_1)-difference set. Then we can omit S_1 and the remaining $t - 1$ sets form a $(v, k, \lambda - \lambda_1)$-difference set family.

Example 2.5.2 Let $G = \mathbb{Z}_9$, $S_1 = \{0, 1, 2, 4\}$, and $S_2 = \{0, 3, 4, 7\}$. Show that S_1, S_2 form a difference set family, and find its parameters.

Solution. The nonzero differences in S_1 are

$$\pm(1 - 0) = 1, 8; \quad \pm(2 - 0) = 2, 7; \quad \pm(4 - 0) = 4, 5;$$
$$\pm(2 - 1) = 1, 8; \quad \pm(4 - 1) = 3, 6; \quad \pm(4 - 2) = 2, 7$$

Likewise, the differences in S_2 are

$$\pm(3 - 0) = 3, 6; \quad \pm(4 - 0) = 4, 5; \quad \pm(7 - 0) = 7, 2;$$
$$\pm(4 - 3) = 1, 8; \quad \pm(7 - 3) = 4, 5; \quad \pm(7 - 4) = 3, 6$$

By direct observation we find that each nonzero element in G occurs three times as a difference. Hence $\{S_1, S_2\}$ is a 2-fold $(9, 4, 3)$-difference set family in the group $\mathbb{Z}_9$. Note that neither of the two sets S_1, S_2 is in itself a difference set in G. In S_1, 1 occurs twice as a difference but 3 occurs only once. In S_2, it is the other way around.

The following theorem is a generalization of Theorem 2.3.3. Its proof is practically the same and is therefore omitted.

THEOREM 2.5.3 If there exists a t-fold (v, k, λ)-difference set family, then

$$tk(k - 1) = \lambda(v - 1)$$

The following theorem, which is a generalization of Theorem 2.3.4, shows that a difference set family in a group G determines a BIBD on the set G.

THEOREM 2.5.4 Let $G = \{g_1, \ldots, g_v\}$ be an additive group, and suppose $\{S_1, \ldots, S_t\}$ is a t-fold (v, k, λ)-difference set family in G. Let

$$B_{ij} = g_i + S_j, \qquad i = 1, \ldots, v; \quad j = 1, \ldots, t$$

Then $D = \{B_{ij} \mid i = 1, \ldots, v; \ j = 1, \ldots, t\}$ is a BIBD on the set G with parameters (v, vt, kt, k, λ).

Proof: It is clear that D consists of vt blocks. Moreover, as shown in the proof of Theorem 2.3.4, each block has k elements. Let $a \in G$. Then for each $x \in S_j$, $j = 1, \ldots, t$, we have $a = (a - x) + x \in (a - x) + S_j$. Hence a occurs in kt blocks of D.

Now let $a, b \in G$, $a \neq b$. Then there are λ triples (x_i, y_i, j_i), $i = 1, \ldots, \lambda$ such that $x_i, y_i \in S_{j_i}$ and $a - b = x_i - y_i$. So $a - x_i = b - y_i = z_i$ (say) and hence $a = z_i + x_i$, $b = z_i + y_i$. Thus $a, b \in z_i + S_{j_i}$, $i = 1, \ldots, \lambda$. Conversely, if a, b are in the same block $z + S_j$, then $a = z + x$, $b = z + y$ for some $x, y \in S_j$. So $a - b = x - y$ and hence $(x, y, z) = (x_i, y_i, j_i)$ for some $i = 1, \ldots, \lambda$. Thus any two elements in G occur together in λ blocks of D. This proves that D is a BIBD on the set G with parameters (v, vt, kt, k, λ).   ∎

Example 2.5.5 Construct the BIBD D induced by the difference set family of Example 2.5.2. What are the parameters of D?

Solution. The blocks of D are given by $i + S_1, i + S_2, i = 0, \ldots, 8$, where $S_1 = \{0, 1, 2, 4\}$, $S_2 = \{0, 3, 4, 7\}$. Hence, on using addition modulo 9 in G, we obtain

$$D = \left\{ \begin{array}{l} \{0, 1, 2, 4\}, \{1, 2, 3, 5\}, \{2, 3, 4, 6\}, \{3, 4, 5, 7\}, \{4, 5, 6, 8\}, \{5, 6, 7, 0\}, \\ \{6, 7, 8, 1\}, \{7, 8, 0, 2\}, \{8, 0, 1, 3\}, \{0, 3, 4, 7\}, \{1, 4, 5, 8\}, \{2, 5, 6, 0\}, \\ \{3, 6, 7, 1\}, \{4, 7, 8, 2\}, \{5, 8, 0, 3\}, \{6, 0, 1, 4\}, \{7, 1, 2, 5\}, \{8, 2, 3, 6\} \end{array} \right\}$$

D has parameters $(9, 18, 8, 4, 3)$.

2.6 CONSTRUCTION OF BIBDs FROM FINITE FIELDS

We shall now prove two theorems that enable us to obtain difference set families from finite fields. We make use of the following two properties of a finite field:

1. The order of every finite field is some power p^r, where p is a prime. Conversely, for every prime power p^r, there exists a finite field of order p^r.
2. The set F^* of nonzero elements in a finite field F is a cyclic group under multiplication. Any generator of the cyclic group F^* is called a *primitive element* of the field F.

THEOREM 2.6.1 Let F be a finite field of order $6t + 1$, and let a be a primitive element in F. Let

$$S_i = \{a^i, a^{2t+i}, a^{4t+i}\}, \qquad i = 0, 1, \ldots, t - 1$$

Then the sets $S_0, \ldots, S_{t-1}$ form a t-fold $(6t + 1, 3, 1)$-difference set family in the additive group F.

Proof: The six nonzero differences in the set S_i are

$$\pm a^i (a^{2t} - 1), \quad \pm a^i (a^{4t} - 1), \quad \pm a^i (a^{4t} - a^{2t})$$

Since F^* is a multiplicative group of order $6t$, we have $a^{6t} = 1$. Hence $(a^{3t} - 1)$ $(a^{3t} + 1) = a^{6t} - 1 = 0$, so $a^{3t} = \pm 1$. But a, being a generator of F^*, is an element of order $6t$, so $a^{3t} \neq 1$. Hence $a^{3t} = -1$. Again, $a^{2t} \neq 1$ and hence $a^{2t} - 1 = a^s$ for some s. Therefore the six nonzero differences in S_i can be expressed as

$$a^i (a^{2t} - 1) = a^{s+i}$$
$$-a^i (a^{2t} - 1) = a^{s+i+3t}$$
$$a^i (1 - a^{4t}) = a^i (a^{6t} - a^{4t}) = a^i (a^{2t} - 1) a^{4t} = a^{s+i+4t}$$
$$-a^i (1 - a^{4t}) = a^{s+i+7t} = a^{s+i+t}$$
$$a^i (a^{4t} - a^{2t}) = a^i (a^{2t} - 1) a^{2t} = a^{s+i+2t}$$
$$-a^i (a^{4t} - a^{2t}) = a^{s+i+5t}$$

Thus the nonzero differences in S_i are

$$a^{s+i+mt}, \qquad m = 0, \ldots, 5$$

Now

$$F^* = \{a^j \mid j = 0, \ldots, 6t - 1\}$$

By the division algorithm, we have $j = mt + i$, where $0 \leq i \leq t - 1$. Hence

$$F^* = \{a^{i+mt} \mid i = 0, \ldots, t - 1; \; m = 0, \ldots, 5\}$$

If each element in F^* is multiplied by a fixed element, the resulting set is again F^*. Hence

$$F^* = a^s F^* = \{a^{s+i+mt} \mid i = 0, \ldots, t - 1; \; m = 0, \ldots, 5\}$$

Thus each nonzero element in the additive group F occurs once as a difference in some set S_i. This proves that the sets $S_0, \ldots, S_{t-1}$ form a t-fold $(6t + 1, 3, 1)$-difference set family in the additive group F. ∎

Using Theorem 2.5.4, and from the existence of a finite field of order p^r for an arbitrary prime power p^r, we get the following corollary from Theorem 2.6.1:

COROLLARY 2.6.2 For any positive integer t such that $6t + 1$ is a prime power, there exists a BIBD with parameters $(6t + 1, t(6t + 1), 3t, 3, 1)$.

THEOREM 2.6.3 Let F be finite field of order $4t + 1$, and let a be a primitive element in F. Let

$$S_i = \{a^i, a^{t+i}, a^{2t+i}, a^{3t+i}\}, \qquad i = 0, 1, \ldots, t - 1$$

Then the sets $S_0, \ldots, S_{t-1}$ form a t-fold $(4t + 1, 4, 3)$-difference set family in the additive group F.

Proof: The nonzero differences in the set S_i are

$$\pm a^i(a^t - 1), \quad \pm a^i(a^{2t} - 1), \quad \pm a^i(a^{3t} - 1),$$
$$\pm a^{t+i}(a^t - 1), \quad \pm a^{t+i}(a^{2t} - 1), \quad \pm a^{2t+i}(a^t - 1)$$

Now F^* is a multiplicative group of order $4t$; hence $a^{4t} = 1$. So $(a^{2t} - 1)(a^{2t} + 1) = 0$. Since a is an element of order $4t$, $a^{2t} \neq 1$ and hence $a^{2t} = -1$. Again, $a^t \neq 1$. So $a^t - 1 = a^u$ and $a^{2t} - 1 = a^v$ for some integers u, v. Hence $a^{3t} - 1 = a^{3t} - a^{4t} = -a^{3t}(a^t - 1) = a^{u+5t} = a^{u+t}$. Therefore the 12 nonzero differences in S_i can be expressed as

$$a^i(a^t - 1) = a^{i+u}, \qquad -a^i(a^t - 1) = a^{i+u+2t}$$
$$a^i(a^{2t} - 1) = a^{i+v}, \qquad -a^i(a^t - 1) = a^{i+v+2t}$$
$$a^i(a^{3t} - 1) = a^{i+u+t}, \qquad -a^i(a^{3t} - 1) = a^{i+u+3t}$$
$$a^{t+i}(a^t - 1) = a^{i+u+t}, \qquad -a^i(a^{3t} - 1) = a^{i+u+3t}$$
$$a^{t+i}(a^{2t} - 1) = a^{i+v+t}, \qquad -a^i(a^{3t} - 1) = a^{i+v+3t}$$
$$a^{2t+i}(a^t - 1) = a^{i+u+2t}, \qquad -a^i(a^{3t} - 1) = a^{i+u}$$

Equivalently, these 12 differences in S_i are given by

$$a^{i+u+mt}, \quad a^{i+u+mt}, \quad a^{i+v+mt}, \qquad m = 0, 1, 2, 3$$

Now

$$F^* = \{a^j \mid j = 0, \ldots, 4t - 1\}$$
$$= \{a^{i+mt} \mid i = 0, \ldots, t - 1; \; m = 0, 1, 2, 3\}$$

Moreover, if each element in F^* is multiplied by a fixed element, the resulting set is again F^*. Hence

$$F^* = a^u F^* = \{a^{i+u+mt} \mid i = 0, \ldots, t - 1; \; m = 0, 1, 2, 3\}$$

and also

$$F^* = a^v F^* = \{a^{i+v+mt} \mid i = 0, \ldots, t - 1; \; m = 0, 1, 2, 3\}$$

Thus each element in F^* occurs three times as a difference in the sets $S_0, \ldots, S_{t-1}$. This proves that the sets $S_0, \ldots, S_{t-1}$ form a t-fold $(4t + 1, 4, 3)$-difference set family in the additive group F.   ∎

Using Theorem 2.5.4, we get the following corollary:

COROLLARY 2.6.4 For any positive integer t such that $4t + 1$ is a prime power, there exists a BIBD with parameters $(4t + 1, t(4t + 1), 4t, 4, 3)$.

Example 2.6.5 Find a 3-fold difference set family in the additive group $\mathbb{Z}_{13}$. What are the parameters of the BIBD induced by it?

Solution. $\mathbb{Z}_{13}$ is a field of order $13 = 4(3) + 1$. By Theorem 2.6.3, the sets S_0, S_1, S_2 constitute a 3-fold $(13, 4, 3)$-difference set family in the additive group $\mathbb{Z}_{13}$, where

$$S_i = \{a^i, a^{3+i}, a^{6+i}, a^{9+i}\}, \qquad i = 0, 1, 2$$

and a is a generator of the cyclic multiplicative group $\mathbb{Z}_{13}^*$. We find by direct computation that 2 is a generator of $\mathbb{Z}_{13}^*$. Since the group $\mathbb{Z}_{13}^*$ is of order 12, the order of 2 must divide 12. Now $2^2 = 4, 2^3 = 8, 2^4 = 3, 2^6 = 12$, and $2^{12} = 12^2 = 1$, so $o(2) = 12$.

Taking $a = 2$, we obtain

$$S_0 = \{2^0, 2^3, 2^6, 2^9\} = \{1, 8, 12, 5\}$$
$$S_1 = \{2^1, 2^4, 2^7, 2^{10}\} = \{2, 3, 11, 10\}$$
$$S_2 = \{2^2, 2^5, 2^8, 2^{11}\} = \{4, 6, 9, 7\}$$

By Theorem 2.5.4, the BIBD induced on the set $\mathbb{Z}_{13}$ by the difference set family (S_0, S_1, S_2) has parameters $(13, 39, 12, 4, 3)$.

Example 2.6.6 Find a 2-fold difference set family in the additive group $\mathbb{Z}_{13}$. What are the parameters of the BIBD induced by it?

Solution. $\mathbb{Z}_{13}$ is a field of order $13 = 6(2) + 1$. By Theorem 2.6.1, the sets S_0, S_1 constitute a 2-fold $(13, 3, 1)$-difference set family in the additive group $\mathbb{Z}_{13}$, where

$$S_i = \{a^i, a^{4+i}, a^{8+i}\}, \qquad i = 0, 1$$

and a is a generator of the cyclic multiplicative group $\mathbb{Z}_{13}^*$. Taking $a = 2$, we obtain

$$S_0 = \{2^0, 2^4, 2^8\} = \{1, 3, 9\}$$
$$S_1 = \{2^1, 2^5, 2^9\} = \{2, 6, 5\}$$

By Theorem 2.5.4, the BIBD induced on the set $\mathbb{Z}_{13}$ by the difference set family (S_0, S_1) has parameters $(13, 26, 6, 3, 1)$.

In the foregoing two examples, we applied Theorems 2.6.1 and 2.6.3 to a field of prime order. We now consider their application to a field of order $p^r, r > 1$.

The first task here is to construct a field of order p^r. Recall from Section 0.8 that for any prime p and positive integer r, there exists a field of order p^r. Moreover, any two fields of the same order p^r are isomorphic. So there is a unique field of order p^r up to isomorphism. It is called the *Galois field* of order p^r and denoted by $GF(p^r)$. In Chapter 0, we gave a construction of this field. Let $p(x)$ be a monic irreducible polynomial of degree r over the field $\mathbb{Z}_p$. Then the field $GF(p^r)$ is given by

$$GF(p^r) = \{a_0 + a_1 t + \cdots + a_{r-1} t^{r-1} \mid a_0, a_1, \ldots, a_{r-1} \in \mathbb{Z}_p\}$$

and t satisfies the relation $p(t) = 0$.

There are many monic irreducible polynomials of degree r over the field $\mathbb{Z}_p$, and any one of these can be chosen to obtain a representation of the field $GF(p^r)$ as given above. In the application of Theorems 2.6.1 and 2.6.3, we need a primitive element in the field F. It would be convenient for our purpose if we find a polynomial whose root t is a primitive element in the field. (It is known that such a monic irreducible polynomial exists.)

Example 2.6.7 Find a 2-fold difference set family in the additive group of the field GF (9). Construct a BIBD with parameters (9, 18, 8, 4, 3).

Solution. It is easily shown that $p(x) = x^2 + 2x + 2$ is an irreducible polynomial in $\mathbb{Z}_3[x]$. Hence the field $F = GF(9)$ can be represented as

$$F = \{x\beta + y \mid x, y = 0, 1, 2\}$$

where β is a root of the polynomial $x^2 + x + 2$. It can be checked that β is a primitive element in the field F. Now $9 = 4(2) + 1$. Hence, by Theorem 2.6.3, the sets

$$S_0 = \{\beta^0, \beta^2, \beta^4, \beta^6\}, \quad S_1 = \{\beta^1, \beta^3, \beta^5, \beta^7\}$$

form a 2-fold (9, 4, 3)-difference set family in the additive group F. Using the relation $\beta^2 = -\beta - 2 = 2\beta + 1$, we obtain

$$S_0 = \{1, 2\beta + 1, 2, \beta + 2\}, \quad S_1 = \{\beta, 2\beta + 2, 2\beta, \beta + 1\}$$

This determines a BIBD D on the set F with parameters (9, 18, 8, 4, 3), consisting of the sets $S_0 + \xi$ and $S_1 + \xi$, where $\xi \in F$. For the sake of convenience, let us use a shorthand notation and write an element $x\beta + y$ in F as xy. Then

$$F = \{00, 01, 02, 10, 11, 12, 20, 21, 22\}$$

and

$$S_0 = \{01, 02, 12, 21\}, \quad S_1 = \{10, 11, 20, 22\}$$

Hence the BIBD D is given by

$$D = \begin{cases} \{01, 02, 12, 21\}, \{02, 00, 10, 22\}, \{00, 01, 11, 20\}, \{11, 12, 22, 01\}, \\ \{12, 10, 20, 02\}, \{10, 11, 21, 00\}, \{21, 22, 02, 11\}, \{22, 20, 00, 12\}, \\ \{20, 21, 01, 10\}, \{10, 11, 20, 22\}, \{11, 12, 21, 20\}, \{12, 10, 22, 21\}, \\ \{20, 21, 00, 02\}, \{21, 22, 01, 00\}, \{22, 20, 02, 01\}, \{00, 01, 10, 12\}, \\ \{01, 02, 11, 10\}, \{02, 00, 12, 11\} \end{cases}$$

EXERCISES 2.6

1. Show that any subset of $\mathbb{Z}_n$ of $n-1$ elements is a difference set in the additive group $\mathbb{Z}_n$.

2. Show that $\{1, 5, 6, 8\}$ is a difference set in the additive group $\mathbb{Z}_{13}$.

3. Show that $\{0, 1, 2, 4, 5, 8, 10\}$ is a difference set in the additive group $\mathbb{Z}_{15}$. What are its parameters?

4. Construct a symmetric BIBD with parameters $(13, 4, 1)$. Hence obtain a BIBD with parameters $(9, 12, 8, 6, 5)$.

5. Construct a symmetric BIBD with parameters $(15, 8, 4)$.

6. Construct a BIBD with parameters $(7, 14, 6, 3, 2)$.

7. Construct a BIBD with parameters $(8, 14, 7, 4, 3)$.

8. Show that if S is a difference set in a group G, then $-S = \{-x \in G \mid x \in S\}$ is also a difference set.

9. Find all quadratic residues modulo 19.

10. Verify that $\{0, 1, 2, 3, 5, 7, 12, 13, 16\}$ is a difference set in $\mathbb{Z}_{19}$. What are its parameters?

11. Show, without direct verification, that $\{0, 1, 2, 3, 5, 7, 12, 13, 16\}$ is a difference set in $\mathbb{Z}_{19}$.

12. Show, without direct verification, that $\{0, 3, 6, 7, 12, 14, 16, 17, 18\}$ is a difference set in $\mathbb{Z}_{19}$.

13. Verify that $\{0, 1, 2, 5, 12, 18, 22, 24, 26, 27, 29, 32, 33\}$ is a difference set in $\mathbb{Z}_{40}$. Find its parameters.

14. Construct a BIBD with parameters $(9, 18, 8, 4, 3)$.

15. Construct a BIBD with parameters $(9, 18, 10, 5, 5)$.

16. Construct a BIBD with parameters $(10, 18, 9, 5, 4)$.

17. Construct a BIBD with parameters $(11, 22, 10, 5, 4)$.

18. Construct a BIBD with parameters $(12, 22, 11, 6, 5)$.

19. Construct a BIBD with parameters $(13, 39, 12, 4, 3)$.

20. Find a difference set in $\mathbb{Z}_{31}$.

21. If p is prime, $p \equiv 1 \pmod 4$, prove that Q_p is not a difference set in $\mathbb{Z}_p$.

22. Show that if a is a primitive element in $\mathbb{Z}_p$, p prime, then a is not a quadratic residue modulo p.

23. Let F be a finite field of order $p^r = 6t + 1$, and let a be a primitive element in F. Let $S_i = \{0, a^i, a^{t+i}\}$, $i = 0, 1, \ldots, t - 1$. Show that the sets $S_0, \ldots, S_{t-1}$ form a $(6t + 1, 3, 1)$-difference set family in the additive group F.

24. Construct a 3-fold difference set family in $\mathbb{Z}_{19}$.

25. Construct a 4-fold difference set family in $\mathbb{Z}_{17}$.

26. Construct a 5-fold difference set family in $\mathbb{Z}_{31}$.

27. Show that a BIBD induced by a difference set family has the following property: The elements in each block can be assigned k different given colors in such a way that each element in the whole set is assigned each color the same number of times.

28. Design an experiment to test nine brands of soaps with 18 subjects, with each subject receiving four brands wrapped in red, yellow, green, and blue paper in such a way that (a) each brand is used by the same number of subjects, (b) each pair of brands is used by the same number of subjects, and (c) each brand is given each color the same number of times.

29. Show that a polynomial of degree 3 is reducible over a field F if and only if it has a root in F.

30. Find all irreducible monic polynomials of degrees 2 and 3 over the field $\mathbb{Z}_2$.

31. Find all irreducible monic polynomials of degrees 2 and 3 over the field $\mathbb{Z}_3$.

32. Find all irreducible monic polynomials of degree 2 over the field $\mathbb{Z}_5$, and identify those whose root is a primitive element in $GF(25)$.

33. Find an irreducible monic polynomial of degree 2 over the field $\mathbb{Z}_7$ whose root is a primitive element in $GF(49)$.

34. Construct a 4-fold difference set family in $GF(25)$.

35. Construct a 6-fold difference set family in $GF(25)$.

2.7 CONSTRUCTION OF BIBDs FROM NEARRINGS

Besides difference sets and difference set families, nearrings are another good source for obtaining BIBDs. A nearring is a generalization of a ring in which two of the postulates—namely, commutativity of addition and left (or right) distributivity of multiplication over addition—are dropped.

Definition 2.7.1 A *nearring* is an algebraic structure $(N, +, \cdot)$ with the following properties.

(a) $(N, +)$ is a group (not necessarily abelian).
(b) $(N, \cdot)$ is a semigroup.

(c)   · is right distributive over +, that is,

$$(a + b) \cdot c = (a \cdot c) + (b \cdot c)$$

for all $a, b, c \in N$.

What we have defined is, strictly speaking, a *right nearring*. If we replace the right distributive property in (c) with left distributive, we obtain the definition of a *left nearring*. In this book, however, we shall always use the term *nearring* to mean a right nearring.

A nearring $(N, +, \cdot)$ is said to be *abelian* if $(N, +)$ is an abelian group.

Note that, contrary to the general convention regarding the use of the symbol + for a commutative binary operation only, we use + here even when the group $(N, +)$ is nonabelian. Consequently, the identity in the group is written 0 and the inverse of an element x is written $-x$.

Clearly, every ring is a nearring. A nearring $(N, +, \cdot)$ is a ring if it is abelian and · is both left and right distributive over +.

To understand the motivation behind the concept of a nearring, let us consider the ring of endomorphisms of an abelian group.

Let $(G, +)$ be an abelian group. Let $R = End(G)$ be the set of all endomorphisms of G—that is, homomorphisms from G to G. We define the sum of two endomorphisms $f, g \in R$ to be the pointwise sum of the mappings $f : G \to G$ and $g : G \to G$—that is, $f + g : G \to G$ such that

$$(f + g)(x) = f(x) + g(x) \quad \text{for all } x \in G$$

Then for all $x, y \in G$,

$$\begin{aligned}
(f + g)(x + y) &= f(x + y) + g(x + y) \\
&= f(x) + f(y) + g(x) + g(y) \\
&= f(x) + g(x) + f(y) + g(y) \\
&= (f + g)(x) + (f + g)(y)
\end{aligned}$$

Hence $f + g$ is an endormorphism of G. Thus R is closed under +.

Given any $f, g \in R$, the composite of the mappings f, g is again a mapping from G to G, and for all $x, y \in G$,

$$\begin{aligned}
(f \circ g)(x + y) &= f(g(x + y)) \\
&= f(g(x) + g(y)) \\
&= f(g(x)) + f(g(y)) \\
&= (f \circ g)(x) + (f \circ g)(y)
\end{aligned}$$

Hence $f \circ g$ is an endomorphism of G. Thus R is closed under the operation $\circ$. We claim that $(R, +, \circ)$ is a ring.

We first show that $(R, +)$ is an abelian group. Let $f, g, h \in R$. Then for all $x \in G$,

$$
\begin{aligned}
((f + g) + h)(x) &= (f + g)(x) + h(x) \\
&= (f(x) + g(x)) + h(x) \\
&= f(x) + (g(x) + h(x)) \\
&= f(x) + (g + h)(x) \\
&= (f + (g + h))(x)
\end{aligned}
$$

Hence $(f + g) + h = f + (g + h)$. Further,

$$(f + g)(x) = f(x) + g(x) = g(x) + f(x) = (g + f)(x)$$

Hence $f + g = g + f$. Thus the operation $+$ in R is commutative and associative.

Let $z : G \to G$ be the zero mapping; that is, $z(x) = 0$ for all $x \in G$. Then for all $x, y \in G$,

$$z(x + y) = 0 = 0 + 0 = z(x) + z(y)$$

Hence z is an endomorphism of G. Now for any $f \in R$,

$$(f + z)(x) = f(x) + z(x) = f(x) + 0 = f(x)$$

Hence $f + z = f$. Similarly $z + f = f$. Thus z is the identity for addition in R.

Given any $f \in R$, we define the mapping $f' : G \to G$ by

$$f'(x) = -f(x) \qquad \text{for all } x \in G$$

Then

$$(f + f')(x) = f(x) + f'(x) = f(x) - f(x) = 0 = z(x)$$

for all $x \in G$. Hence $f + f' = z$. Similarly $f' + f = z$. Thus f' is the inverse of f under addition. This proves that $(R, +)$ is an abelian group with identity z. As usual, we may write 0 for the identity element z and $-f$ for f'.

The composition of mappings is associative in general. Hence $(R, \circ)$ is a semigroup.

We now show that the operation $\circ$ is distributive over $+$. Let $f, g, h \in R$. Then for all $x \in G$,

$$
\begin{aligned}
((f + g) \circ h)(x) &= (f + g)(h(x)) \\
&= f(h(x)) + g(h(x)) \\
&= (f \circ h)(x) + (g \circ h)(x) \\
&= ((f \circ h) + (g \circ h))(x)
\end{aligned}
$$

Hence $(f + g) \circ h = (f \circ h) + (g \circ h)$, so $\circ$ is right distributive. Further, for all $x \in G$,

$$
\begin{aligned}
(h \circ (f + g))(x) &= h((f + g)(x)) \\
&= h(f(x) + g(x))
\end{aligned}
$$

$$= h(f(x)) + h(g(x))$$
$$= (h \circ f)(x) + (h \circ g)(x)$$
$$= ((h \circ f) + (h \circ g))(x)$$

Hence $h \circ (f + g) = (h \circ f) + (h \circ g)$. Thus $\circ$ is both right and left distributive over $+$ in R. This proves that $(R, +, \circ)$ is a ring. It is called the *ring of endomorphisms* of G.

Now suppose, instead of endomorphisms of an abelian group, we consider the set $N = Map(G, G)$ of all mappings from G to G, where G is a group (not necessarily abelian). As before, we write the group operation in G as $+$ and define the sum $f + g$ of any two mappings $f, g \in N$ to be pointwise. It is clear that N is closed under $+$ as well as under the composition $\circ$. If we go through the proof given above for $(R, +, \circ)$ to be a ring, we see that all the arguments there hold equally for $(N, +, \circ)$, with two exceptions:

1. If the group G is not abelian, then the addition in N is not commutative.
2. In proving the left distributive property in R, we made use of the equality

$$h(f(x) + g(x)) = h(f(x)) + h(g(x))$$

which is a consequence of the fact that h is a homomorphism. This equality does not necessarily hold if h is an arbitrary mapping from G to G.

Thus we conclude that $(N, +, \circ)$ satisfies all the postulates for a ring, except the commutativity of addition and the left distributivity of $\circ$ over $+$. Hence $(N, +, \circ)$ is a nearring. It is called the *nearring of mappings* of the group G.

We record these results in the following theorem.

THEOREM 2.7.2

(a) If $(G, +)$ is an abelian group, then $(End(G), +, \circ)$ is a ring.
(b) If $(G, +)$ is a group (not necessarily abelian), then $(Map(G, G), +, \circ)$ is a nearring.

Since a nearring lacks the symmetry of the distributive law of a ring, it is natural to expect that some of the results that hold symmetrically in a ring may fail to do so in a nearring. For example, for all a, b in a ring, $0a = 0 = a0$ and $(-ab) = -(ab) = a(-b)$. But in a nearring, in general, only the first of the two equalities holds in each case, as shown by the following theorem.

THEOREM 2.7.3 Let N be a nearring. Then

(a) $0a = 0$ for all $a \in N$.
(b) $(-a)b = -(ab)$ for all $a, b \in N$.

Proof:

(a)   For all $a \in N$,

$$0 + 0a = 0a = (0 + 0)a = 0a + 0a$$

Hence, by right cancellation in the group $(N, +)$, $0 = 0a$.

(b)   For all $a, b \in N$,

$$(-a)b + ab = (-a + a)b = 0b = 0$$

Hence $(-a)b = -(ab)$.   ∎

The following example shows that $a0$ is not necessarily equal to 0 in a nearring, in general: Let $(N = Map(G, G), +, \circ)$ be the nearring of mappings of a nontrivial group G. Let a be a nonzero element in G, and let $f \in N$ such that $f(0) = a$ and $f(x) = 0$ for all $x \neq 0$. Let z denote the zero mapping—that is, $z(x) = 0$ for all $x \in G$. Then $(f \circ z)(x) = f(z(x)) = f(0) = a$ for all $x \in G$, so $f \circ z \neq z$.

Given a nearring $(N, +, \cdot)$, let M be a nonempty subset of N. If M itself is a nearring under the operations $+$ and $\cdot$ (restricted to M), then M is called a *subnearring* of N.

Clearly, if M is a subnearring of N, then $(M, +)$ is a subgroup of $(N, +)$ and M is closed under $\cdot$. Conversely, if these conditions hold, then $(M, +, \cdot)$ satisfies all the conditions for a nearring and hence M is a subnearring of N.

Example 2.7.4 Given a group $(G, +)$, let $M(G)$ denote the set $Map(G, G)$ and

$$M_0(G) = \{f \in M(G) \mid f(0) = 0\}$$

Show that $M_0(G)$ is a subnearring of the nearring $(M(G), +, \circ)$.

Solution. Let $f, g \in M_0(G)$. Then $(f - g)(0) = f(0) - g(0) = 0 - 0 = 0$, so $f - g \in M_0(G)$. Thus $M_0(G)$ is a subgroup of $(M(G), +)$. Further, $(f \circ g)(0) = f(g(0)) = f(0) = 0$, so $f \circ g \in M_0(G)$. Hence $M_0(G)$ is closed under the operation $\circ$. This proves that $M_0(G)$ is a subnearring of $M(G)$.

2.8 PLANAR NEARRINGS

We now consider a special class of nearrings known as planar nearrings. To define a planar nearring, we first introduce an equivalence relation in a nearring.

Definition 2.8.1 Let N be a nearring. For $a, b \in N$, we define $a \sim b$ to mean that

$$na = nb \qquad \text{for all } n \in N$$

It is clear that $\sim$ is an equivalence relation in N. We refer to an equivalence class under $\sim$ as a $\sim$-class and denote the set of all $\sim$-classes in N by $N/\sim$. Thus $N/\sim$ is a partition of the set N. If $a \sim b$ does not hold, we write $a \not\sim b$.

The following theorem shows that $\sim$ is a right congruence with respect to multiplication in N.

THEOREM 2.8.2 Let N be a nearring, and let $a, b \in N$. Then

$$a \sim b \Rightarrow ac \sim bc \qquad \text{for all } c \in N$$

Proof: Let $a \sim b$. Then for all $n \in N$, $na = nb$. Hence for any $c \in N$, $n(ac) = (na)c = (nb)c = n(bc)$ for all $n \in N$. So $ac \sim bc$. ∎

Definition 2.8.3 A nearring N is said to be *planar* if

(a) $|N/\sim| \geq 3$; that is, there are at least three $\sim$-classes in N.
(b) For all $a, b, c \in N$, with $a \not\sim b$, the equation

$$xa = xb + c$$

has a unique solution for x in N.

Example 2.8.4 For $u, v \in \mathbb{C}$, define $u * v = u|v|$. Show that $(\mathbb{C}, +, *)$ is a planar nearring.

Solution. For all $u, v, w \in \mathbb{C}$,

$$(u * v) * w = u|vw| = u|(v|w|)| = u * (v * w)$$

and

$$(u + v) * w = (u + v)|w| = u * w + v * w$$

Hence $(\mathbb{C}, +, *)$ is a nearring. Further, $u \sim v$ holds if and only if $|u| = |v|$. So there are infinitely many equivalence classes in $\mathbb{C}$. Given $u, v \in \mathbb{C}$ with $|u| \neq |v|$, the equation $x * u = x * v + w$ has the unique solution $x = w(|u| - |v|)^{-1}$. This proves that $(\mathbb{C}, +, *)$ is a planar nearring.

We now prove some basic properties of a planar nearring.

THEOREM 2.8.5 Let N be a planar nearring. Then for every $a \in N$,

$$0a = 0 = a0$$

Proof: The equality $0a = 0$ was proved in Theorem 2.7.3. To prove the second equality, let $b \in N$ such that $b \not\sim 0$. Then, by the definition of a planar nearring, the equation

$$xb = x0 + 0$$

has a unique solution. Now $x = 0$ is obviously a solution because $0b = 0 = 00 + 0$. But $x = a0$ is also a solution because

$$(a0)b = a(0b) = a0 = a(00) = (a0)0 = (a0)0 + 0$$

Hence, by the uniqueness of the solution, it follows that $a0 = 0$.   ∎

A nearring N is said to be *zero symmetric* if $0a = 0 = a0$ for all $a \in N$. Thus every planar nearring is zero symmetric. The following result is an immediate consequence of Theorem 2.8.5.

THEOREM 2.8.6 Let N be a planar nearring. Then for all $a, b \in N$, with $a \not\sim 0$, the equation $xa = b$ has a unique solution.

Proof: Since $x0 = 0$ for all $x \in N$, the equation $xa = b$ can be rewritten as $xa = x0 + b$, which has a unique solution because $a \not\sim 0$.   ∎

THEOREM 2.8.7 Let N be a planar nearring and let $a, b \in N$. Then

(a)   If $ca = cb$ for some nonzero $c \in N$, then $a \sim b$. Consequently, $a \not\sim b \Rightarrow na \neq nb$ for all $n \in N$, $n \neq 0$.
(b)   If $ca \sim cb$ for some $c \in N$, $c \not\sim 0$, then $a \sim b$. Consequently, $a \not\sim b \Rightarrow na \not\sim nb$ for all $n \in N$, $n \not\sim 0$.

Proof:

(a)   Suppose $ca = cb$ for some $c \neq 0$ but $a \not\sim b$. Then the equation $xa = xb$ is satisfied by $x = 0$ as well as by $x = c$. Therefore $c = 0$, a contradiction. Hence $a \sim b$.
(b)   Suppose $ca \sim cb$ for some $c \not\sim 0$. Then for any nonzero $n \in N$, $nc \neq 0$ and $(nc)a = n(ca) = n(cb) = (nc)b$, hence $(nc)a = (nc)b$. Therefore $a \sim b$.   ∎

Let us write N^* to denote the set of all elements in a nearring N that are not equivalent to 0 under $\sim$; that is,

$$N^* = \{a \in N \mid a \not\sim 0\}$$

THEOREM 2.8.8 Let N be a planar nearring, and let $a \in N^*$. Then

(a)   There exists a unique element $1_a \in N^*$ such that $1_a a = a$.
(b) $1_a 1_a = 1_a$.
(c) 1_a is a right identity for multiplication in N; that is, $n1_a = n$ for all $n \in N$.
(d)   There exists a unique element $a' \in N^*$ such that $a'a = 1_a$.

Proof:

(a) Since $a \not\sim 0$, the equation $xa = a$ has a unique solution $x = 1_a \in N$, so $1_a a = a$. If $1_a \sim 0$, then $1_a a \sim 0a$ and hence $a \sim 0$, a contradiction. Hence $1_a \in N^*$.

(b) $(1_a 1_a)a = 1_a(1_a a) = 1_a a = a$, so $x = 1_a 1_a$ is also a solution of the equation $xa = a$. Hence, by the uniqueness of solution, $1_a 1_a = 1_a$.

(c) Let b be any element in N. Then the equation $x1_a = b1_a$ has a unique solution. Now $x = b$ is obviously a solution, but $x = b1_a$ is also a solution because $(b1_a)1_a = b(1_a 1_a) = b1_a$. Hence $b1_a = b$.

(d) The equation $xa = 1_a$ has a unique solution $x = a'$, so $a'a = 1_a$. If $a' \sim 0$, then $1_a = a'a \sim 0$, a contradiction. Hence $a' \in N^*$. ∎

Definition 2.8.9 A nearring N is said to be *integral* if for all $a, b \in N$,
$$ab = 0 \Rightarrow a = 0 \text{ or } b = 0.$$

It follows from the definition that in an integral nearring N, $a \sim 0$ if and only if $a = 0$. Hence
$$N^* = \{a \in N \mid a \neq 0\}$$

THEOREM 2.8.10 Let N be an integral nearring, and let $a, b, \in N$. Then

(a) If $ac = bc$ for some $c \neq 0$, then $a = b$. Consequently, $a \neq b \Rightarrow an \neq bn$ for all nonzero $n \in N$.

(b) If $ac \sim bc$ for some $c \neq 0$, then $a \sim b$. Consequently, $a \not\sim b \Rightarrow an \not\sim bn$ for all nonzero $n \in N$.

Proof:

(a) Suppose $ac = bc$. Then $(a - b)c = ac + (-b)c = ac - bc = 0$. Since $c \neq 0$, it follows that $a - b = 0$. Hence $a = b$.

(b) Suppose $ac \sim bc$ for some $c \neq 0$. Then for all $n \in N$, $n(ac) = n(bc)$; hence $(na)c = (nb)c$. Therefore $na = nb$ and hence $a \sim b$. ∎

The following theorem shows that every nonzero element a in an integral planar nearring has a two-sided inverse with respect to its left identity 1_a.

THEOREM 2.8.11 Let N be an integral planar nearring, and let $a \in N^*$. Then there exists a unique element $a^{-1} \in N^*$ such that $a^{-1}a = 1_a = aa^{-1}$.

Proof: By Theorem 2.8.8, there exists a unique element $a' \in N^*$ such that $a'a = 1_a$, so
$$(aa')a = a(a'a) = a1_a = a = 1_a a$$

Hence, by Theorem 2.8.10, $aa' = 1_a$. On writing $a' = a^{-1}$, we have $a^{-1}a = 1_a = aa^{-1}$.

■

2.9 FINITE INTEGRAL PLANAR NEARRINGS AND BIBDs

We now obtain some results for finite integral planar nearrings, leading to the main result that every finite integral planar nearring N yields a BIBD on the set N. As usual, we write Na and aN to mean

$$Na = \{na \mid n \in N\}, \quad aN = \{an \mid n \in N\}$$

THEOREM 2.9.1 Let N be a finite integral nearring. Then for every nonzero $a \in N$, $Na = N$.

Proof: By Theorem 2.8.10, for all $x, y \in N$, $x \neq y \Rightarrow xa \neq ya$. Hence $|Na| = |N|$. Since Na is a subset of a finite set N, it follows that $Na = N$.   ■

By a *set of $\sim$-representatives* in N^* we mean a subset U of N^* consisting of exactly one representative element from each $\sim$-class in N^*. Obviously, for any such set U, $|U| = |N^*/\sim|$.

THEOREM 2.9.2 Let N be a finite integral planar nearring, and let $|N^*/\sim| = k$. Let U be a set of $\sim$-representatives in N^*. Let $a \in N^*$. Then

(a) aU is a set of $\sim$-representatives in N^*.
(b) $aN^* = aU$.
(c) $abN^* = aN^*$ for all $b \in N^*$.

Proof:

(a) U has k elements, and no two of them belong to the same $\sim$-class. By Theorem 2.8.7, for any $x, y \in N$, $x \not\sim y \Rightarrow ax \not\sim ay$. Hence aU has k elements, and no two of them are in the same $\sim$-class. Thus aU consists of exactly one element from each $\sim$-class in N^*.
(b) Every element $n \in N^*$ lies in some $\sim$-class in N^*. So $n \sim u$ for some $u \in U$ and hence $an = au$. Therefore $aN^* \subset aU$. Conversely, $U \subset N^*$, so $aU \subset aN^*$. Hence $aN^* = aU$.
(c) If $b \in N^*$, then $bN^* = bU$, and bU is a set of $\sim$-representatives in N^*. Hence $abN^* = a(bN^*) = a(bU) = aN^*$.   ■

THEOREM 2.9.3 Let N be a finite integral planar nearring. Then the sets aN^*, $a \in N^*$, form a partition of N^*.

Consequently, the number of distinct sets of the form aN^* is $(v-1)/k$, where $v = |N|, k = |N^*/\sim|$.

Proof: By Theorem 2.8.8, for any $b \in N^*$, $b = 1_b b$, and hence $b \in 1_b N^*$. Thus each element in N^* lies in some set aN^*.

Let $a, b \in N^*$. We claim that the sets aN^* and bN^* are either equal or disjoint. Suppose they have a common element c. Then $c = am = bn$ for some $m, n \in N^*$. Hence, by Theorem 2.9.2, $aN^* = amN^* = bnN^* = bN^*$. Hence any two sets aN^* and bN^* are either equal or disjoint. This proves that the sets aN^*, $a \in N^*$, form a partition of N^*.

Let r be the number of distinct sets of the form aN^*. By Theorem 2.9.2, $|aN^*| = k$ for each $a \in N^*$, and $|N^*| = v - 1$. It follows that $rk = v - 1$; hence $r = (v-1)/k$. ∎

COROLLARY 2.9.4 Let N be a finite integral planar nearring, with $v = |N|, k = |N^*/\sim|$. Then $k \mid (v-1)$.

THEOREM 2.9.5 Let N be a finite integral planar nearring, and let $a, b \in N^*$. If $aN^* = bN^*$, then either $a = b$ or $a \not\sim b$.

Proof: Suppose $c \in aN^* = bN^*$. Then $c = am = bn$ for some $m, n \in N^*$. If $m \sim n$, then $am = bn = bm$. Hence, by Theorem 2.8.10, $a = b$. Now suppose $m \not\sim n$. By Theorem 2.8.7, $xm \neq xn$ for all $x \in N^*$. So for any $t \in N^*$, $am = bn \Rightarrow tam = tbn \Rightarrow ta \neq tb \Rightarrow a \not\sim b$. This proves that either $a = b$ or $a \not\sim b$. ∎

THEOREM 2.9.6 Let N be a finite integral planar nearring, and let $|N| = v$, $|N^*/\sim| = k$. Then

(a) Each $\sim$-class C in N^* contains exactly $(v-1)/k$ elements.
(b) For any $\sim$-class C in N^*, the sets cN^*, $c \in C$, form a partition of N^*.
(c) Given any $a \in N^*$, there are exactly k elements $a_1, \ldots, a_k$ in N^* such that $a_i N^* = aN^*$ for each $i = 1, \ldots, k$.

Proof:

(a) Suppose C consists of r elements $c_1, \ldots, c_r$. Then, by Theorem 2.9.5, $c_1 N^*, \ldots, c_r N^*$ are r distinct and hence disjoint sets. By Theorem 2.9.3, the total number of distinct sets aN^*, $a \in N^*$, is $(v-1)/k$. Hence $r \leq (v-1)/k$. Since there are k $\sim$-classes and they form a partition of N^*, it follows that each $\sim$-class C in N^* contains exactly $r = (v-1)/k$ elements.
(b) This follows immediately from (a).
(c) Let $C_1, \ldots, C_k$ be the $\sim$-classes in N^*. For each i, the sets cN^*, $c \in C_i$, form a partition of N^*. Hence, given any set aN^*, C_i has a unique element a_i such that $a_i N^* = aN^*$. ∎

We now consider subsets of N^* of the form $aN^* + b$, where $a, b \in N, a \neq 0$.

THEOREM 2.9.7 Let N be a finite integral planar nearring. Let $a_1, a_2, b_1, b_2 \in N$, with $a_1, a_2 \neq 0$. Then

$$a_1 N^* + b_1 = a_2 N^* + b_2 \quad \Leftrightarrow \quad a_1 N^* = a_2 N^*, \qquad b_1 = b_2$$

Proof: It is clear that $a_1 N^* + b_1 = a_2 N^* + b_2$ holds if and only if $a_1 N^* = a_2 N^* + (b_2 - b_1)$. So we need only prove that $aN^* = bN^* + c$ if and only if $aN^* = bN^*$, $c = 0$.

Suppose $aN^* = bN^* + c$ and $a, b, c \neq 0$. Since $|cN^*| = k \geq 2$, we can choose $d \in N^*$ such that $cd \neq c$. By Theorem 2.9.1, $N^*d = N^*$. Hence

$$bN^* + c = aN^* = a(N^*d) = (aN^*)d = (bN^* + c)d = bN^*d + cd = bN^* + cd$$

Therefore $bN^* = bN^* + t$, where $t = cd - c \neq 0$. It follows that $bN^* = bN^* + it$ for any integer i. Let $T = [t]$ be the subgroup of $(N, +)$ generated by t, and let q be the order of T. So $T = \{it \mid i = 0, \ldots, q - 1\}$. Hence

$$bN^* = \bigcup_{i=0}^{q-1}(bN^* + it) = \{bn + it \mid n \in N^*, i = 0, \ldots, q - 1\} = \bigcup_{n \in N^*} (bn + T)$$

Now every coset of T has q elements and hence

$$k = |bN^*| = \left| \bigcup_{n \in N^*} (bn + T) \right| = mq$$

for some integer m, so $q \mid k$. By Theorem 2.9.3, $k \mid (v - 1)$ and hence $q \mid (v - 1)$, where $v = |N|$. By the Lagrange theorem, $q \mid v$; hence $q = 1$. But this implies $t = 0$, a contradiction. We conclude that $aN^* = bN^* + c$ implies $c = 0$ and hence also $aN^* = bN^*$. The converse is obvious.   ∎

Now we prove the main result.

THEOREM 2.9.8 Let N be a finite integral planar nearring, and let $|N| = v$, $|N^*/ \sim| = k$. Then

$$D = \{aN^* + b \mid a, b \in N, a \neq 0\}$$

is a BIBD on the set N with parameters (v, b, r, k, λ), where

$$b = \frac{v(v - 1)}{k}, \quad r = v - 1, \quad \lambda = k - 1$$

Proof: It is clear that every block $aN^* + b$ in D consists of k elements. By Theorem 2.9.3, there are exactly $(v - 1)/k$ distinct sets $aN^*, a \in N^*$. Hence, by Theorem 2.9.7, D has exactly $v(v - 1)/k$ distinct blocks.

Given $c \in N$, let b be any element in N, $b \neq c$. Then $c - b \neq 0$. Hence, by Theorem 2.9.3, there is exactly one set of the form aN^*, say $a_b N^*$, that contains $c - b$, so $c = (c - b) + b \in a_b N^* + b$. Now b can be chosen in $v - 1$ ways.

Hence there are exactly $v - 1$ blocks of D that contain c—namely, $a_b N^* + b$, with $b \in N, b \neq c$.

Finally, we show that every two elements of N occur together in exactly $k - 1$ blocks of D. Let $c, d \in N$, $c \neq d$. It is clear that $c, d \in aN^* + b$ if and only if $0, d - c \in aN^* + (b - c)$. Hence we need only show that for any $c \neq 0$, there are exactly $k - 1$ blocks of D that contain both 0 and c.

Let $U = \{u_1, \ldots, u_k\}$ be a set of $\sim$-representatives in N^*. By Theorem 2.9.2, $aN^* = aU$. If $0, c \in aN^* + b$, then $0 = au_i + b$ and $c = au_j + b$ for some u_i, $u_j \in U$, $i \neq j$. Hence $b = -au_i$ and $c = au_j - au_i$. The last equation can be rewritten as

$$(-a)u_i = (-a)u_j + c$$

Conversely, given any $u_i, u_j \in U$, $i \neq j$, the equation

$$xu_i = xu_j + c \tag{1}$$

has the unique solution $x = -a_{ij}$. Then, taking $b_{ij} = -a_{ij}u_i$, we have $0 = a_{ij}u_i + b_{ij}$ and $c = a_{ij}u_j + b_{ij}$, so $0, c \in a_{ij}N^* + b_{ij}$. Thus the blocks of D that contain $0, c$ are of the form

$$a_{ij}N^* - a_{ij}u_i, \qquad i, j = 1, \ldots, k; \quad i \neq j$$

where $-a_{ij}$ is the unique solution of equation (1). We claim that there are exactly $k - 1$ distinct blocks among these.

It is easily seen that for any fixed i, the elements a_{ij} are all distinct. If $a_{ij} = a_{is}$, then

$$(-a_{ij})u_j + c = (-a_{ij})u_i = (-a_{is})u_i = (-a_{is})u_s + c = (-a_{ij})u_s + c$$

So $(-a_{ij})u_j = (-a_{ij})u_s$, which implies that $u_j \sim u_s$ and hence $j = s$. Thus $j \neq s$ implies $a_{ij} \neq a_{is}$. In particular, it follows that $a_{12}, \ldots, a_{1k}$ are all distinct. Hence $a_{1j}N^* - a_{1j}u_1$, $j = 2, \ldots, k$, are $k - 1$ distinct blocks containing $0, c$.

On the other hand, there are only $k(k - 1)$ ordered pairs (u_i, u_j), $i \neq j$. It is easily seen that the $k(k - 1)$ ordered pairs $(a_{ij}, -a_{ij}u_i)$ are all distinct. By Theorem 2.9.6, for any given set aN^*, there are exactly k elements α_i such that $aN^* = \alpha_i N^*$, $i = 1, \ldots, k$. So the number of distinct blocks among $a_{ij}N^* - a_{ij}u_i$ is equal to $\dfrac{k(k - 1)}{k} = k - 1$. Hence there are exactly $k - 1$ blocks in D that contain both 0 and c.

This proves that D is a BIBD with parameters $(v, v(v - 1)/k, v - 1, k, k - 1)$.   ∎

We give below another proof, which is interesting for its own sake, for the assertion that there are only $k - 1$ distinct blocks among $a_{ij}N^* - a_{ij}u_i$. We show directly that any such block is equal to one of the blocks $a_{1j}N^* - a_{1j}u_1$, $j = 2, \ldots, k$. Since $-a_{ij}$ is the solution of equation (1), we have

$$(-a_{ij})u_i = (-a_{ij})u_j + c$$

By Theorem 2.8.11, for every $a \in N^*$, $aa^{-1} = 1_a = a^{-1}a$ is a right identity in N. Hence the above equation can be rewritten as

$$(-a_{ij})u_i(u_1^{-1}u_1) = (-a_{ij})(u_i(u_1^{-1}u_1)u_i^{-1})u_j + c$$

or

$$(-a_{ij}u_iu_1^{-1})u_1 = (-a_{ij}u_iu_1^{-1})(u_1u_i^{-1}u_j) + c$$

Now $u_1u_i^{-1}u_j \sim u_s$ for some $u_s \in U$, so we have

$$(-a_{ij}u_iu_1^{-1})u_1 = (-a_{ij}u_iu_1^{-1})u_s + c$$

If $s = 1$, then $c = 0$, which is not true. Hence $s > 1$, so $u_1 \not\sim u_s$. Therefore $-a_{ij}u_iu_1^{-1}$ is the unique solution of the equation $xu_1 = xu_s + c$. But the solution of this equation, by hypothesis, is $x = -a_{1s}$, so $a_{1s} = a_{ij}u_iu_1^{-1}$. Hence

$$a_{1s}N^* - a_{1s}u_1 = a_{ij}u_iu_1^{-1}N^* - a_{ij}u_iu_1^{-1}u_1 = a_{ij}N^* - a_{ij}u_i$$

Hence every block $a_{ij}N^* - a_{ij}u_i$ is equal to some block $a_{1s}N^* - a_{1s}u_1$, $s = 2, \ldots, k$.

   We observe that Theorem 2.9.8 gives a trivial result when $k = 2$. Then $b = v(v - 1)/2$, so D is the set of all subsets of order 2 of N, which we already know from Theorem 2.1.3 to be a BIBD. Thus if the nearring N has only three equivalence classes, the theorem doesn't yield anything new.

2.10 FINITE FIELDS AND PLANAR NEARRINGS

We now show how a finite integral planar nearring and hence a BIBD can be obtained from a finite field.

THEOREM 2.10.1 Let F be a finite field of order $q = p^r$ (p prime), and let β be a generator of the cyclic multiplicative group F^*. Suppose $q - 1 = sk$ is a nontrivial factorization. Define a new multiplication $*$ in F as follows: For any $a, b \in F$, $a * b = 0$ if $a = 0$ or $b = 0$. Otherwise, if $a = \beta^i, b = \beta^j$, then

$$a * b = \beta^{i+j-j'}$$

where $j' = res_s(j)$ is the residue on dividing j by s. Then $(F, +, *)$ is a finite integral planar nearring with $|F^*/\sim| = k$.

Proof: Let $a, b, c \in F$. If any of them is zero, the equalities $(a * b) * c = a * (b * c)$ and $(a + b) * c = a * c + b * c$ hold trivially. So let $a = \beta^i, b = \beta^j, c = \beta^r$, and suppose $i = us + i', j = vs + j', r = ws + r'$. Then

$$(a * b) * c = \beta^{i+vs} * \beta^r = \beta^{i+vs+ws}$$
$$a * (b * c) = \beta^i * \beta^{j+ws} = \beta^{i+vs+ws}$$

So $(a * b) * c = a * (b * c)$. Further

$$(a + b) * c = (\beta^i + \beta^j) * \beta^r$$
$$= (\beta^i + \beta^j)\beta^{ws}$$
$$= \beta^{i+ws} + \beta^{j+ws}$$
$$= a * c + b * c$$

Thus $*$ is associative and right distributive over $+$. Hence $(F, +, *)$ is a nearring.

With $a = \beta^i$, $b = \beta^j$ as before, the equality $c * a = c * b$, $c \in F^*$, holds if and only if $c\beta^{us} = c\beta^{vs}$. Thus $a \sim b$ if and only if $u = v$. Hence the equivalence class containing a is $\{\beta^{us+t} \mid t = 0, 1, \ldots, s - 1\}$. Thus every equivalence class has s elements. Since F^* has $q - 1$ elements, it follows that the number of equivalence classes in F^* is equal to $(q - 1)/s = k$, so $|F^*/\sim| = k$. By hypothesis, $q - 1 = sk$ is a nontrivial factorization of $q - 1$; hence $k > 1$. So there are at least three equivalence classes in F.

Consider the equation $x * a = x * b + c$. This can be rewritten as $x\beta^{us} = x\beta^{vs} + c$. If $a \not\sim b$, then $u \neq v$ and hence $\beta^{us} - b^{vs} \neq 0$. The equation has a unique solution $x = c(\beta^{us} - b^{vs})^{-1}$.

This proves that $(F, +, *)$ is a finite integral planar nearring with $|F^*/\sim| = k$. ∎

The applicability of Theorem 2.10.1 depends on the factorization of $q - 1$. If p is an odd prime, then $p^r - 1$ is even, so there are at least two ways of factorizing $q - 1$. We can take $s = 2$ or $(q - 1)/2$, but if $p = 2$, then $q - 1$ is odd. The theorem is then applicable if and only if $q - 1$ is not prime.

Example 2.10.2 Let F be a field of order 11. Use Theorem 2.10.1 to construct a planar nearring $(F, +, *)$ with $|F^*/\sim| = 5$. Obtain the BIBD from this nearring.

Solution. We take $F = \mathbb{Z}_{11}$. Then $11 - 1 = 10 = 2.5$. We take $s = 2, k = 5$. It is easily checked that 2 is a primitive element in F, so we define the multiplication $*$ in F^* by the rule

$$2^i * 2^j = 2^{i+j-j'}$$

where j' is the remainder on dividing j by 2. The equivalence classes in F^* are given by $\{2^{2t}, 2^{2t+1}\}$, $t = 0, 1, 2, 3, 4$, so we have the following five equivalence classes:

$$\{1, 2\}, \{4, 8\}, \{5, 10\}, \{9, 7\}, \{3, 6\}$$

Using the rule for $*$, we get this multiplication table:

$*$	1	2	3	4	5	6	7	8	9	10
1	1	1	3	4	5	3	9	4	9	5
2	2	2	6	8	10	6	7	8	7	10
3	3	3	9	1	4	9	5	1	5	4
4	4	4	1	5	9	1	3	5	3	9
5	5	5	4	9	3	4	1	9	1	3
6	6	6	7	2	8	7	10	2	10	8
7	7	7	10	6	2	10	8	6	8	2
8	8	8	2	10	7	2	6	10	6	7
9	9	9	5	3	1	5	4	3	4	1
10	10	10	8	7	6	8	2	7	2	6

By Theorem 2.9.8, the BIBD D determined by the planar nearring $(F^*, +, *)$ consists of the blocks $a * F^* + b$, where $a, b \in F$, $a \neq 0$. To obtain all the distinct blocks, we need to consider a in some fixed equivalence class only, say $\{1,2\}$. So D consists of the blocks $1 * F^* + b$ and $2 * F^* + b$, where $b = 0, 1, \ldots, 10$. Now

$$1 * F^* = \{1, 3, 4, 5, 9\}, \quad 2 * F^* = \{2, 6, 7, 8, 10\}$$

Hence the blocks of D are

$$1 * F^* + 0 = \{1, 3, 4, 5, 9\}, \quad 2 * F^* + 0 = \{2, 6, 7, 8, 10\}$$
$$1 * F^* + 1 = \{2, 4, 5, 6, 10\}, \quad 2 * F^* + 1 = \{3, 7, 8, 9, 0\}$$
$$1 * F^* + 2 = \{3, 5, 6, 7, 0\}, \quad 2 * F^* + 2 = \{4, 8, 9, 10, 1\}$$
$$1 * F^* + 3 = \{4, 6, 7, 8, 1\}, \quad 2 * F^* + 3 = \{5, 9, 10, 0, 2\}$$
$$1 * F^* + 4 = \{5, 7, 8, 9, 2\}, \quad 2 * F^* + 4 = \{6, 10, 0, 1, 3\}$$
$$1 * F^* + 5 = \{6, 8, 9, 10, 3\}, \quad 2 * F^* + 5 = \{7, 0, 1, 2, 4\}$$
$$1 * F^* + 6 = \{7, 9, 10, 0, 4\}, \quad 2 * F^* + 6 = \{8, 1, 2, 3, 5\}$$
$$1 * F^* + 7 = \{8, 10, 0, 1, 5\}, \quad 2 * F^* + 7 = \{9, 2, 3, 4, 6\}$$
$$1 * F^* + 8 = \{9, 0, 1, 2, 6\}, \quad 2 * F^* + 8 = \{10, 3, 4, 5, 7\}$$
$$1 * F^* + 9 = \{10, 1, 2, 3, 7\}, \quad 2 * F^* + 9 = \{0, 4, 5, 6, 8\}$$
$$1 * F^* + 10 = \{0, 2, 3, 4, 8\}, \quad 2 * F^* + 10 = \{1, 5, 6, 7, 9\}$$

D is a BIBD with parameters $(11, 22, 10, 5, 4)$.

An interesting feature of D is worth noting. The sets $1 * F^*$ and $2 * F^*$ are both difference sets in the group $\mathbb{Z}_{11}$, so each of them determines a symmetric BIBD. Thus D is the union of two symmetric BIBDs. As a matter of fact, $1 * F^*$ is equal to the set Q_{11} of quadratic residues modulo 11, and $2 * F^* = -Q_{11} = \{-x \in \mathbb{Z}_{11} \mid x \in Q_{11}\}$.

EXERCISES 2.10

1. Let $(M(G), +, \circ)$ be the nearring of mappings of a group G, and let $S \subset End(G)$. Show that $M_S(G) = \{f \in M_0(G) \mid f \circ s = s \circ f \text{ for all } s \in S\}$ is a subnearring of $M(G)$.

2. Given a nearring N, show that:
   (a) $N_0 = \{a \in N \mid a0 = 0 = 0a\}$ is a subnearring of N.
   (b) $N0 = \{a0 \mid a \in N\}$ is a subnearring of N.
   (c) $N = N0 + N_0$.

3. Given a nearring N, $a \in N$, show that
   (a) $Na = \{na \mid n \in N\}$ is a subgroup of $(N, +)$.
   (b) $\text{Ann}(a) = \{n \in N \mid na = 0\}$ is a normal subgroup of $(N, +)$.

4. For all $u, v \in \mathbb{C}$, define

$$u * v = \begin{cases} u \operatorname{Re} v & \text{if } \operatorname{Re} v \neq 0 \\ u \operatorname{Im} v & \text{if } \operatorname{Re} v = 0 \end{cases}$$

Show that $(\mathbb{C}, +, *)$ is an integral planar nearring.

5. For all $u, v \in \mathbb{C}$, define

$$u * v = \begin{cases} 0 & \text{if } v = 0 \\ \dfrac{uv}{|v|} & \text{if } v \neq 0 \end{cases}$$

Show that $(\mathbb{C}, +, *)$ is an integral planar nearring.

6. For all $u, v \in \mathbb{C}$, define

$$u * v = u |(\operatorname{Re} v)^2 - (\operatorname{Im} v)^2|^{1/2}$$

Show that $(\mathbb{C}, +, *)$ is a nonintegral planar nearring.

7. Let N be a planar nearring. For $a, b \in N^*$, let $a \sim_1 b$ mean that $1_a = 1_b$. Show that $\sim_1$ is an equivalence relation in N^*, and every equivalence class under $\sim_1$ is a subgroup of $(N, \cdot)$.

8. Construct an integral planar nearring N, with $|N| = 13$, $|N^*/\sim| = 4$. Obtain the BIBD determined by it.

9. Construct a BIBD with parameters $(17, 34, 16, 8, 7)$ from a suitable planar nearring.

10. Construct a BIBD with parameters $(19, 57, 18, 6, 5)$ from a suitable planar nearring.

CHAPTER
3

ALGEBRAIC CRYPTOGRAPHY

3.1 SUBSTITUTION CIPHERS

Cryptography is the study of methods of transforming a secret message in such a way that it can be understood only by an authorized recipient who has been provided with a secret *key* for deciphering it. If the message is intercepted by an unauthorized recipient (let us call him the *Enemy*), he should not be able to interpret it.

Cryptanalysis deals with the reverse problem of devising ways (used by the Enemy) to decipher an intercepted secret message without possessing the secret key used by the sender. This is referred to as "breaking the code." Cryptography and cryptanalysis are the two main components of the science of *cryptology*. In this book we confine ourselves to cryptography.

Basically there are two ways in cryptography of transforming a message. One of them, called *transposition,* is to rearrange the characters in the message according to some rule. The other method, called *substitution,* is to replace each character by another according to some rule. In this book we mainly consider substitution methods, especially those that employ algebraic techniques and make use of algebraic systems like rings and fields. We will see an application of the transposition method in the block ciphers discussed in Section 3.3.

The general principle involved in the substitution method is to select a permutation f of the set of letters of the alphabet (in which the message is written) and replace each letter x in the message with $f(x)$. We refer to this permutation f as the *key.*

Suppose we wish to send this message to a friend:

ALGEBRA IS GREAT FUN

Since we don't want the Enemy to know this secret, we decide to send the message in a disguised form. Let us take f to be the permutation given by the table below. For each letter x in the top row, $f(x)$ is the letter directly below x.

A	B	C	D	E	F	G	H	I	J	K	L	M	N	O	P	Q	R	S	T	U	V	W	X	Y	Z
N	F	P	R	K	S	C	E	L	A	J	Q	G	T	B	I	M	D	H	O	X	Z	V	Y	U	W

Replacing each letter x in the message with $f(x)$, and suppressing the blank spaces, we transform the original message to

NQCKFDNLHCDKNOSXT

We send this disguised message to our friend, who has been provided in advance with the secret key f. She can recover the original message by replacing each letter y in the received message with $f^{-1}(y)$. But an unauthorized recipient, not knowing the secret key, cannot find out the original message.

It might seem that the Enemy could try all possible permutations of the alphabet until he finds the one that generates a meaningful message from the disguised message he intercepted. Such a procedure for breaking the code is known as an *exhaustive key search*. But the number of permutations of the set of 26 letters in the English alphabet is equal to

$$26! = 403291461126605635584000000 \approx 4 \times 10^{26}$$

This number is so large that it is not feasible to try all possible permutations. If our Enemy takes one second to check each permutation, it will take him more that 10^{19} years to go through all possible permutations. Even if he is a wizard and can check a million permutations in one second, it will still take him more than ten thousand billion years. So an exhaustive key search is not a feasible option.

Moreover, by including lowercase letters, punctuation marks, numerals, and so on, we can have a much larger set of characters, and the number of permutations of this set becomes mind-boggling. It is thus clear that it is computationally infeasible to break the code by the brute-force method of trying out all possible permutations of the alphabet.

Now let us introduce some terminology. The original message in readable form is called the *plaintext* message (or simply plaintext). The transformed messsage is called the *ciphertext* (or *cryptogram*). In our example, the plaintext is "ALGEBRA IS GREAT FUN," and the ciphertext is "NQCKFDNLHCDKNOSXT." The process of transforming the plaintext into the ciphertext is called *enciphering* (or *encryption*). The reverse process of recovering the original plaintext from the received ciphertext is called *deciphering*. A particular system of enciphering is called a *cipher* (or *cryptosystem*). A system of enciphering that uses the method of substitution is called a *substitution cipher*. The set of characters used in writing the plaintext message is called the *plaintext alphabet*. It may consist of the letters of the English alphabet alone, or it may also include other characters like punctuation marks, numerals, and so on. Likewise, the set of characters used for the ciphertext is called the *cipher alphabet*. The plaintext alphabet and the cipher alphabet may be the same (as in our example) or may be different. For example, the plaintext alphabet may consist of capital letters A, B, . . . , Z, but the cipher alphabet may be the set of numbers 1, 2, . . . ,26.

Let A and B denote the plaintext alphabet and the cipher alphabet, respectively. Let $f : A \rightarrow B$ be an injective mapping. (If A and B have the same number of elements, then f is bijective.) If the plaintext is transformed into the ciphertext by replacing each character x in it with $f(x)$, then the mapping f is called the *enciphering key* (or simply the *key*). Since f is injective, there is an inverse mapping g from Im $f \subset B$ to A such that $g(f(x)) = x$ for all $x \in A$. This mapping g is called the *deciphering key*. The original plaintext can be recovered from the ciphertext by replacing each character y in it with $g(y)$.

When a fixed key f is used to encipher every letter in the plaintext, as described above, we refer to the enciphering procedure as a *simple substitution cipher* or a *monoalphabetic cipher*. It suffers from the following disadvantage: It is known that the letters in the English alphabet occur with definite varying frequencies. These frequencies can be found by statistical analysis of long texts chosen at random. For example, the letter E has been observed to occur most frequently, followed by T. If we use the same mapping f to encipher all the letters in the plaintext, the frequency distribution of different letters is preserved in the ciphertext. If the received ciphertext is sufficiently long, it is possible to break the code by observing the relative frequencies of different characters in the ciphertext and comparing these with the known frequencies of letters in the English language. For example, if we find that the letter S occurs most frequently in the ciphertext, then we can conclude that S corresponds to E in the plaintext. Of course, this method of breaking the code does not work unless the ciphertext is long enough. For instance, in the example at the beginning of this chapter, the letter that occurs most frequently in the plaintext is A instead of E.

This problem can be overcome in two ways. One is to use different keys for enciphering successive letters. Such a system is called a *stream cipher* or a *polyalphabetic cipher*. A *periodic stream cipher with period* p is defined as follows: As before, let A and B denote the plaintext alphabet and the cipher alphabet, respectively. Let p be a positive integer greater than 1. Let $f_i :\rightarrow B$, $i = 1, \ldots, p$, be injective mappings (not all equal, but not necessarily all distinct). The plaintext is transformed into the ciphertext by using the mappings $f_1, \ldots, f_p$ in cyclic repetition to encipher successive characters in the plaintext. We take up the first p characters in the plaintext and encipher them by using the mappings $f_1, \ldots, f_p$. Then we do the same for the next p characters, and so on. In short, if the plaintext message is

$$m = a_1 a_2 \ldots a_p a_{p+1} \ldots a_{2p} a_{2p+1} \ldots$$

where $a_1, a_2, \ldots \in A$, then the ciphertext message is

$$m' = f_1(a_1)f_2(a_2) \ldots f_p(a_p)f_1(a_{p+1}) \ldots f_p(a_{2p})f_1(a_{2p+1}) \ldots$$

A periodic substitution cipher conceals the frequencies of different letters in the plaintext because the same character in the ciphertext can come from different characters in the plaintext. Therefore it is not possible to break the code by simply analyzing the ciphertext for the frequency distribution of the characters. However, if the period p is known, one can consider subsequences of the ciphertext obtained by taking every pth character. From a frequency analysis of these subsequences, one may succeed in discovering the enciphering keys f_i. The period p itself can be discovered by observing the presence of repeated blocks in the ciphertext.

The ultimate generalization of a periodic cipher is to have a randomly generated, indefinitely long sequence of keys $f_i : A \to B$, $i = 1, 2, \ldots$. The ith letter a_i in the plaintext is enciphered as $f_i(a_i)$. Once this particular sequence of keys has been used, it is discarded and not used again. Such a system is called *one-time pad substitution cipher.*

The second method of enciphering that conceals the relative frequencies of letters in the plaintext is *block substitution*. Instead of enciphering one letter at a time, we encipher a block of letters. Let r be a positive integer, and let A^r denote the set of all r-tuples of elements of the plaintext alphabet A. Let $f : A^r \to A^r$ be a permutation of the set A^r. (More generally, we can have an injective mapping $f : A^r \to B$.) The plaintext is divided into blocks of r characters each. If the blank space is not an element of A, then the blank spaces in the plaintext are suppressed. If the last block falls short of r characters, it is stuffed with some prefixed character to make up the shortage. Enciphering is done by replacing each block $x = (x_1, \ldots, x_r) \in A^r$ with $f(x)$. In other words, we treat the set A^r as an enlarged plaintext alphabet.

These two methods can be combined to obtain a *periodic block substitution*. We select injective mappings $f_i : A^r \to B$, $i = 1, \ldots, p$, and use them in cyclic repetition to encipher successive blocks of the plaintext.

These two methods, either alone or combined, make it difficult to break the code, but the enciphering process becomes harder to carry out. Consider the simplest case, $r = 2$. If the plaintext alphabet has 26 characters (the bare minimum in English), then the set A^2 has $26^2 = 676$ elements. So we need a permutation f of a set of 676 elements to use as an enciphering key. Moreover, it can be taken for granted that the Enemy is devilishly smart and well versed in all the tricks of cryptanalysis and, sooner or later, will be able to break the code. Therefore the only safeguard lies in changing the secret key frequently. Further the key has to be provided to the authorized recipient in advance, so we must have a secure method of communicating it. Both these tasks, selection and secure transmission of the secret key, pose a serious problem.

The solution is to devise an algorithm for generating enciphering keys—that is, permutations of the set A^r. Such an algorithm is called an *enciphering algorithm.*

Generally, the algorithm involves one or several parameters, say $k_1, \ldots, k_s$. For every choice of suitable values for these parameters, the algorithm determines a permutation f of the set A^r (or an injective mapping $f : A^r \to B$). That is, for every $x \in A^r$, we use the algorithm to find $f(x)$. In the context of a given algorithm of this kind, we use the term *key* to refer to the s-tuple $k = (k_1, \ldots, k_s)$. This algorithm must be reversible, by which we mean that for any given key k, we can find the deciphering algorithm that converts the ciphertext to the original plaintext.

The use of an enciphering algorithm makes it easier to perform the task of enciphering and deciphering. Moreover, the general form of the algorithm is given to the authorized recipient only once. Whenever the key is changed, it is only the new key k that has to be communicated. Since the key k consists of a small number of parameters, it is relatively easy to communicate. We may, in fact, take it for granted that the Enemy also knows the general form of the enciphering algorithm being used. It is only the key k that has to be kept secret.

The disadvantage of using an enciphering algorithm is that it reduces the number of available keys. The set of all possible values of the key k is called the *key space* of the cryptosystem. If the key space is not sufficiently large, the code can be broken by an exhaustive key search.

It is in order to devise such enciphering algorithms that we make use of abstract algebraic systems like rings and fields. Before considering these algebraic methods, however, we describe a very simple enciphering algorithm that does not require high-brow abstract algebra. It is also of some historical interest, since it was used in ancient times by the Roman emperor, Julius Caesar. Imagine the letters of the alphabet arranged around a circle in their usual order. If we shift each letter forward by k places around the circle, the resulting permutation of the alphabet is called a *cyclic shift* of length k or simply a *k-shift*. Enciphering done by using a cyclic shift is called a *shift cipher* (or a *Caesar cipher*). The actual cipher used by Caesar was a cyclic shift with $k = 3$. In this case, the letter A shifts to D, B to E, and so on. The table below shows the complete permutation given by the Caesar cipher with $k = 3$.

A	B	C	D	E	F	G	H	I	J	K	L	M	N	O	P	Q	R	S	T	U	V	W	X	Y	Z
D	E	F	G	H	I	J	K	L	M	N	O	P	Q	R	S	T	U	V	W	X	Y	Z	A	B	C

Example 3.1.1

(a) Use the Caesar cipher with $k = 3$ to encipher

ATTACK IS IMMINENT

(b) Decipher the following cryptogram, assuming it has been enciphered by using the Caesar cipher with $k = 3$.

SURFHHGWRURPH

Solution.

(a) Using the table above and suppressing blank spaces, we obtain

DWWDFNLVLPPLQHQW

(b) To decipher, we use the table in the reverse order; that is, we replace each letter in the bottom row with the one directly above. The result is

PROCEED TO ROME

The Caesar cipher is fairly easy to break, although nobody did it during Caesar's time, probably because nobody thought of it. If it is known that a ciphertext has been obtained by using a shift cipher, the code can be broken if just one correspondence between a ciphertext letter and a plaintext letter can be found. If the letter y in the ciphertext is known to correspond to x in the plaintext, then k is the number of steps in going from x to y. Moreover, the key k can have only 25 values—namely, $1, 2, \ldots, 25$. So the code can be broken by an exhaustive key search.

Example 3.1.2 The following cryptogram has been obtained by using a shift cipher. Decipher it by doing an exhaustive key search.

UIBPMUIBQKAQAMIAG

Solution. If the key used for enciphering is k, then we obtain the plaintext by replacing each letter in the ciphertext by the one that is k places before it in the cyclic order of the alphabet. We try successively $k = 1, 2, \ldots$ and continue until we come upon a meaningful message.

THAOLTHAPJZPZLHZF
SGZNKSGZOIYOYKGYE
RFYMJRFYNHXNXJFXD
QEXLIQEXMGWMWIEWC
PDWKHPDWLFVLVHDVB
OCVJGOCVKEUKUGCUA
NBUIFNBUJDTJTFBTZ
MATHEMATICSISEASY

Thus we have found the original message: MATHEMATICS IS EASY. The key used for enciphering is $k = 8$.

We now describe a periodic substitution cipher composed of shift ciphers. It is called a *Vigenère cipher,* named after Claise de Vigenère, a 16th-century cryptologist. A Vigenère cipher with period p and key sequence $(k_1, \ldots, k_p)$ consists of a sequence of p shift ciphers with lengths $k_1, \ldots, k_p$, respectively. It is more common in this case to identify a shift cipher by the letter to which it moves the letter A. We refer to it as the *key letter* of the cipher. For example, the key letter of the shift cipher with $k = 3$ is D. The word formed by the key letters of the shift ciphers that constitute a Vigenère cipher is called its *key word*. Given the key word of a Vigenère cipher, we can make a table to represent the cipher. For each letter in the key word, the table contains a row beginning with that letter. The letters below A in the first column form the key word of the cipher. The table below represents the Vigenère cipher with the key word PARIS.

A	B	C	D	E	F	G	H	I	J	K	L	M	N	O	P	Q	R	S	T	U	V	W	X	Y	Z
P	Q	R	S	T	U	V	W	X	Y	Z	A	B	C	D	E	F	G	H	I	J	K	L	M	N	O
A	B	C	D	E	F	G	H	I	J	K	L	M	N	O	P	Q	R	S	T	U	V	W	X	Y	Z
R	S	T	U	V	W	X	Y	Z	A	B	C	D	E	F	G	H	I	J	K	L	M	N	O	P	Q
I	J	K	L	M	N	O	P	Q	R	S	T	U	V	W	X	Y	Z	A	B	C	D	E	F	G	H
S	T	U	V	W	X	Y	Z	A	B	C	D	E	F	G	H	I	J	K	L	M	N	O	P	Q	R

Example 3.1.3 Use the Vigenère cipher with the key word *PARIS* to **(a)** encipher SEND MORE MONEY and **(b)** decipher XADKCBPCMLTLPJJDKV.

Solution.

(a) The period of the given cipher is 5. We encipher the first five letters in the plaintext—namely, SENDM—by using the shift ciphers with keys P, A, R, I, S, respectively. The result is HEELE. Likewise, the next five letters OREMO encipher into DRVUG. Finally, the last three letters NEY become CEP. Thus, we get the cryptogram HEELEDRVUGCEP.

(b) Using the table in the reverse direction, we decipher the first five letters XADKC as IAMCO. Similarly, the remaining blocks of letters are deciphered as MPLET, ELYBR, and OKE. Inserting blank spaces at appropriate places, we get the message I AM COMPLETELY BROKE.

EXERCISES 3.1

1. Encipher this message by using a shift cipher with $k = 7$:

MISSION IMPOSSIBLE

2. Decipher the cryptogram, assuming it was enciphered by using a shift cipher with $k = 10$:

WSCCSYXMYWZVODON

3. Decipher the cryptogram by an exhaustive key search, assuming it was enciphered by using a shift cipher:

RZYPHTESESPHTYO

4. Encipher this message by using a Vigenère cipher with the key word THUNDER:

SILENCE IS GOLDEN

5. Decipher this cipher message by using a Vigenère cipher with the key word HOPE:

LHTVUOAPPTT.

3.2 ALGEBRAIC ENCIPHERING ALGORITHMS AND CLASSICAL CRYPTOSYSTEMS

In this section we consider how an algebraic system like a ring or a field can be used to devise an enciphering algorithm. The general principle involved is as follows: Let A be the common alphabet for the plaintext and the ciphertext. Let S be a finite algebraic system (ring or field) that has the same number of elements as A. Choose a fixed bijective mapping $p : A \to S$. Usually we take the mapping p that is determined in a natural way by the usual order of letters in the alphabet. Then, for any permutation σ of S, the composite mapping $f = p^{-1}\sigma p : A \to A$ is bijective and hence a permutation of A. Thus every permutation of S provides an enciphering key. The advantage in going from A to A via S is that we can exploit the algebraic structure of S to create an algorithm for generating permutations of S. The following diagram represents this enciphering scheme:

$$A \xrightarrow{\ p\ } S \xrightarrow{\ \sigma\ } S \xrightarrow{\ p^{-1}\ } A$$

More generally, if the plaintext alphabet and the ciphertext alphabet are different, we have the following scheme:

$$A \xrightarrow{\ p\ } S \xrightarrow{\ \sigma\ } S \xrightarrow{\ q\ } B$$

where p and q are injective mappings. The mappings p and q are fixed, but σ is variable and determined by the values of the parameters in the algorithm. Clearly, the sets A, B and the mappings p, q have no bearing on the algorithm. In fact, we may treat S itself as the plaintext alphabet as well as the ciphertext alphabet.

Now we describe examples of algebraic enciphering algorithms of this kind.

3.2.1 Modular Enciphering and Affine Cipher

Let n be the number of characters in the plaintext alphabet A. Let $S = \mathbb{Z}_n$ be the ring of integers modulo n. An enciphering that makes use of the algebraic operations in $\mathbb{Z}_n$ is called a *modular enciphering*. The simplest example of a modular enciphering is an *affine cipher.*

Let $a, b \in \mathbb{Z}_n$ and suppose a is relatively prime to n. Then a is an invertible element in the ring $\mathbb{Z}_n$. Hence the mapping $\sigma : \mathbb{Z}_n \to \mathbb{Z}_n$ given by

$$\sigma(x) = ax + b$$

is bijective. In the notation of the usual addition and multiplication operations in $\mathbb{Z}$, the mapping σ is given by

$$\sigma(x) = (ax + b) \bmod n$$

(Recall that $x \bmod n$ denotes the remainder left on dividing x by n.) We refer to σ, or the enciphering determined by σ, as an *affine cipher*. When $a = 1$, σ represents a shift cipher described in the foregoing section.

The inverse of the mapping σ is given by

$$\sigma^{-1}(y) = a^{-1}(y - b) = a^{-1}y - a^{-1}b$$

for every $y \in \mathbb{Z}_n$. So σ^{-1} also represents an affine cipher, with parameters a^{-1} and $-a^{-1}b$.

If the plaintext alphabet A is the usual set of letters A, B, ..., Z, then $n = 26$. We take the preliminary mapping $p : A \rightarrow \mathbb{Z}_{26}$ as given in the table below.

A	B	C	D	E	F	G	H	I	J	K	L	M
1	2	3	4	5	6	7	8	9	10	11	12	13
N	O	P	Q	R	S	T	U	V	W	X	Y	Z
14	15	16	17	18	19	20	21	22	23	24	25	0

Example 3.2.1 Use the mapping p given above and the affine cipher

$$\sigma(x) = (7x + 4) \bmod 26$$

(a) to encipher ALGEBRA and (b) to decipher AMEQMNZW.

Solution. We note that 7 is relatively prime to 26; hence σ is bijective.

(a) We replace each letter in the plaintext with its corresponding number x as given in the table above and apply the mapping σ. Then we write the letter corresponding to $\sigma(x)$. The table below shows the complete enciphering process.

plaintext	A	L	G	E	B	R	A
x	1	12	7	5	2	18	1
$7x + 4$	11	88	53	39	18	130	11
$(7x + 4) \bmod 26$	11	10	1	13	18	0	11
ciphertext	K	J	A	M	R	Z	K

(b) To decipher, we use the inverse mapping σ^{-1}. If $7x + 4 = y$, then $x = 7^{-1}(y - 4)$. By trial and error, or by using the method of generalized Euclidean algorithm explained in Chapter 0, we find that in $\mathbb{Z}_{26}$, $7^{-1} = 15$. [Indeed, we can check that $7 \times 15 = 105 \equiv 1 \pmod{26}$.] So the deciphering is done by using the mapping $d(y) = 15y - 60 = 15y + 18$. The table below gives the deciphering process.

ciphertext	A	M	E	Q	M	N	Z	W
y	1	13	5	17	13	14	0	23
$15y + 18$	33	213	93	273	213	228	18	363
$(15y + 18) \bmod 26$	7	5	15	13	5	20	18	25
plaintext	G	E	O	M	E	T	R	Y

As we noted, the shift cipher is a special case of the affine cipher where $a = 1$. We may therefore use the method shown in Example 3.2.1 to encipher with a shift cipher or, more generally, a Vigenère cipher.

Example 3.2.2 Use a Vigenère cipher with the key word ATHENS to **(a)** encipher SOCRATES and **(b)** decipher *PEHXB*.

Solution.

(a) The values of k in the shift ciphers corresponding to the letters in the key word ATHENS are, respectively, $k = 0, 19, 7, 4, 13, 18$. We apply the mappings $\sigma(x) = (x + k) \bmod 26$, where k takes these values in cyclic repetition.

plaintext	S	O	C	R	A	T	E	S
x	19	15	3	18	1	20	5	19
k	0	19	7	4	13	18	0	19
$(x + k) \bmod 26$	19	8	10	22	14	12	5	12
ciphertext	S	H	J	V	N	L	E	L

(b) For deciphering, we use the inverse mapping $\sigma^{-1}(y) = (y - k) \bmod 26$.

ciphertext	P	E	H	X	B
y	16	5	8	24	2
k	0	19	7	4	13
$(y - k) \bmod 26$	16	12	1	20	15
plaintext	P	L	A	T	O

If n is prime, then $\mathbb{Z}_n$ is a field, so every nonzero element in $\mathbb{Z}_n$ is invertible. For example, let us take $n = 29$. Suppose the plaintext alphabet A contains the usual letters A to Z and three additional characters: , (comma), . (period), and # (blank space). The mapping $p : A \to \mathbb{Z}_{29}$ is given in the table:

A	B	C	...	X	Y	Z	,	.	(blank)
1	2	3	...	24	25	26	27	28	0

Example 3.2.3 Use the mapping $p : A \rightarrow \mathbb{Z}_{29}$ given above and the affine cipher

$$\sigma(x) = (4x + 10) \bmod 29$$

(a) to encipher WAR AND PEACE and **(b)** to decipher *CL .CLW.*

Solution. We show the work in the tables.

(a)

plaintext	W	A	R		A	N	D		P	E	A	C	E
x	23	1	18	0	1	14	4	0	16	5	1	3	5
$4x + 10$	102	14	82	10	14	66	26	10	74	30	14	22	30
$(4x + 10) \bmod 29$	15	14	24	10	14	8	26	10	16	1	14	22	1
ciphertext	O	N	X	J	N	H	Z	J	P	A	N	V	A

(b) In $\mathbb{Z}_{29}$, $4^{-1} = 22$. Hence we use the mapping $\sigma^{-1}(y) = 22(y - 10) = 22y + 12$ for deciphering.

ciphertext	C	L		.	C	L	W
y	3	12	0	28	3	12	23
$(22y + 12) \bmod 29$	20	15	12	19	20	15	25
plaintext	T	O	L	S	T	O	Y

The affine cipher $\sigma(x) = ax + b$ contains two parameters a and b, where $a, b \in \mathbb{Z}_n$ and a is relatively prime to n. The number of possible values for a is $\varphi(n)$, where φ denotes Euler's phi-function. The second parameter b can be assigned any value in $\mathbb{Z}_n$. So the total number of possible keys (a, b) is $\varphi(n)n$. For example, if $n = 26$, then the number of keys is $12 \times 26 = 312$. If n is prime, then $\varphi(n)n = n(n - 1)$. The number of keys is not large enough to make it a secure cipher; the code can be broken easily by an exhaustive key search. Moreover, since there are only two parameters, the code can be broken even more readily if we know just two associations between the plaintext and ciphertext characters. Suppose we have found out that x_1, x_2 are enciphered as y_1, y_2, respectively. Then $ax_1 + b = y_1$ and $ax_2 + b = y_2$. We can find a, b by solving these equations.

We have observed that the variations in the relative frequencies of letters in the plaintext can be ironed out by using a variable key for enciphering. The variable key may be periodic, as in a Vigenère cipher, for example, or we may have a stream cipher where we have some rule for determining the key k_i, $i = 1, 2, \ldots$, for enciphering the ith letter in the plaintext. A possible way of doing this is to use the plaintext itself to determine k_i. For example, we can design a variable affine cipher as follows: Select $a, b, x_0 \in \mathbb{Z}_n$, with $\gcd(a, n) = 1$. Let $x_i \in \mathbb{Z}_n$ represent the ith letter in the plaintext message. We encipher x_i by using the affine mapping

$\sigma_i : \mathbb{Z}_n \to \mathbb{Z}_n$ given by

$$\sigma_i(x) = ax + bx_{i-1}, \qquad i = 1, 2, \ldots$$

In other words, x_i is enciphered as $ax_i + bx_{i-1}$. We refer to a stream cipher of this kind as *text-induced*.

Example 3.2.4 Use the text-induced stream affine cipher

$$\sigma_i(x) = (5x + 9x_{i-1}) \bmod 26, \qquad i = 1, 2, \ldots$$

with $x_0 = 1$ to **(a)** encipher HAPPINESS and **(b)** decipher ZLQRCP.

Solution.

(a) We use the mapping σ_i to encipher the ith letter. The following table shows the work:

plaintext	H	A	P	P	I	N	E	S	S
i	1	2	3	4	5	6	7	8	9
x_i	8	1	16	16	9	14	5	19	19
$(5x_i + 9x_{i-1}) \bmod 26$	23	25	11	16	7	21	21	10	6
ciphertext	W	Y	K	P	G	U	U	J	F

(b) Let y_i represent the ith letter in the ciphertext. So $y_i = (5x_i + 9x_{i-1})$ mod 26 and hence

$$x_i = 5^{-1}(y_i - 9x_{i-1}) \bmod 26 = (21y_i + 19x_{i-1}) \bmod 26$$

The following table shows the complete deciphering:

ciphertext	Z	L	Q	R	C	P
i	1	2	3	4	5	6
y_i	0	12	17	18	3	16
$x_i = (21y_i + 19x_{i-1}) \bmod 26$	19	15	18	18	15	23
plaintext	S	O	R	R	O	W

The affine cipher can be extended to block-wise substitution. Let r be a positive integer and n the number of letters in the plaintext alphabet A. We can map the set A^r onto $\mathbb{Z}_{n^r}$ in a natural way as follows: Let $p : A \to \mathbb{Z}_n$ be a bijective mapping determined by the usual order of letters in the alphabet. Then p can be extended to the mapping $p : A^r \to (\mathbb{Z}_n)^r$ given by $p(a_1, \ldots, a_r) = (p(a_1), \ldots, p(a_r))$. Further there is a natural way in which $(\mathbb{Z}_n)^r$ can be mapped onto $\mathbb{Z}_{n^r}$. We define $h : (\mathbb{Z}_n)^r \to \mathbb{Z}_{n^r}$ by

$$h(x_1, x_2, \ldots, x_r) = x_1 n^{r-1} + x_2 n^{r-2} + \cdots + x_r$$

The mapping h is easily seen to be bijective. Given $y \in \mathbb{Z}_{n^r}$, by repeated division algorithm, we can express y uniquely as $y = x_1 n^{r-1} + x_2 n^{r-2} + \cdots + x_r$, where

$x_1, x_2, \ldots, x_r \in \mathbb{Z}_n$. Hence h is bijective. Thus we have a bijective mapping $f = h \circ p : A^r \to \mathbb{Z}_{n^r}$. Now we can have an affine mapping $\sigma : \mathbb{Z}_{n^r} \to \mathbb{Z}_{n^r}$ given by $\sigma(x) = (ax + b) \bmod n^r$, where $a, b \in \mathbb{Z}_{n^r}$, with a relatively prime to n. The following diagram represents the enciphering scheme:

$$A^r \xrightarrow{\ p\ } (\mathbb{Z}_n)^r \xrightarrow{\ h\ } \mathbb{Z}_{n^r} \xrightarrow{\ \sigma\ } \mathbb{Z}_{n^r} \xrightarrow{\ h^{-1}\ } (\mathbb{Z}_n)^r \xrightarrow{\ p^{-1}\ } A^r$$

The number of keys for the affine cipher $\sigma : \mathbb{Z}_{n^r} \to \mathbb{Z}_{n^r}$ is $n^r \varphi(n^r)$. For example, if $n = 26$ and $r = 2$, then the number of keys is $26^2 \varphi(26^2) = 26^2 \varphi(2^2 \cdot 13^2) = 676 \times 2 \times 156 = 210,912$.

Example 3.2.5 Use blocks of two letters and the affine cipher

$$\sigma(x) = (35x + 120) \bmod 676$$

to **(a)** encipher HELLO and **(b)** decipher IUOZDG.

Solution.

(a) We break the message into blocks of two letters each, adding an extra letter Z at the end to complete the last block.

plaintext	HE	LL	OZ
x_1, x_2	8, 5	12, 12	5, 0
$x = 26x_1 + x_2$	213	324	130
$y = (35x + 120) \bmod 676$	139	644	614
y_1, y_2	5, 9	24, 20	23, 16
ciphertext	EI	XT	WP

(b) If $y = (35x + 120) \bmod 676$, then

$$x = 35^{-1}(y - 120) \bmod 676 = (367y + 576) \bmod 676$$

We decipher as shown in the table:

ciphertext	IU	OZ	DG
y_1, y_2	9, 21	15, 0	4, 7
$y = 26y_1 + y_2$	255	390	111
$x = (367y + 576) \bmod 676$	197	394	77
x_1, x_2	7, 15	15, 4	2, 25
plaintext	GO	OD	BY

In our discussion of the affine cipher, we considered an affine mapping $x \mapsto ax + b$ on a ring $\mathbb{Z}_n$. If n is prime, then $\mathbb{Z}_n$ is a field. More generally, if n is a power of a prime, we can use an affine mapping on a field of n elements. We saw in Chapter 0 that for every prime p and every positive integer r, there exists a field $F = \mathbb{F}_{p^r}$ (unique up to isomorphism) of p^r elements. This field is constructed as follows: Let $g(x) = x^r + c_1 x^{r-1} + \cdots + c_r$ be a monic irreducible polynomial over the field $\mathbb{Z}_p$. Then

$$\mathbb{F}_{p^r} = \{a_1 t^{r-1} + \cdots + a_{r-1} t + a_r \mid a_1, \ldots, a_r \in \mathbb{Z}_p\}$$

where t satisfies the relation $g(t) = t^r + c_1 t^{r-1} + \cdots + c_r = 0$.

Let n be the number of characters in the alphabet A, and suppose $n = p^r$, where p is prime and $r > 1$ (for example, $n = 27, 32, 49$, or 64). We can map A onto F as follows. As usual, we start with a preliminary mapping $p : A \to \mathbb{Z}_n$ induced by the usual order of letters in the alphabet. Now, given $x \in \mathbb{Z}_n$, by the division algorithm, x can be expressed uniquely as

$$x = x_1 p^{r-1} + \cdots + x_{r-1} p + x_r$$

where $x_1, x_2, \ldots, x_r \in \mathbb{Z}_p$. Thus we have a bijective mapping $h : \mathbb{Z}_{p^r} \to \mathbb{F}_{p^r}$ given by

$$h(x_1 p^{r-1} + \cdots + x_{r-1} p + x_r) = x_1 t^{r-1} + \cdots + x_{r-1} t + x_r$$

We can now use an affine mapping $\sigma : \mathbb{F}_{p^r} \to \mathbb{F}_{p^r}$ given by $\sigma(u) = au + b$, where $a, b \in \mathbb{F}_{p^r}$ and $a \neq 0$. The following diagram shows the enciphering scheme:

$$A \xrightarrow{\;p\;} \mathbb{Z}_{p^r} \xrightarrow{\;h\;} \mathbb{F}_{p^r} \xrightarrow{\;\sigma\;} \mathbb{F}_{p^r} \xrightarrow{\;h^{-1}\;} \mathbb{Z}_{p^r} \xrightarrow{\;p^{-1}\;} A$$

We illustrate the above procedure by taking the case $n = 27 = 3^3$. Consider the monic polynomial $g(x) = x^3 + 2x + 1$. It is easily checked that none of the elements $0, 1, 2$ in $\mathbb{Z}_3$ is a root of $g(x)$. Hence $g(x)$ is irreducible over $\mathbb{Z}_3$. Therefore the field $\mathbb{F}_{27}$ is given by

$$\mathbb{F}_{27} = \{at^2 + bt + c \mid a, b, c \in \mathbb{Z}_3\}$$

where t satisfies the relation $t^3 + 2t + 1 = 0$. To find the product of two elements in the field $\mathbb{F}_{27}$, we make use of this relation $t^3 + 2t + 1 = 0$—that is, $t^3 = t + 2$.

By computing successive powers of t, we find that $o(t) = 26$. So t is a primitive element and hence every nonzero element in the field is equal to some power of t. This property provides a convenient method for finding the inverse of an element $\alpha \in \mathbb{F}_{27}$. If $\alpha = t^m$, then $\alpha^{-1} = t^{26-m}$.

Let A be the plaintext alphabet consisting of the letters A to Z and the blank space. Let the mapping $p : A \to \mathbb{Z}_{27}$ take the blank space to 0 and the letters A, $\ldots$, Z to 1, $\ldots$, 26. The following table shows for each character in A its numerical value in $\mathbb{Z}_{27}$, the corresponding element in $\mathbb{F}_{27}$, and its expression as a power of t:

A	1	1	t^0	J	10	t^2+1	t^{21}	S	19	$2t^2+1$	t^{25}
B	2	2	t^{13}	K	11	t^2+2	t^{12}	T	20	$2t^2+2$	t^8
C	3	t	t^1	L	12	t^2+t	t^{10}	U	21	$2t^2+t$	t^{17}
D	4	$t+1$	t^9	M	13	t^2+t+1	t^6	V	22	$2t^2+t+1$	t^{20}
E	5	$t+2$	t^3	N	14	t^2+t+2	t^{11}	W	23	$2t^2+t+2$	t^5
F	6	$2t$	t^{14}	O	15	t^2+2t	t^4	X	24	$2t^2+2t$	t^{23}
G	7	$2t+1$	t^{16}	P	16	t^2+2t+1	t^{18}	Y	25	$2t^2+2t+1$	t^{24}
H	8	$2t+2$	t^{22}	Q	17	t^2+2t+2	t^7	Z	26	$2t^2+2t+2$	t^{19}
I	9	t^2	t^2	R	18	$2t^2$	t^{15}	0	0	0	

Example 3.2.6 Use the representation of the alphabet by $\mathbb{F}_{27}$ described above and the affine cipher $\sigma : \mathbb{F}_{27} \to \mathbb{F}_{27}$ given by

$$\sigma(u) = (t+1)u + t^2 + 2$$

to (a) encipher FIELD and (b) decipher FVC A.

Solution.

(a) We write for each letter in the plaintext its representation $u \in \mathbb{F}_{27}$, and find $\sigma(u) = (t+1)u + t^2 + 2$. The table shows the work:

plaintext	F	I	E	L	D
$u \in \mathbb{F}_{27}$	$2t$	t^2	$t+2$	t^2+t	$t+1$
$\sigma(u)$	$2t+2$	$2t^2+t+1$	$2t^2+1$	$2t+1$	$2t^2+2t$
ciphertext	H	V	S	G	X

(b) From the table given above, we find that $t+1 = t^9$. Hence $(t+1)^{-1} = t^{17} = 2t^2 + t$. [Check $(t+1)(2t^2+t) = 1$.] Hence the inverse mapping σ^{-1} is given by

$$\sigma^{-1}(v) = (2t^2+t)(v - (t^2+2)) = (2t^2+t)v + 2t + 1$$

ciphertext	F	V	C		A
$v \in \mathbb{F}_{27}$	$2t$	$2t^2+t+1$	t	0	1
$\sigma^{-1}(v)$	$2t^2$	t^2	t^2+t+2	$2t+1$	$2t^2+1$
plaintext	R	I	N	G	S

3.2.2 Hill Cipher

We now consider a generalization of the affine cipher known as the *Hill cipher*. Let n be the number of characters in the plaintext alphabet A, and let r be a positive integer greater than 1. As explained above, the mapping $p : A \to \mathbb{Z}_n$ can be extended to $p : A^r \to (\mathbb{Z}_n)^r$ by defining $p(a_1, \ldots, a_r) = (p(a_1), \ldots, p(a_r))$.

Let us write the elements of $(\mathbb{Z}_n)^r$ as columns, and let $(\mathbb{Z}_n)^{r \times r}$ denote the set of all $r \times r$ matrices with entries in $\mathbb{Z}_n$. Let $K \in (\mathbb{Z}_n)^{r \times r}$. If K is invertible, then $\det K \det K^{-1} = \det(K K^{-1}) = \det I = 1$. Hence $\det K$ is an invertible element in

$\mathbb{Z}_n$. Conversely, if det K is an invertible element in $\mathbb{Z}_n$, then K^{-1} is given by $K^{-1} = (\det K)^{-1} adj\, K$. Thus we see that a square matrix K over $\mathbb{Z}_n$ is invertible if and only if det K is relatively prime to n.

Let K be an invertible $r \times r$ matrix over $\mathbb{Z}_n$ and $D \in (\mathbb{Z}_n)^r$. Then the mapping $\sigma : (\mathbb{Z}_n)^r \to (\mathbb{Z}_n)^r$ given by

$$\sigma(X) = KX + D$$

for all $X \in (\mathbb{Z}_n)^r$ is called a *Hill cipher.* The following diagram shows the scheme of the cipher:

$$A^r \xrightarrow{p} (\mathbb{Z}_n)^r \xrightarrow{\sigma} (\mathbb{Z}_n)^r \xrightarrow{p^{-1}} A^r$$

To decipher the ciphertext, we use the inverse mapping

$$\sigma^{-1}(Y) = K^{-1}(Y - D)$$

Example 3.2.7 Use the Hill cipher, with $n = 26, r = 2$, given by

$$\sigma\left(\begin{bmatrix} x_1 \\ x_2 \end{bmatrix}\right) = \begin{bmatrix} 8 & 5 \\ 5 & 6 \end{bmatrix} \begin{bmatrix} x_1 \\ x_2 \end{bmatrix} + \begin{bmatrix} 7 \\ 9 \end{bmatrix} \pmod{26}$$

to **(a)** encipher MOUNTAIN and **(b)** decipher AJGLLRKE.

Solution.

(a) We verify that $\det \begin{bmatrix} 8 & 5 \\ 5 & 6 \end{bmatrix} = 23$ is relatively prime to 26. To encipher, we divide the plaintext message into blocks of two letters, represent each block by a column of numbers given by the mapping p, and then apply the mapping σ.

plaintext	MO	UN	TA	IN
$\begin{bmatrix} x_1 \\ x_2 \end{bmatrix}$	$\begin{bmatrix} 13 \\ 15 \end{bmatrix}$	$\begin{bmatrix} 21 \\ 14 \end{bmatrix}$	$\begin{bmatrix} 20 \\ 1 \end{bmatrix}$	$\begin{bmatrix} 9 \\ 14 \end{bmatrix}$
$\sigma\left(\begin{bmatrix} x_1 \\ x_2 \end{bmatrix}\right)$	$\begin{bmatrix} 4 \\ 8 \end{bmatrix}$	$\begin{bmatrix} 11 \\ 16 \end{bmatrix}$	$\begin{bmatrix} 16 \\ 11 \end{bmatrix}$	$\begin{bmatrix} 19 \\ 8 \end{bmatrix}$
ciphertext	DH	KP	PK	SH

(b) In $\mathbb{Z}_{26}$, $(\det K)^{-1} = 23^{-1} = 17$. Hence $K^{-1} = 17 \begin{bmatrix} 6 & -5 \\ -5 & 8 \end{bmatrix} = \begin{bmatrix} 24 & 19 \\ 19 & 6 \end{bmatrix}$. To decipher, we use the mapping

$$\sigma^{-1}\left(\begin{bmatrix} y_1 \\ y_2 \end{bmatrix}\right) = \begin{bmatrix} 24 & 19 \\ 19 & 6 \end{bmatrix} \left(\begin{bmatrix} y_1 \\ y_2 \end{bmatrix} - \begin{bmatrix} 7 \\ 9 \end{bmatrix}\right)$$

$$= \begin{bmatrix} 24 & 19 \\ 19 & 6 \end{bmatrix} \begin{bmatrix} y_1 \\ y_2 \end{bmatrix} + \begin{bmatrix} 25 \\ 21 \end{bmatrix}$$

ciphertext	AJ	GL	LR	KE
$\begin{bmatrix} y_1 \\ y_2 \end{bmatrix}$	$\begin{bmatrix} 1 \\ 10 \end{bmatrix}$	$\begin{bmatrix} 7 \\ 12 \end{bmatrix}$	$\begin{bmatrix} 12 \\ 18 \end{bmatrix}$	$\begin{bmatrix} 11 \\ 5 \end{bmatrix}$
$\sigma^{-1}\left(\begin{bmatrix} y_1 \\ y_2 \end{bmatrix}\right)$	$\begin{bmatrix} 5 \\ 22 \end{bmatrix}$	$\begin{bmatrix} 5 \\ 18 \end{bmatrix}$	$\begin{bmatrix} 5 \\ 19 \end{bmatrix}$	$\begin{bmatrix} 20 \\ 0 \end{bmatrix}$
plaintext	EV	ER	ES	TZ

Evidently, the last letter Z was stuffed in to complete the block. Hence the deciphered message is EVEREST.

As explained earlier for an affine cipher, we can generate a text-induced sequence of Hill ciphers. We select $r \times r$ matrices K, C over $\mathbb{Z}_n$, where K is invertible, and fix $X_0 \in (\mathbb{Z}_n)^r$. Let $X_i \in (\mathbb{Z}_n)^r$ represent the ith block of r letters in the plaintext. We define $\sigma_i : (\mathbb{Z}_n)^r \to (\mathbb{Z}_n)^r$ by $\sigma_i(X) = KX + CX_{i-1}$ for all $X \in (\mathbb{Z}_n)^r$. The ith block X_i is enciphered by using the Hill cipher σ_i.

EXERCISES 3.2

1. Use the affine cipher $\sigma(x) = (3x + 14) \bmod 26$ to (a) encipher REPUBLICAN and (b) decipher ZCAGWPQV.

2. Use the affine cipher $\sigma(x) = (5x + 11) \bmod 29$ to (a) encipher ALBERT EINSTEIN and (b) decipher XVG.NTK.LKNGMPX, E, XT.

3. Use a text-induced stream affine cipher with $x_0 = 5$ and $\sigma_i(x) = (7x + 4x_{i-1})$ $\bmod 26$ to (a) encipher ABSTRACT and (b) decipher AJSKHDA.

4. Use blocks of two letters and the affine cipher $(23x + 314) \bmod 26^2$ to (a) encipher HEAVEN and (b) decipher RMLR.

5. Use the representation of the alphabet by the field $\mathbb{F}_{27}$ (as in Example 3.2.6) and the affine cipher $\sigma(u) = (t^2 + 1)u + t + 1$ to (a) encipher NEW and (b) decipher ITJ.

6. Use the Hill cipher $\sigma(X) = KX + D$, with $K = \begin{bmatrix} 7 & 3 \\ 8 & 7 \end{bmatrix}$, $D = \begin{bmatrix} 4 \\ 5 \end{bmatrix}$, over $\mathbb{Z}_{26}$ to (a) encipher HAPPY and (b) decipher JHFK.

7. Use the Hill cipher $\sigma(X) = KX + D$, with $K = \begin{bmatrix} 17 & 23 & 4 \\ 17 & 12 & 7 \\ 11 & 15 & 25 \end{bmatrix}$, $D = \begin{bmatrix} 4 \\ 5 \\ 9 \end{bmatrix}$, over $\mathbb{Z}_{26}$ to (a) encipher LINEAR and (b) decipher RYLWMYGOA.

8. Show that if a square matrix K over $\mathbb{Z}_p$ (p prime) is involutory (or self-inverse), then $\det K = \pm 1$. (An $n \times n$ matrix K is called *involutory* if K is invertible and $K^{-1} = K$.)

9. Show that a 2×2 matrix $K = \begin{bmatrix} a & b \\ c & d \end{bmatrix}$ over $\mathbb{Z}_n$, with $\det K = -1$, is involutory if and only if $a + d = 0$. (See Exercise 8 for the definition of an involutory matrix.)

3.3 BLOCK CIPHERS AND ADVANCED ENCRYPTION STANDARD

The discussion of various enciphering algorithms in Section 3.2 provides a synoptic view of the historical development of cryptography from antiquity to about 1970. We have seen how algebraic concepts and algebraic systems like rings and fields can be used to devise enciphering algorithms. But the cryptosystems described by us, now referred to as classical cryptosystems, have become obsolete and are of historical and theoretical interest only. With advances in the science of cryptanalysis and the availability of fast computers in modern times, these cryptosystems are no longer secure. The size of the key space in them is so small that an exhaustive key search can be performed easily.

We now give a brief description of the latest cryptosystem approved for general use by the National Institute of Standards and Technology (NIST). It is called the *Advanced Encryption Standard* (AES) and was adopted, effective May 26, 2002, as the official Federal Information Processing Standard (FIPS) to be used by all U.S. government organizations to protect sensitive information. It is also expected to be used by other organizations, institutions, and individuals all over the world.

The AES was developed to replace two currently existing standards: DES (Data Encryption Standard) and Triple DES (a variation of DES). DES was developed by IBM and was adopted by NIST (then called the National Bureau of Standards) on January 15, 1977. It soon became the most widely used cryptosystem in the world. However, from the very beginning, DES attracted criticism for not having a sufficiently large key space to make it really secure. The size of the key space in DES is 2^{56} (approximately 7.2×10^{16}). From early on, attempts were made to build a special-purpose machine devoted exclusively to the task of breaking the DES code. In 1998 a massive parallel network computer, called "DES Cracker," was built by Electronic Frontier Foundation that could search 88 billion DES keys per second. It succeeded in finding a DES secret key in 56 hours. In 1999, working in conjunction with a worldwide network of 100,000 computers, the DES Cracker could search 245 billion keys per second and succeeded in finding a secret DES key in a little more than 22 hours. It was thus clear that DES was no longer a secure cryptosystem. Therefore it was necessary to phase out the DES and adopt a more secure encryption standard.

The size of the key space in AES is 2^{128}, or approximately 3.4×10^{38}. This number is so large that the fastest "cracker machine" available at present would take trillions of years to crack the AES code by doing an exhaustive key search. It is therefore expected that AES will remain a secure cryptosystem for many years to come.

The enciphering algorithm in AES was designed by two Belgian cryptographers, Dr. Joan Daeman and Dr. Vincent Rijmen. It was given the name *Rijndael* (pronounced "rhine dahl"). The basic structure of the Rijndael algorithm is that of an *iterated block cipher,* but with some additional features. Before considering the Rijndael algorithm, we will explain the general scheme of an iterated block cipher.

Nowadays all data are stored in a computer and transmitted electronically to other computers in digital form—that is, as a stream of 1's and 0's. (The symbols 1 and 0 are in fact notations for high voltage and low voltage in the computer's electronic circuit.) The digits 1 and 0 are referred to as *bits,* short for *binary digits.* A positive integer n is represented as a string of bits by expressing n in binary form; that is, the string $a_k a_{k-1} \ldots a_1 a_0$ (where each a_i is 1 or 0) represents the integer $a_k 2^k + a_{k-1} 2^{k-1} + \cdots + a_1 2 + a_0$. A string of 8 bits is called a *byte.* The numerical value of a byte lies between 0 ($= 00000000$) and 255 ($= 11111111$). When we write the numerical value of a byte, it is more convenient to use the hexadecimal notation than the decimal notation used in ordinary arithmetic. The hexadecimal system uses 16 digits, consisting of the ordinary decimal digits 0 to 9 and the letters A, B, C, D, E, F, which represent the numbers 10, 11, 12, 13, 14, 15, respectively. The numerical value of a 4-bit string (called a *nibble*) lies between 0 ($= 0000$) and 15 ($= 1111$). Hence a nibble is represented by a single hexadecimal digit. The following table gives the hexadecimal representations of the 16 nibbles:

0000	0001	0010	0011	0100	0101	0110	0111
0	1	2	3	4	5	6	7
1000	1001	1010	1011	1100	1101	1110	1111
8	9	A	B	C	D	E	F

Since a byte consists of two nibbles, it is represented by two hexadecimal digits. For example, the byte 11001111 is written CF in hexadecimal notation and represents the number 207 in decimal notation.

Readable English text is converted to a stream of bits by using some fixed scheme of representation. The most commonly used scheme is *ASCII,* an acronym for *American Standard Code for Information Interchange.* It represents each character (a letter of the alphabet or a punctuation mark) as a byte. There are $2^8 = 256$ distinct bytes, so it is possible to represent 256 different characters by bytes. But ASCII uses a set of 128 characters only. Of these, only 95 characters are printable. The remaining 33 characters are nonprintable and are used by the computer's operating system for its internal working. The 95 printable characters include the capital and lowercase letters of the alphabet, the numerals 0 to 9, and various punctuation marks. To see how the ASCII representation works, we number these 128 characters from 0 to 127. Each of these $128 = 2^7$ numbers is expressed in binary form as a string of 7 bits. An extra bit equal to 0 is then added in the leftmost position to make up a byte. For example, the capital letters A to Z are numbered 65 to 90, and the lowercase letters a to z are numbered 97 to 122. To get the ASCII representation of A, we express the number 65 in binary form—that is, 1000001. So the ASCII representation of A is 01000001. When you press A on the keyboard, it sends the binary string 01000001 to the computer's CPU (central processing unit).

In addition to ASCII, there are other schemes for converting text to binary form. For example, EBCDIC (Extended Binary Coded Decimal Interchange Code) uses a set of 256 characters (of which 72 are nonprintable). For the purpose of the cryptosystems that we are about to discuss, it is immaterial which particular scheme is used for converting text to binary form. We consider the binary string itself as the plaintext. The enciphering algorithm of the cryptosystem converts it into another binary string. Thus the plaintext and the ciphertext are both binary strings.

In the cryptosystems discussed in Section 3.2, we saw that an enciphering algorithm determines for every key k belonging to some set K, an injective mapping from the set $M = A^r$ (where A is the plaintext alphabet) to a set C (ciphertext alphabet). We can formulate a general definition of a cryptosystem as follows: A cryptosystem consists of sets M, C, K, and an algorithm for a mapping $E : M \times K \to C$ such that for every $k \in K$, the induced mapping $f_k : M \to C$, given by $f_k(m) = E(m, k)$, is injective. The sets M, C, and K are called the *message space, cipher space,* and *key space,* respectively. The mapping E is the enciphering algorithm. The corresponding deciphering algorithm is a mapping $D : C \times K \to M$ such that $D(E(m, k), k) = m$ for all $m \in M, k \in K$.

In an iterated block cipher, $M = C = (\mathbb{Z}_2)^n$ and $K = (\mathbb{Z}_2)^q$, where n and q are some specified positive integers. The numbers n and q are called the *block length* and the *key length* of the cipher, respectively. The block length and the key length may be the same. The numbers n and q are usually multiples of 8, so that the plaintext and the key both consist of an integral number of bytes. For the cipher to be secure, q should be sufficiently large, say 64 or larger. We further assume that n is a multiple of s, where s is a relatively small integer; typically, $s = 4, 6,$ or 8. (We will demonstrate the relevance of this number s later when we discuss the s-bit substitution.) The complete enciphering algorithm consists of a specified number of iterations of a basic algorithm, called the *round function.* Let r be the number of iterations, which are referred to as *rounds.* There is a built-in algorithm in the cryptosystem that, for any given key $k \in K$, generates r *round keys* $k_1, \ldots, k_r \in (\mathbb{Z}_2)^n = M$. In other words, we have an algorithm for a mapping $\phi : K \to M^r$, with $k \mapsto (k_1, \ldots, k_r)$. This mapping is called the *key schedule.* The round function uses the round key k_i in the ith round. Let $g_{k_i} : M \to M$ denote the round function with the key k_i. Then the complete enciphering is given by the function $f_k : M \to C$, where

$$f_k(m) = g_{k_r} \cdots g_{k_1}(m)$$

Typically, the round function g_{k_i} consists of three operations: (1) round key addition, (2) s-bit substitution, and (3) permutation. Only the first of these operations is variable and uses the round key. The other two operations are the same in each round. For the s-bit substitution, there is a fixed permutation $\sigma : (\mathbb{Z}_2)^s \to (\mathbb{Z}_2)^s$. For the third operation, there is a fixed permutation $\pi : \{1, \ldots, n\} \to \{1, \ldots, n\}$. We now explain these operations in detail. We write $m \in M$ to denote the current state of the message at the start of the operation under consideration, and m' denotes the result of the operation on m.

1. **Round key addition.** In the ith round, the round key k_i is added to the current state $m \in (\mathbb{Z}_2)^n$; that is, $m \mapsto m' = m + k_i$. Note that this addition is in the ring $(\mathbb{Z}_2)^n$. In terms of bits, if $m = (a_1, \ldots, a_n)$ and $k_i = (b_1, \ldots, b_n)$, then

$$m' = (a_1 + b_1, \ldots, a_n + b_n)$$

   Recall that addition in $\mathbb{Z}_2$ is modulo 2, which, in the terminology of computer science, is the operation XOR.

2. **s-bit substitution.** As stated above, we assume $n = st$, where s is a suitably small number. The n-bit string m is divided into t strings of s bits each, say $m = (u_1, \ldots, u_t)$. Each s-bit string u_i is subjected to a fixed bijective mapping $\sigma : (\mathbb{Z}_2)^s \to (\mathbb{Z}_2)^s$. Thus $m = (u_1, \ldots, u_t) \mapsto m'$, where

$$m' = (\sigma(u_1), \ldots, \sigma(u_t))$$

3. **Permutation.** The n bits in the string $m = (a_1, \ldots, a_n)$ are permuted by using a fixed permutation $\pi : \{1, \ldots, n\} \to \{1, \ldots, n\}$. Thus $m = (a_1, \ldots, a_n) \mapsto m'$, where

$$m' = (a_{\pi(1)}, \ldots, a_{\pi(n)})$$

In computer science jargon, the permutation $\sigma : (\mathbb{Z}_2)^s \to (\mathbb{Z}_2)^s$ used for substitution in operation 2 is called an *S-box* (S for substitution). [You may imagine a black box in which you input an s-bit string u and the box outputs $\sigma(u)$.] The reason for performing substitution separately on each s-bit part of the string m, instead of doing so on the whole string itself, is to save on computer memory. The substitution operation is usually performed by using a look-up table that gives $\sigma(u)$ for each s-bit u. The total number of s-bits is 2^s. Hence the memory requirement for storing 2^s s-bit strings is $s2^s$ bits. For example, if $s = 8$, the memory requirement is 8×2^8 bits—that is, 256 bytes. This requirement can be met easily. But if we decided to use substitution on the entire 64-bit string m, then the memory requirement for the look-up table would be 64×2^{64} bits—that is, 2^{37} gigabytes. This requirement is impossible to meet.

3.3.1 Rijndael Algorithm

We now give a brief description of the Rijndael algorithm in AES. We will give only the basic features that are of algebraic interest. The technical details regarding its implementation can be found in the NITS publication FIPS 197.

There are several versions of Rijndael in AES; they differ only in the block/key length and the number of rounds. The general scheme is the same in all versions. The possible values for the block length and the key length are 128, 192, and 256 bits. The number of rounds can be 8, 10, or 12. In the description below, we have taken the simplest case where the block length and the key length are both 128 bits. Thus we assume $M = C = K = (\mathbb{Z}_2)^{128}$.

As mentioned earlier, the general scheme of Rijndael is that of an iterated block cipher. But in addition to the three operations of key addition, substitution, and permutation that we have described, the round function in Rijndael has a fourth operation called column mix. As we explain these four operations, we write $m \in M$ to denote the current state.

1. **Round key addition.** If k is the round key, then $m \mapsto m + k$.
2. **Byte substitution.** Substitution is performed on an 8-bit string by using a fixed permutation $\sigma : (\mathbb{Z}_2)^8 \to (\mathbb{Z}_2)^8$. Since m is a 128-bit string, m consists of 16 bytes. Let $m = (u_1, \ldots, u_{16})$, where $u_1, \ldots, u_{16} \in (\mathbb{Z}_2)^8$. Then

$$m = (u_1, \ldots, u_{16}) \mapsto (\sigma(u_1), \ldots, \sigma(u_{16}))$$

The permutation $\sigma : (\mathbb{Z}_2)^8 \to (\mathbb{Z}_2)^8$ used in the Rijndael algorithm is obtained as follows. Each element in the set $(\mathbb{Z}_2)^8$ is identified with a corresponding element in the finite field; $\mathbb{F}_{2^8} = \mathbb{F}_{256}$. To obtain a representation of the field $\mathbb{F}_{256}$, Rijndael uses the polynomial $x^8 + x^4 + x^3 + x + 1$ (which is irreducible over the field $\mathbb{Z}_2$). Thus the field $\mathbb{F}_{256}$ is given by

$$\mathbb{F}_{256} = \{x_7 t^7 + \cdots + x_1 t + x_0 \mid x_0, x_1, \ldots, x_7 \in \mathbb{Z}_2\}$$

where t satisfies the relation $t^8 + t^4 + t^3 + t + 1 = 0$. So we have a natural one-to-one correspondence between $(\mathbb{Z}_2)^8$ and $\mathbb{F}_{256}$ given by

$$(x_7, \ldots, x_1, x_0) \longleftrightarrow x_7 t^7 + \cdots + x_1 t + x_0$$

With this identification of $(\mathbb{Z}_2)^8$ with $\mathbb{F}_{256}$, the mapping $\sigma : \mathbb{F}_{256} \to \mathbb{F}_{256}$ is defined as a composite of two bijective mappings: $\sigma = h \circ f$. The first mapping $f : \mathbb{F}_{256} \to \mathbb{F}_{256}$ is given by $f(0) = 0$ and $f(u) = u^{-1}$ for all nonzero $u \in \mathbb{F}_{256}$. (u^{-1} denotes the inverse of u in the field $\mathbb{F}_{256}$.) The second mapping

$h : \mathbb{F}_{256} \to \mathbb{F}_{256}$ is defined as follows: If $u = \sum_{i=0}^{7} x_i t^i$, then $h(u) = \sum_{i=0}^{7} y_i t^i$, where

$$
\begin{bmatrix} y_0 \\ y_1 \\ y_2 \\ y_3 \\ y_4 \\ y_5 \\ y_6 \\ y_7 \end{bmatrix}
=
\begin{bmatrix}
1 & 0 & 0 & 0 & 1 & 1 & 1 & 1 \\
1 & 1 & 0 & 0 & 0 & 1 & 1 & 1 \\
1 & 1 & 1 & 0 & 0 & 0 & 1 & 1 \\
1 & 1 & 1 & 1 & 0 & 0 & 0 & 1 \\
1 & 1 & 1 & 1 & 1 & 0 & 0 & 0 \\
0 & 1 & 1 & 1 & 1 & 1 & 0 & 0 \\
0 & 0 & 1 & 1 & 1 & 1 & 1 & 0 \\
0 & 0 & 0 & 1 & 1 & 1 & 1 & 1
\end{bmatrix}
\begin{bmatrix} x_0 \\ x_1 \\ x_2 \\ x_3 \\ x_4 \\ x_5 \\ x_6 \\ x_7 \end{bmatrix}
+
\begin{bmatrix} 1 \\ 1 \\ 0 \\ 0 \\ 0 \\ 1 \\ 1 \\ 0 \end{bmatrix}
$$

The mapping h can be expressed more succinctly as

$$ h(u) = (t^4 + t^3 + t^2 + t + 1) * u + (t^6 + t^5 + t + 1) $$

where $*$ represents multiplication modulo $(t^8 + 1)$. Note that this operation is different from the multiplication in the field $\mathbb{F}_{256}$, which is multiplication modulo $(t^8 + t^4 + t^3 + t + 1)$.

For the next two operations in the round function—namely, permutation (or row shift) and column mix—we impose a matrix structure on elements of M. We rewrite $m = (u_1, \ldots, u_{16})$ as a 4×4 matrix:

$$
m =
\begin{bmatrix}
u_1 & u_5 & u_9 & u_{13} \\
u_2 & u_6 & u_{10} & u_{14} \\
u_3 & u_7 & u_{11} & u_{15} \\
u_4 & u_8 & u_{12} & u_{16}
\end{bmatrix}
$$

Changing the notation and introducing double subscripts, we write

$$
m =
\begin{bmatrix}
m_{00} & m_{01} & m_{02} & m_{03} \\
m_{10} & m_{11} & m_{12} & m_{13} \\
m_{20} & m_{21} & m_{22} & m_{23} \\
m_{30} & m_{31} & m_{32} & m_{33}
\end{bmatrix}
$$

This matrix representation of m is used in the next two operations.

3. **Row shift.** The third operation in the round function—that is, permutation—is called a *row shift* and is performed as follows: The row i $(i = 0, 1, 2, 3)$ in the matrix m is shifted cyclically by i places to the left. Thus the top row remains unaltered, row 1 is shifted one place to the left, row 2 is shifted two places to the left, and row 3 is shifted three places to the left. The result of the operations is

$$
m =
\begin{bmatrix}
m_{00} & m_{01} & m_{02} & m_{03} \\
m_{10} & m_{11} & m_{12} & m_{13} \\
m_{20} & m_{21} & m_{22} & m_{23} \\
m_{30} & m_{31} & m_{32} & m_{33}
\end{bmatrix}
\longmapsto
\begin{bmatrix}
m_{00} & m_{01} & m_{02} & m_{03} \\
m_{11} & m_{12} & m_{13} & m_{10} \\
m_{22} & m_{23} & m_{20} & m_{21} \\
m_{33} & m_{30} & m_{31} & m_{32}
\end{bmatrix}
$$

4. **Column mix.** In this operation, the matrix m is multiplied on the left by a fixed matrix c given by

$$c = \begin{bmatrix} c_0 & c_3 & c_2 & c_1 \\ c_1 & c_0 & c_3 & c_2 \\ c_2 & c_1 & c_0 & c_3 \\ c_3 & c_2 & c_1 & c_0 \end{bmatrix}$$

where $c_0, c_1, c_2, c_3 \in \mathbb{F}_{256}$. These elements c_0, c_1, c_2, c_3 are chosen so that the matrix c is invertible. The result of the operation is given by

$$m = \begin{bmatrix} m_{00} & m_{01} & m_{02} & m_{03} \\ m_{10} & m_{11} & m_{12} & m_{13} \\ m_{20} & m_{21} & m_{22} & m_{23} \\ m_{30} & m_{31} & m_{32} & m_{33} \end{bmatrix} \mapsto \begin{bmatrix} c_0 & c_3 & c_2 & c_1 \\ c_1 & c_0 & c_3 & c_2 \\ c_2 & c_1 & c_0 & c_3 \\ c_3 & c_2 & c_1 & c_0 \end{bmatrix} \begin{bmatrix} m_{00} & m_{01} & m_{02} & m_{03} \\ m_{10} & m_{11} & m_{12} & m_{13} \\ m_{20} & m_{21} & m_{22} & m_{23} \\ m_{30} & m_{31} & m_{32} & m_{33} \end{bmatrix}$$

Recall that the entries in the matrix m are treated as elements of $\mathbb{F}_{256}$, so the product of the matrices c and m is obtained by using multiplication in the field $\mathbb{F}_{256}$.

The actual values of the entries in the matrix c used in Rijndael are

$$c_0 = t, \quad c_1 = 1, \quad c_2 = 1, \quad c_3 = t + 1$$

The inverse of the matrix c is

$$d = \begin{bmatrix} d_0 & d_3 & d_2 & d_1 \\ d_1 & d_0 & d_3 & d_2 \\ d_2 & d_1 & d_0 & d_3 \\ d_3 & d_2 & d_1 & d_0 \end{bmatrix}$$

where

$$d_0 = t^3 + t^2 + t, \quad d_1 = t^3 + 1, \quad d_2 = t^3 + t^2 + 1, \quad d_3 = t^3 + t + 1$$

In describing the last two operations in the round function—namely, row shift and column mix—we treated the elements of M as 4×4 matrices with entries in the field $\mathbb{F}_{256}$. We can obtain a simpler expression for these operations by imposing yet another structure on M. Let $\mathbb{F}_{256}[x, y]$ denote the ring of polynomials in indeterminates x and y over the field $\mathbb{F}_{256}$. Let $\langle x^4 + 1, y^4 + 1 \rangle$ denote the ideal generated by $\{x^4 + 1, y^4 + 1\}$, and let R be the quotient ring

$$R = \frac{\mathbb{F}_{256}}{\langle x^4 + 1, y^4 + 1 \rangle}$$

Then every element in R is a polynomial of the form $\sum_{i=0}^{3} \sum_{j=0}^{3} a_{ij} x^i y^j$, where $a_{ij} \in \mathbb{F}_{256}$. To find the product of two elements of R, we make use of the relations $x^4 + 1 = 0$, $y^4 + 1 = 0$—that is, $x^4 = 1$, $y^4 = 1$. For example,

$$x^2(a + bx + cx^2 + dx^3) = ax^2 + bx^3 + c + dx$$

Now we have a natural one-to-one correspondence between M and R given by

$$m = \begin{bmatrix} m_{00} & m_{01} & m_{02} & m_{03} \\ m_{10} & m_{11} & m_{12} & m_{13} \\ m_{20} & m_{21} & m_{22} & m_{23} \\ m_{30} & m_{31} & m_{32} & m_{33} \end{bmatrix} \longleftrightarrow \sum_{i=0}^{3} \sum_{j=0}^{3} m_{ij} x^i y^j$$

With this identification of M with R, the row shift operation is given by

$$m = \sum_{i=0}^{3} \sum_{j=0}^{3} m_{ij} x^i y^j \longmapsto \sum_{i=0}^{3} \sum_{j=0}^{3} m_{ij} x^i y^{j-i}$$

The column mix operation is given by

$$m = \sum_{i=0}^{3} \sum_{j=0}^{3} m_{ij} x^i y^j \longmapsto \sum_{i=0}^{3} \sum_{j=0}^{3} m_{ij} x^i y^j (c_0 + c_1 x + c_2 x^2 + c_3 x^3)$$

where $c_0, c_1, c_2, c_3 \in \mathbb{F}_{256}$ have the same meaning as before.

If we combine all four operations, the round function $g_k : M \to M$ is given by

$$g_k(m) = \sum_{i=0}^{3} \sum_{j=0}^{3} \sigma(m_{ij} + k_{ij}) x^i y^{j-1} (c_0 + c_1 x + c_2 x^2 + c_3 x^3)$$

3.4 PUBLIC-KEY CRYPTOSYSTEMS

In the cryptosystems discussed in the foregoing sections, we saw that the deciphering key can be easily obtained from the enciphering key. In essence, therefore, there is a single key, which must be kept secret and known only to the sender and the authorized recepient. Therefore such cryptosystems are known as *single-key* (or *secret-key*) *cryptosystems.* The requirement of keeping the key secret can pose a serious problem if many persons are using the key, as is often the case in large organizations.

All cryptosystems in use before 1976 were of the single-key type. Since 1976, cryptosystems of a different type, called *public-key cryptosystems,* have been developed. In these sytems it is not possible to find the deciphering key even when the enciphering key is known (unless some additional information is provided). Therefore it is not necessary to keep the enciphering key secret. In fact, the enciphering key is made public, so that anyone interested can have access to it. But the deciphering key is secret and known only to the recipient. The system works in the following manner.

Suppose a person R (receiver) decides to set up a cryptosystem so that anyone in the world can send her a secret message. She chooses a public-key cryptosystem and selects an enciphering key E together with the corresponding deciphering key D. The enciphering key E is made public (by publishing it in a directory or on her Website, possibly), but the deciphering key D is secret and known only to herself. Anyone who wishes to send a secret message to R can use this enciphering key E. When R receives the enciphered message, she uses the deciphering key D to decipher it. But nobody else who intercepts the message can decipher it, even though he may know the key E used to encipher it, because it is not possible to derive the deciphering key D from the enciphering key E.

The essential element in a public-key cryptosystem is a special kind of mapping, known as a trapdoor one-way function.

Definition 3.4.1 An injective mapping $f : A \to B$ is called a *trapdoor one-way function* if:

(a) There is an algorithm E by which we can find $f(x)$ for every $x \in A$.
(b) There is an algorithm D by which, given $y \in \text{Im } f$, we can find $x \in A$ such that $f(x) = y$.
(c) It is not possible to discover the algorithm D from knowledge of the algorithm E (unless some additional information is provided).

The term *trapdoor* refers to the additional information that enables us to find the algorithm D from the algorithm E. If only the algorithm E for finding $f(x)$ is known, and we do not have the additional information by which D may be discovered, then we refer to f as a *one-way function*.

By the term *not possible* in Definition 3.4.1 we mean computationally infeasible. Of course, what is computationally infeasible is relative to our existing theoretical knowledge and the speed and capacity of the computers available at present. It is quite possible that what is considered computationally infeasible at present may become feasible at some time in the future as a result of advances in our theoretical knowledge or the availability of much faster computers.

The idea of a one-way function may appear strange at first sight. One may argue that if we can compute $f(x)$ for every $x \in A$, then we can prepare a look-up table that shows $f(x)$ against x for all x in the domain of the function. Then, given any y in the range of f, we can look up y in the $f(x)$ column in the table and find the corresponding x against it. This argument, however, begs the question. The preparation of such a table is itself computationally infeasible if the size of the set A is very large, say of the order 10^{100}. It is thus implied in the concept of a one-way function that the domain of the function must be so large that the brute-force method of preparing a look-up table is computationally infeasible.

Apart from the advantage of not having to keep the enciphering key secret, the public-key cryptosystems have some other interesting applications. One of them is the technique of "digital signature," which we describe next.

Suppose Alice and Bob exchange secret messages and both have their own public-key cryptosystems. Let E_A, D_A be the public and private keys, respectively, of Alice, and E_B, D_B those of Bob. When Alice sends a message to Bob, she enciphers it by using Bob's public key E_B. When Bob receives the message, he deciphers it by using the secret key D_B (known to himself alone). But how can Bob make sure that the message is really from Alice? Because Bob's enciphering key E_B is public, anybody pretending to be Alice could have sent a fake message. The solution to the problem is to use a digital signature. Let m be the plaintext signature used by Alice. She uses her own secret key D_A to obtain her digital signature $s = D_A(m)$. When she sends a secret message to Bob, she appends her digital signature to the main message and then enciphers the whole thing by using Bob's public key E_B. When Bob deciphers the message and finds the digital signature s included in it, he uses Alice's public key and obtains $E_A(s) = E_A(D_A(m)) = m$. Bob recognizes Alice's signature m and therefore feels assured that the message is indeed from Alice. Because the secret key D_A is known to Alice alone, nobody else could have faked the digital signature $D_A(m)$.

We describe below two public-key cryptosystems.

3.4.1 Knapsack Cryptosystem

A knapsack is a canvas or leather bag commonly worn on the back of hikers or soldiers to carry miscellaneous stuff. Suppose we have a knapsack containing a large number of rods of different lengths. Consider the following problem: Given a length l, how can we pick rods from the knapsack so that, joined end to end, they add up exactly to the length l? This is known as a *knapsack problem.*

In a simpler mathematical version of the knapsack problem, let K be a finite set consisting of positive integers. Given an arbitrary positive integer n, how can we find a subset S of K whose elements add up to n (or show that no such S exists)? In general, for an arbitrary set K, the problem does not have an easy answer. For example, if K has 100 elements, then the number of subsets of K is $2^{100} \approx 10^{30}$. The brute-force approach of checking all subsets of K is computationally infeasible. Thus we have a one-way function here—namely, $f : \mathcal{P}(K) \to \mathbb{N}$, where $\mathcal{P}(K)$ denotes the power set of K, and $f(S)$ is the sum of all elements in S. Given any $S \in \mathcal{P}(K)$, we can easily find $f(S)$, but we do not have an algorithm by which, given a positive integer n, we can find S such that $f(S) = n$. However, we can construct a trapdoor that will enable us to find this algorithm.

A set $L = \{b_1, b_2, \ldots, b_m\}$ is called a *simple knapsack* if

$$b_{i+1} > b_1 + \cdots + b_i \qquad \text{for all } i = 1, \ldots, m - 1$$

In this case, given a positive integer y, we can easily find a unique subset S of L whose elements add up to y, or we can conclude that no such S exists. If $y < b_1$ or $y > b_1 + \cdots + b_m$, then obviously no such S exists. Suppose $b_1 \le y \le b_1 + \cdots + b_m$, and let i_1 be the greatest integer such that $b_{i_1} \le y$. If $b_{i_1} = y$, we are done. Otherwise, we write $y_1 = y - b_{i_1}$. If $y_1 < b_1$ or $y_1 > b_1 + \cdots + b_{i_1 - 1}$, there is no solution. Otherwise, let i_2 be the greatest integer such that $b_{i_2} \leq y_1$. If $b_{i_2} = y_1$, then $y = b_{i_1} + b_{i_2}$, and we are done. Otherwise, we write $y_2 = y_1 - b_{i_2}$ and apply the same argument to y_2. Continuing in this manner, either we find a unique subset $S = \{b_{i_1}, b_{i_2}, \ldots\}$ of L such that $y = b_{i_1} + b_{i_2} + \cdots$, or we arrive at the conclusion that no such subset exists.

The general principle in constructing a knapsack cryptosystem is to start with a simple knapsack L and transform it into a trapdoor knapsack K, as follows: Let $L = (b_1, b_2, \ldots, b_m)$ be a simple knapsack. Select two positive integers q, k such that $q > b_1 + \cdots + b_m$ and k is less than and relatively prime to q. We write

$$a_i = (kb_i) \bmod q$$

for all $i = 1, \ldots, m$. Then $K = (a_1, a_2, \ldots, a_m)$ is a trapdoor knapsack. Since k is relatively prime to q, k is an invertible element in $\mathbb{Z}_q$. Hence we have $b_i = (k^{-1}a_i) \bmod q$ for all $i = 1, \ldots, m$. Therefore, if k and q are both known, L can be obtained from K. But if only q is given and k is unknown, it is not possible to derive L from K.

In a knapsack cryptosystem, the knapsack vector $K = (a_1, a_2, \ldots, a_m)$ and the modulus integer q are made public but the key k is kept secret. The message space M and the cipher space C are $(\mathbb{Z}_2)^m$ and $\mathbb{Z}_q$, respectively. The enciphering algorithm E is given by the mapping

$$f : (\mathbb{Z}_2)^m \to \mathbb{Z}_q$$

where

$$f(x_1, \ldots, x_m) = (x_1 a_1 + \cdots + x_m a_m) \bmod q$$

for all $(x_1, \ldots, x_m) \in (\mathbb{Z}_2)^m$. In brief, using the scalar product notation for vectors, we have

$$f(x) = (x \cdot K) \bmod q$$

Note that each x_i is 1 or 0. Hence $x \cdot K$ is the sum of elements in the subset $S = \{a_i \in K \mid x_i = 1\}$.

Deciphering is done by using the secret key k as follows: Let $y \in \mathbb{Z}_q$ be the received ciphertext. Suppose $y = f(x_1, \ldots, x_m)$. Then, working in $\mathbb{Z}_q$ and using the relation $b_i = (k^{-1}a_i) \bmod q$, we have

$$k^{-1}y = k^{-1}(x_1 a_1 + \cdots + x_m a_m)$$
$$= x_1 b_1 + \cdots + x_m b_m$$

Since $L = (b_1, \ldots, b_m)$ is a simple knapsack, we can find $x_1, \ldots, x_m$ from the value of $k^{-1}y$, as explained above.

To use the knapsack cryptosystem, we first have to transform the plaintext in English into a string of bits. If each character in the alphabet is represented by t bits, then m is chosen to be a multiple of t, say $m = rt$. Then every block of r

characters in the plaintext transforms into an m-tuple $(x_1, \ldots, x_m) \in (\mathbb{Z}_2)^m$. Thus we have a preliminary injective mapping $p : A^r \to (\mathbb{Z}_2)^m$. The enciphering scheme is represented by the diagram

$$A^r \xrightarrow{\ p\ } (\mathbb{Z}_2)^m \xrightarrow{\ f\ } \mathbb{Z}_q$$

In ASCII, for example, $t = 8$, but we may use a smaller value for t if a smaller alphabet is used. In general, if the alphabet has n characters, we can take t to be smallest integer such that $2^t \geq n$. For example, suppose the alphabet consists of only 26 letters A to Z. Then $t = 5$. We represent the letters A, ..., Z by the numbers $1, \ldots, 26$, respectively, and then write each of these numbers in binary form by using 5 bits. The following table shows the complete representation:

A	B	C	D	E	F	G	H	I
00001	00010	00011	00100	00101	00110	00111	01000	01001
J	K	L	M	N	O	P	Q	R
01010	01011	01100	01101	01110	01111	10000	10001	10010
S	T	U	V	W	X	Y	Z	
10011	10100	10101	10110	10111	11000	11001	11010	

In the following example, we take $t = 5$ and $m = 10$.

Example 3.4.2 (a) Use the knapsack system (K, q) with

$$K = (302, 453, 906, 1812, 784, 1417, 145, 290, 580, 1305), \quad q = 2991$$

to encipher REST.

(b) Given that the secret key is $k = 151$, decipher 2773 747.

Solution.

(a)

plaintext	RE	ST
$x \in (\mathbb{Z}_2)^{10}$	1, 0, 0, 1, 0, 0, 0, 1, 0, 1	1, 0, 0, 1, 1, 1, 0, 1, 0, 0
$x \cdot K$	$302 + 1812 + 290 + 1305$	$302 + 1812 + 784 + 1417 + 290$
$(x \cdot K) \bmod q$	$3709 \bmod 2991 = 718$	$4605 \bmod 2991 = 1614$
ciphertext	718	1614

(b) We first use the secret key to discover the simple knapsack L from which K was obtained. In the ring $\mathbb{Z}_{2991}$, $k^{-1} = 151^{-1} \bmod 2991 = 2476$. Hence L is given by

$$\begin{aligned} L &= k^{-1}K \\ &= 2476(302, 453, 906, 1812, 784, 1417, 145, 290, 580, 1305) \bmod 2991 \\ &= (2, 3, 6, 12, 25, 49, 100, 200, 400, 900) \end{aligned}$$

Deciphering is done as shown in the table:

ciphertext $y \in \mathbb{Z}_{2991}$	2773	747
$k^{-1}y = 2476y \bmod 2991$	1603	1134
$k^{-1}y = \sum x_i b_i$	$3 + 100 + 200 + 400 + 900$	$3 + 6 + 25 + 200 + 900$
x	0, 1, 0, 0, 0, 0, 1, 1, 1, 1	0, 1, 1, 0, 1, 0, 0, 1, 0, 1
plaintext	HO	ME

3.4.2 RSA Cryptosystem

We now describe the RSA cryptosystem, named after Rivest, Shamir, and Adelman, which is the most commonly used public-key cryptosystem. The following result in number theory is the basis of the RSA system. Recall that $\varphi(n)$ denotes Euler's phi-function, which gives the number of positive integers not exceeding n and relatively prime to n.

THEOREM 3.4.3 Let p, q be distinct primes. Let $n = pq$ and $m = \varphi(n) = (p-1)(q-1)$. Let $e, d \in \mathbb{Z}_m$ such that $ed \equiv 1 \pmod{m}$. Then for every integer x,

$$x^{ed} \equiv x \pmod{n}$$

Proof: Since $ed \equiv 1 \pmod{m}$, there is an integer t such that $ed = 1 + tm = 1 + t(p-1)(q-1)$. If $x \equiv 0 \pmod{p}$, then trivially $x^{ed} \equiv x \pmod{p}$. Suppose $x \not\equiv 0 \pmod{p}$. By Fermat's theorem, $x^{p-1} \equiv 1 \pmod{p}$. Hence

$$x^{ed} = x^{1+t(p-1)(q-1)} = x(x^{p-1})^{t(q-1)} \equiv x \pmod{p}$$

and $x^{ed} \equiv x \pmod{p}$ for all x. By a similar reasoning, $x^{ed} \equiv x \pmod{q}$ for all x. So p, q both divide $x^{ed} - x$, and hence pq divides $x^{ed} - x$. Therefore $x^{ed} \equiv x \pmod{n}$. ∎

To construct an RSA system, the first step is to select two large primes p, q. We write $n = pq, m = (p-1)(q-1)$. Next we pick a positive integer e less than and relatively prime to m. So e is invertible in the ring $\mathbb{Z}_m$, and hence we can find $d \in \mathbb{Z}_m$ such that $ed \equiv 1 \pmod{m}$. The numbers n and e constitute the public key, but p, q, m, d are all kept secret. We take the message space and the cipher space to be $M = C = \mathbb{Z}_n$. The enciphering algorithm is given by the mapping $f : \mathbb{Z}_n \to \mathbb{Z}_n$, where

$$f(x) = x^e \bmod n$$

Deciphering is done by using the inverse mapping $g : \mathbb{Z}_n \to \mathbb{Z}_n$ given by

$$g(y) = y^d \bmod n$$

By Theorem 3.4.3, $g(f(x)) = (x^e)^d \bmod n = x$. Hence g is the inverse of f. The numbers e and d are called the enciphering and deciphering exponents, respectively.

The security of an RSA system lies in the difficulty of factoring a very large number. As of now, an RSA system is considered secure if the numbers p and q each have 100 or more digits, so $n = pq$ has more than 200 digits. It is estimated that it would take the today's fastest computer thousands of years to factor a 200-digit number. Therefore it would be computationally infeasible to find the deciphering key d from knowledge of the public key (n, e).

The preliminary transformation of the plaintext in English into $\mathbb{Z}_n$ can be done by adopting some suitable scheme. For example, suppose we use an alphabet A consisting of only 26 letters A, B, ..., Z. We represent these letters by two-digit strings 01, 02, ..., 26, respectively. Then a block of r letters transforms into a string of $2r$ digits, which we interpret as a number in $\mathbb{Z}_n$. We choose the number r such that $2r$ does not exceed the number of digits in n. The enciphering scheme is represented by the diagram $A^r \to \mathbb{Z}_n \xrightarrow{f} \mathbb{Z}_n$.

In the example shown below, $r = 1$.

Example 3.4.4 Construct an RSA cryptosystem by taking $p = 3, q = 11$. Use this system to **(a)** encipher PUBLIC and **(b)** decipher 28 26 27 24 26 14.

Solution. Here $n = 3 \times 11 = 33$ and $m = 2 \times 10 = 20$. The numbers relatively prime to 20 are 1, 3, 7, 9, 11, 13, 17, and 19. We can take any of these (except 1) as the enciphering exponent e. Let $e = 13$. Then the public key is $(n = 33, e = 13)$. In $\mathbb{Z}_{20}$, the inverse of 13 is $13^{-1} \bmod 20 = 17$. Hence the secret private key is $d = 17$. Enciphering is done by using the mapping $x \mapsto x^{13} \bmod 33$ and deciphering by $y \mapsto y^{17} \bmod 33$.

(a)

plaintext $x \in \mathbb{Z}_{33}$	P	U	B	L	I	C
	16	21	02	12	09	03
ciphertext $y = x^{13} \bmod 33$	04	21	08	12	03	27

(b)

ciphertext $y \in \mathbb{Z}_{33}$	28	26	27	24	26	14
$y^{17} \bmod 33$	19	05	03	18	05	20
plaintext	S	E	C	R	E	T

The following example shows how the secret key can be found if the modulus n can be factored.

Example 3.4.5 An RSA cryptosystem has the public key $(n = 2747, e = 2051)$. Find the secret key d, and decipher the cryptogram 1795 2483 1922 2313.

Solution. We can find the secret key if we can factor n. This is easily done: $n = 2747 = 41 \times 67$. So $m = \varphi(n) = 40 \times 66 = 2640$ and hence

$$d = e^{-1} \bmod m = 2051^{-1} \bmod 2640 = 251$$

We use the key $d = 251$ to decipher the cryptogram:

ciphertext $y \in \mathbb{Z}_{2747}$	1795	2483	1922	2313
$y^{251} \bmod 2747$	2305	1212	0415	1405
plaintext	WE	LL	DO	NE

EXERCISES 3.4

1. Transform the simple knapsack $L = (3, 5, 9, 18, 36)$ into a trapdoor knapsack with modulus $q = 97$ and private key $k = 29$.

2. Use the knapsack system with public key

$$K = (45, 60, 22, 44, 88), \quad q = 98$$

and private key $k = 15$ to (a) encipher AMERICAN and (b) decipher 66 89 12 28 34 60.

3. Use the knapsack system with public key

$$K = (73, 146, 93, 186, 173), \quad q = 199$$

and private key $k = 73$ to (a) encipher PROBLEM and (b) decipher 173 27 34 127 67 60.

4. Use the knapsack system with public key

$$K = (2013, 356, 41, 753, 520, 1711, 2436, 887, 1815, 42), \quad q = 2999$$

and private key $k = 671$ to (a) encipher DIVINE and (b) decipher 455 1846 1231.

5. An RSA system has the public key $n = 35$, $e = 19$. Find the private key d.

6. An RSA system has the public key $n = 323$, $e = 35$. Find the private key d.

7. An RSA system has the public key $n = 3337$, $e = 17$. Find the private key d.

8. Use the RSA cryptosystem with public key $n = 3337$, $e = 17$, and blocks of two letters to (a) encipher GOLD and (b) decipher 369 3029.

9. Use the RSA cryptosystem with public key $n = 2881$, $e = 65$, and blocks of two letters to (a) encipher SILVER and (b) decipher 2690 109 2103.

CHAPTER

4

CODING THEORY

4.1 INTRODUCTION TO ERROR-CORRECTING CODES

Coding theory deals with the problem of errors that occur when a message is transmitted through a communication channel. The channel may be a telephone line, a radio or television link, a recording device, or whatever. The error may be caused by thermal noise, atmospheric disturbance, faulty equipment, or human negligence. A channel prone to transmission errors is called a *noisy* channel. An error-correcting code is a scheme of encoding the message in such a way that the correct message may be recovered even when errors have taken place during transmission. The general principle of an error-correcting code is to add redundancy to the message so that the errors can be detected and corrected in most cases.

Let us consider an artificially simple example. Suppose the only messages we wish to send are "YES" and "NO." (Think of an examination where each question requires a yes/no answer.) We have a digital communication channel through which the symbols 1 and 0 can be transmitted. Let us decide to represent the message YES by 1 and NO by 0. We assume this scheme of representation is known also to the recipient of the message. If the channel is not noisy, there is no problem. When we wish to send the message YES, we transmit 1. If no error occurs in the transmission, the recipient receives the message 1 and interprets it to mean YES. But if the channel is noisy, it is possible that when we transmit 1, the message received may be 0, which she interprets to mean NO. (Because of a noisy channel, you may get an F on the exam even when you answered all questions correctly.)

An obvious (though not the most efficient) method for dealing with the problem of transmission errors is to repeat the message. Let us represent YES as 11 and NO as 00. We refer to this representation as *encoding* and refer to 11 and 00 as *codewords*. The set $C = \{11, 00\}$ is called a *code*. Suppose we wish to send the message YES, and therefore we transmit 11. If no error takes place, the received message is 11, which is correctly interpreted as YES. If an error occurs in one of the two symbols, the received message is 10 or 01. Because neither of these is a codeword, the receiver concludes that an error has occurred, but she has no way of determining whether the original message was 11 or 00. Further, if an error occurs in both the digits, the received message is 00, which she interprets as NO. In this case, she gets the wrong message. We thus see that we have an encoding scheme in which one error in the message can be detected (but not corrected), but if two errors occur, they remain undetected. Since the probability of two errors occurring is less than the probability of one error, the chances of a wrong message being accepted as correct are less now than in the previous scheme where we represented YES and NO by single symbols.

We observe that the ability to detect an error in the received message is a result of the redundancy that we introduced in the codewords by using two symbols in place of one. Let us see what happens if we further increase redundancy. Let us represent YES as 111 and NO as 000. As before, suppose we transmit 111 to send the message YES. Now, if one of the three digits is received in error, the received message is 110, 101, or 011. If two errors occur, the received message is 100, 010, or 001. Because none of these is a codeword, the receiver concludes that an error has occurred. But if three errors occur, the received message is 000, which, being a codeword, is wrongly accepted as the message NO. We thus see that with this code we can detect up to two errors in the received message.

In fact, with this code $C = \{111, 000\}$, we can do more than just detect up to two errors. We can recover the correct message if only one error has occurred in the received message. If the received message is 110, 101, or 011 and we assume that only one error has occurred, then the original message must have been 111 (=YES). So we adopt the following rule: If the received message is 111, 110, 101, or 011, we decode it as 111. If, on the other hand, the received message is 000, 100, 010, or 001, we decode it as 000. This is called *nearest neighbor decoding* or *maximum-likelihood decoding*. Thus we see that this code can detect up to two errors and, with the nearest neighbor decoding procedure, it can correct one error. Of course, if more than one error has occurred in the received message, then the nearest neighbor decoding rule will give a wrong result. But the chances of two or three errors occurring are less than the chance of one error, so on the whole we are in a better situation than before.

To get a better understanding of the error-correcting ability of the code $C = \{111, 000\}$, let us do some computation and find the probability of a received message being decoded correctly. We assume that for each transmitted symbol (0 or 1), there is the same probability $p(<\frac{1}{2})$ of its being received in error. (A transmission channel with this property is called a *symmetric binary channel*.) Consequently, the probability of a symbol being received correctly is $1 - p$. We refer to p as the *symbol error probability* of the transmission channel. Suppose we transmit the codeword 111. We have seen that the nearest neighbor decoding gives a correct result if the received word is 111, 110, 101, or 011. The probability of 111 being received as 111 is $(1 - p)^3$. The probability that 111 is received as 110 is $(1 - p)^2 p$. The same holds for the probability of 101 and 011. Hence the total probability of 111 being received as 111, 110, 101, or 011 is equal to

$$P_{\text{corr}}(C) = (1 - p)^3 + 3(1 - p)^2 p = (1 + 2p)(1 - p)^2$$

This is the probability that the decoding gives the correct result when the codeword 111 is transmitted. Clearly, the probability of correct decoding is the same when the codeword 000 is transmitted. Hence the probability that a received vector is decoded wrongly is $P_{\text{err}}(C) = 1 - P_{\text{corr}}(C)$. We refer to $P_{\text{err}}(C)$ as the *word error probability* of the code C. Note that the symbol error probability p is a property of the channel and does not depend on the code C, but the word error probability $P_{\text{err}}(C)$ is a characteristic of the code C and is a function of p.

Thus we have shown that the word error probability of the code $C = \{111, 000\}$ is given by

$$P_{\text{err}}(C) = 1 - P_{\text{corr}}(C) = (3 - 2p)p^2$$

For example, suppose $p = 0.1$. (In most real situations, p may be less than this.) Then $P_{\text{err}}(C) = (3 - 0.2)(0.1)^2 = 0.028$ and $P_{\text{corr}}(C) = 0.972$. If $p = 0.01$, then $P_{\text{err}}(C) = 0.000298$ and $P_{\text{corr}}(C) = 0.999702$.

By further increasing the number of repetitions of the symbol in a codeword— for example, by taking $C = \{11111, 00000\}$—we can obtain a lower word error probability $P_{\text{err}}(C)$. But on the whole, mere repetition does not give an efficient code. The number of codewords is only two, which is not enough for any practical application. Also, too much redundancy of symbols in a codeword increases the time and cost of transmission. It is therefore desirable to find more efficient encoding schemes.

We now give a general definition of a code. Let A be a finite set of q (> 1) symbols that can be transmitted through the communication channel. We refer to A as the *alphabet* of transmission. Let $V = A^n$ denote the set of all n-tuples of elements of A, where n is some specified positive integer greater than 1. So V consists of q^n elements, which are called *words* or *vectors*. For the sake of convenience, an n-tuple $(v_1, v_2, \ldots, v_n) \in V$ is commonly written $v_1 v_2 \ldots v_n$. Let C be a nonempty subset of V. Then C is called a *q-ary code* of *length n* over A. In particular, if $q = 2$, the code C is called a *binary* code; if $q = 3$, it is called *ternary*. The members of C are called *codewords*. If C contains only one codeword, or if $C = A^n$, then C is called a *tivial code*. Further, C is called a *repetition code* if every codeword in C is a vector of the form $aa \ldots a$ for some $a \in A$. Thus a q-ary repetition code (of any length) contains exactly q codewords. For example, the code $C = \{111, 000\}$ discussed above is a binary repetition code of length 3.

To use a code C, we must have a fixed scheme (known to both sender and receiver) of representing different messages by codewords in C. This is referred to as *encoding.* For example, if C has at least 26 codewords, we can represent each letter of the English alphabet by a codeword. Then any message written in English can be converted into a sequence of codewords. However, for the purpose of studying the error-correcting properties of a code, which is our objective, the choice of an encoding scheme is immaterial. Therefore we need not concern ourselves with it.

To study the error-detecting and error-correcting properties of a code, we introduce the concept of distance between two vectors in V.

Definition 4.1.1 Let $x, y \in A^n$, $x = x_1 x_2 \ldots x_n$, $y = y_1 y_2 \ldots y_n$. The *Hamming distance* between the vectors x and y, denoted by $d(x, y)$, is defined to be the number of subscripts i such that $x_i \neq y_i$; that is,

$$d(x, y) = |\{i \mid x_i \neq y_i\}|$$

For example, in $\{0, 1\}^3$, $d(110, 011) = 2$. In $\{0, 1, 2\}^4$, $d(1202, 1102) = 1$.

The Hamming distance satisfies the three usual conditions of a distance function:

1. $d(x, y) = 0$ if and only if $x = y$.
2. $d(x, y) = d(y, x)$ for all $x, y \in A^n$.
3. $d(x, z) \leq d(x, y) + d(y, z)$ for all $x, y, z \in A^n$ (*triangle inequality*).

The first two conditions are obvious from the definition. To prove the triangle inequality, let $I = \{i \mid x_i \neq y_i\}$ and $J = \{i \mid y_i \neq z_i\}$. Then for all $i \notin I \cup J$, $x_i = y_i = z_i$. Hence

$$d(x, z) \leq |I \cup J| \leq |I| + |J| = d(x, y) + d(y, z)$$

Consider the problem of decoding a received message. Suppose a codeword x (unknown to the receiver) is transmitted and the vector received is y (which may or may not be the same as x). If y is also a codeword, then it is accepted (possibly wrongly) as the transmitted codeword. If y is not a codeword, then it is clear that one or more errors have occurred during transmission. In this case, we decode y as the codeword $x' \in C$ whose distance from y is the least. If the number of errors that have occurred is sufficiently small, then $x' = x$, and decoding gives the correct result. This procedure is known as *nearest neighbor decoding* or *maximum-likelihood decoding* (as explained in the example discussed above). A code C is said to correct up to t errors, or is called a t-*error correcting* code, if this decoding procedure gives the correct result whenever t or fewer errors have occurred in the transmission of any codeword. This number t depends on the minimum distance of the code, which we define below.

Definition 4.1.2 The *minimum distance* of a code C, denoted by $d(C)$, is the smallest distance between any two distinct codewords in C; that is,

$$d(C) = \min\{d(x, y) \mid x, y \in C, \ x \neq y\}$$

For example, the minimum distance of the binary repetition code $C = \{111, 000\}$ is 3. The minimum distance of $C = \{0000, 1110, 0111\}$ is 2.

The minimum distance of a code determines its error-detecting and error-correcting capabilities, as given by the following theorem. For any real number x, we write $\lfloor x \rfloor$ to denote the greatest integer $\leq x$.

THEOREM 4.1.3 Let C be a code with minimum distance d. Let $t = \left\lfloor \dfrac{d-1}{2} \right\rfloor$.
Then

(a) C can detect up to $d - 1$ errors in any transmitted codeword.
(b) C can correct up to t errors in any transmitted codeword.

Proof: Suppose a codeword $x \in C$ is transmitted and the vector received is y ($\neq x$). Then $d(x, y)$ is equal to the number of errors that have occurred in transmission.

(a) If the number of errors is less than or equal to $d - 1$, then $d(x, y) < d$. Hence y cannot be a codeword because the minimum distance between any two distinct codewords is d. The vector y is recognized as being in error, so up to $d - 1$ errors are detected. If $d(x, y) \geq d$, then y may be (not necessarily) a codeword. So it is possible that more than $d - 1$ errors may go undetected.

(b) Suppose the number of errors is less than or equal to t. Then $d(x, y) \leq t$. We claim that x is the unique codeword whose distance from y is less than or equal to t. Let $x' \in C$ such that $d(x', y) \leq t$. Then, by the triangle inequality, $d(x, x') \leq d(x, y) + d(y, x') \leq 2t \leq d - 1$. So $d(x, x') < d$ and hence $x' = x$. Thus, by the nearest neighbor decoding rule, y is correctly decoded as x. But if more than t errors occur, then a codeword x' (different from x) may be the nearest codeword from y, in which case the decoding will give a wrong result.   ∎

Note that if C is a t-error correcting code, then its minimum distance is $2t + 1$ or $2t + 2$.

A code of length n, minimum distance d, and consisting of M codewords is called an (n, M, d)-code. The desirable properties of a "good" code are to have small length n, large size M, and large minimum distance d. The smaller the length n, the faster the transmission and lower the cost; the larger the size M, the greater the variety of messages that can be sent; the larger the minimum distance d, the greater the number of errors that can be corrected. But these three properties are mutually conflicting. For example, if we keep n fixed, then a larger M requires a smaller d. The main problem of coding theory is this: for given values of n and d, to find the code with the largest possible M. We denote by $A_q(n, d)$ the largest value of M such that there exists a q-ary (n, M, d)-code. In the next theorem we will find an upper bound for $A_q(n, d)$.

Given a vector $u \in A^n$ and a nonnegative integer $r \le n$, let $S_r(u)$ denote the set of all vectors in A^n at a distance $\le r$ from u; that is,

$$S_r(u) = \{v \in A^n \mid d(u, v) \le r\}$$

The set $S_r(u)$ is called the *sphere of radius r* with center u. A simple counting argument gives the number of elements in $S_r(u)$. Let A be an alphabet set of q symbols. Consider the vectors $v \in A^n$ whose distance from the given vector u is equal to m $(0 \le m \le r)$. Let $v = v_1 \ldots v_n$ and $u = u_1 \ldots u_n$. Then there are m subscripts i such that $v_i \ne u_i$. These m subscripts can be chosen in $\binom{n}{m}$ ways. For each such subscript i, v_i can be chosen in $(q-1)$ ways. Hence the number of vectors v such that $d(u, v) = m$ is equal to $\binom{n}{m}(q-1)^m$. It follows immediately that the number of vectors v such that $d(u, v) \le r$ is equal to

$$|S_r(u)| = \sum_{m=0}^{r} \binom{n}{m}(q-1)^m$$

Let $C \subset A^n$ be a code of minimum distance d, and let $t = \lfloor (d-1)/2 \rfloor$. Then the spheres of radius t whose centers are codewords in C are pairwise disjoint. Suppose a vector v lies in both $S_t(u)$ and $S_t(u')$, where $u, u' \in C$. Then $d(u, u') \le d(u, v) + d(v, u') \le 2t < d$, a contradiction. It follows that the union of all these spheres consists of $M |S_t(u)|$ vectors, where M is the number of codewords in C. Since the total number of vectors in A^n is q^n, we have the following result:

THEOREM 4.1.4 Let C be a q-ary (n, M, d)-code, and let $t = \lfloor (d-1)/2 \rfloor$. Then

$$M \sum_{m=0}^{t} \binom{n}{m}(q-1)^m \le q^n$$

This inequality is known as the *sphere-packing bound*. It gives an upper bound for the number $A_q(n, d)$ for given values of q, n, and d; that is,

$$A_q(n, d) \le \frac{q^n}{\sum_{m=0}^{t} \binom{n}{m}(q-1)^m}$$

In most codes the actual number M of codewords is less than the sphere-packing bound given by Theorem 4.1.4. A code is called *perfect* when the upper bound for M is attained. In that case, every vector in A^n is at a distance $\le t$ from some codeword. It follows that the minimum distance of a perfect code must be an odd integer. Suppose C is a perfect code with $d(C) = 2t + 2$, and let $x \in C$. Pick $y \in A^n$ such that $d(x, y) = t + 1$. Then y is at distance greater than t from every codeword. (Prove!) Thus we can define a perfect code as follows:

Definition 4.1.5 Let $C \subset A^n$ be a code with minimum distance $d(C) = 2t + 1$. If for every vector $y \in A^n$, there exists $x \in C$ such that $d(x, y) \le t$, then C is called a *perfect* code.

For example, the binary repetition code $\{00\ldots0, 11\ldots1\}$ of length n, where n is any odd integer, is a perfect code with minimum distance n. It is clear that every vector $y \in \{0, 1\}^n$ is at a distance $\leq t = (n-1)/2$ from either $00\ldots0$ or $11\ldots1$. At the other extreme, if $C = A^n$, then C is a perfect code with minimum distance 1. We refer to these as *trivial* perfect codes.

For future reference, we state below the condition for a perfect code that follows from Theorem 4.1.4.

THEOREM 4.1.6 Let C be a q-ary (n, M, d)-code with $d = 2t + 1$. Then C is a perfect code if and only if

$$M \sum_{m=0}^{t} \binom{n}{m}(q-1)^m = q^n$$

In particular, a binary $(n, M, 2t+1)$-code is perfect if and only if

$$M \sum_{m=0}^{t} \binom{n}{m} = 2^n$$

Example 4.1.7 Show that a binary $(7, 16, 3)$-code (if it exists) is perfect.

Solution. Here $n = 7$, $M = 16$, and $t = 1$. Hence

$$M \sum_{m=0}^{t} \binom{n}{m} = 16 \left\{ \binom{7}{0} + \binom{7}{1} \right\} = 16(1+7) = 2^7$$

which proves that the code is perfect.

[We will show in the next section that a binary $(7, 16, 3)$-code does exist.]

EXERCISES 4.1

1. Show that the word error probability of the binary repetition code of length 5 is $(10 - 15p + 6p^2)p^3$, where p is the symbol error probability of the channel.

2. Suppose C is a code with minimum distance $2t + 2$. If a vector a is at a distance $t + 1$ from some codeword, show that a is at a distance $> t$ from every codeword.

3. Show that a ternary code of length 3 and minimum distance 2 cannot have more than 9 codewords. Show that a ternary $(3, 9, 2)$-code does exist.

4. Show that if C is a q-ary $(3, M, 2)$-code, then $M \leq q^2$. Show that a q-ary $(3, q^2, 2)$-code does exist.

5. Show that if there exists a binary (n, M, d)-code, then there exists a binary $(n-1, M', d)$-code with $M' \geq M/2$.

6. If d is an odd integer, show that a binary (n, M, d)-code exists if and only if a binary $(n+1, M, d+1)$-code exists.

7. Show that $A_q(n, 1) = q^n$ and $A_q(n, n) = q$.

8. Show that $A_q(3, 2) = q^2$ for any $q \geq 2$.

9. Show that $A_2(n, d) \leq 2A_2(n - 1, d)$.

10. Show that the binary repetition code $\{0 \ldots 0, 1 \ldots 1\}$ of any odd-length n satisfies the condition of Theorem 4.1.6 for being perfect.

11. Show that a binary $(15, 2048, 3)$-code (if it exists) is perfect.

12. Show that, for any positive integer $r > 1$, a binary $(2^r - 1, 2^{2^r - r - 1}, 3)$-code (if it exists) is perfect.

13. Show that a binary $(23, 4096, 7)$-code (if it exists) is perfect.

14. Show that a ternary $(4, 9, 3)$-code (if it exists) is perfect.

15. Show that a ternary $(13, 59049, 3)$-code (if it exists) is perfect.

16. Show that a ternary $(11, 729, 5)$-code (if it exists) is perfect.

4.2 LINEAR CODES

The definition of a code given in Section 4.1 is too general. We now confine our attention to codes of a special kind, known as *linear codes*. These codes are more amenable to algebraic treatment and have many interesting algebraic properties. We take the alphabet of symbols to be a finite field F. The set $V = F^n$ is then an n-dimensional vector space over F. A nonempty subset C of V is a subspace if for all $x, y \in C$ and $\alpha \in F$, we have $x + y \in C$ and $\alpha x \in C$.

Definition 4.2.1 Let F be a finite field and n a positive integer. Let C be a subspace of the vector space $V = F^n$. Then C is called a *linear code* over F. If C is a subspace of dimension k, then C is called an $[n, k]$-code. Further, if the code C has minimum distance d, we say C is an $[n, k, d]$-code.

For example, the binary repetition code $\{000, 111\}$ is a linear $[3, 1, 3]$-code over the field $F = \{0, 1\}$. The binary code $\{000, 100, 111\}$ is not linear because $100 + 111 = 011$, but 011 is not a codeword.

Note the difference between the notation (n, M, d) for a code of M words used in Section 4.1 and the notation $[n, k, d]$ for a linear code of dimension k.

Let F be a field of q elements. Let C be a linear $[n, k]$-code over F. Then, being a subspace of F^n of dimension k, C has a basis of k elements, say $(b_1, \ldots, b_k)$. So every element $x \in C$ can be uniquely expressed as a linear combination of the basis elements; that is,

$$x = \alpha_1 b_1 + \cdots + \alpha_k b_k$$

where $\alpha_1, \ldots, \alpha_k \in F$. Since each α_i can be chosen in q ways, it follows that there are exactly q^k elements in C. Thus we see that a q-ary $[n, k]$-code consists of q^k codewords. In particular, a binary $[n, k]$-code consists of 2^k codewords.

In the sequel, F denotes a finite field. A field of q elements is denoted by $\mathbb{F}_q$. (Recall that the number of elements in a finite field is a power of some prime p.) We treat the elements of F^n as row vectors. Depending on the context, we write an element $x \in F^n$ as $(x_1, \ldots, x_n)$ or $[x_1 \ldots x_n]$ or simply $x_1 \ldots x_n$. We write x^T to denote the transpose of x. The symbol 0 denotes the zero element in F as well as the zero vector $(0, 0, \ldots, 0) \in F^n$.

4.2.1 Generator and Parity-Check Matrices

There are two matrices associated with a linear code that play an important role in coding theory.

Definition 4.2.2 Let C be a linear $[n, k]$-code. Let G be a $k \times n$ matrix whose rows form a basis of C. Then G is called a *generator matrix* of the code C.

A linear code is completely determined by a generator matrix. Let G be a generator matrix of an $[n, k]$-code C over F. Then every element $x \in C$ is a linear combination of the rows of G—that is, $x = u_1 G_1 + \cdots + u_k G_k$, where $u_1, \ldots, u_k \in F$ and $G_1, \ldots, G_k$ are the rows of G. In other words, C is the *row space* of the matrix G. Thus we have the following result:

THEOREM 4.2.3 Let C be an $[n, k]$-code over F. Let G be a generator matrix of C. Then

$$C = \{uG \mid u \in F^k\}$$

This representation of C provides a scheme for *encoding*. Let G be a generator matrix of an $[n, k]$-code C over $F = \mathbb{F}_q$. The matrix G determines a bijective mapping $e : F^k \to C$ given by $u \mapsto uG$. We use this mapping to represent q^k distinct messages by codewords. First, we adopt some fixed scheme by which the q^k vectors in F^k are identified with the q^k messages, and then we use the encoding mapping e. For our purpose, however, the way in which the elements of F^k are identified with the actual messages is immaterial. So we may consider F^k itself as the set of messages. We refer to the elements of F^k as *message words*. Thus every k-tuple message word u is encoded as an n-tuple codeword uG. We refer to the number $n - k$ as the *redundancy* of the code C, and k/n as its *transmission rate*.

Consider the homogeneous linear system $GX = 0$—that is, the system of k linear equations in n unknowns:

$$g_{11}x_1 + \cdots + g_{1n}x_n = 0$$

$$\vdots$$

$$g_{k1}x_1 + \cdots + g_{kn}x_n = 0$$

Let N be the set of all solutions. Since the rows of the matrix G are linearly independent, *rank* $G = k$. Hence, by the rank-nullity theorem in linear algebra, N is a subspace of F^n of dimension $n - k$. N is called the *null space* of G.

The matrix G is not a unique generator matrix of C, since a basis of C can be chosen in many ways. But we show that the null space N is determined uniquely by C and is independent of the particular choice of a generator matrix G. Let $y = [y_1 \ldots y_n] \in N$. Then

$$G_i y^T = g_{i1} y_1 + \cdots + g_{in} y_n = 0$$

for each $i = 1, \ldots, k$. Hence $G y^T = 0$. Therefore, for every $x = uG \in C$, $u \in F^k$, we have $x y^T = uG y^T = 0$. Conversely, let $y \in F^n$ such that $x y^T = 0$ for every $x \in C$. Then, in particular, $G_i y^T = 0$ for each $i = 1, \ldots, k$. Hence $y \in N$. Thus $N = \{ y \in F^n \mid x y^T = 0 \text{ for all } x \in C \}$. This proves that N is independent of the matrix G chosen to generate C. Since N is a subspace of F^n of dimension $n - k$, N is a linear $[n, n - k]$-code over F. We call N the *dual code* of C and denote it by $C^\perp$.

For any vectors $x, y \in F^n$, let $x \cdot y$ denote their inner product; that is,

$$x \cdot y = x_1 y_1 + \cdots + x_n y_n = x y^T = y x^T = y \cdot x$$

Then we can define the dual code as follows:

Definition 4.2.4 Let C be an $[n, k]$-code over F. Then the *dual code* of C is defined to be

$$C^\perp = \{ y \in F^n \mid x \cdot y = 0 \text{ for all } x \in C \}$$

Two vectors $x, y \in F^n$ are said be *orthogonal* if $x \cdot y = 0$. Thus each vector in $C^\perp$ is orthogonal to each vector in C. A linear code C is called *self-orthogonal* if every vector in C is orthogonal to itself and to every other vector in C. In other words, C is self-orthogonal if $C \subset C^\perp$.

We now show that the relation between C and $C^\perp$ is symmetric; that is, the dual code of $C^\perp$ is C. Let $x \in C$. Then $x \cdot y = 0$ for each $y \in C^\perp$. Hence $x \in (C^\perp)^\perp$, so $C \subset (C^\perp)^\perp$. We have seen above that if C is a subspace of dimension k, then $C^\perp$ is a subspace of dimension $n - k$. Hence $(C^\perp)^\perp$ has dimension $n - (n - k) = k$. Thus $\dim C = k = \dim(C^\perp)^\perp$. This proves that $C = (C^\perp)^\perp$. We record this result in the following theorem.

THEOREM 4.2.5 Let C be an $[n, k]$-code. Then $C^\perp$ is an $[n, n - k]$-code and $(C^\perp)^\perp = C$.

Example 4.2.6 Find the dual of the binary code $C = \{0000, 1111\}$, and show that C is self-orthogonal.

Solution. C is a $[4, 1]$-code over $\mathbb{F}_2$. Checking the inner product of vectors in $(\mathbb{F}_2)^4$ with every vector in C, we find that

$$C^\perp = \{0000, 1100, 1010, 1001, 0110, 0101, 0011, 1111\}$$

We verify that $C^\perp$ is a $[4, 3]$-code. Clearly, $C \subset C^\perp$, so C is self-orthogonal.

We now define the second of the two matrices associated with a linear code, to which we referred above.

Definition 4.2.7 Let C be an $[n, k]$-code. Let H be a generator matrix of the dual code $C^\perp$. Then H is called a *parity-check matrix* of the code C.

It follows from the definition that if G is a generator matrix of a code C, then G is a parity-check matrix of the dual code $C^\perp$. Since the dual of an $[n, k]$-code is an $[n, n - k]$-code, it follows that a parity-check matrix of an $[n, k]$-code C is an $(n - k) \times n$ matrix H whose rows form a basis of $C^\perp$. We have seen above that if G is a generator matrix of C, then the null space of G is $C^\perp$. Now H is a generator matrix of $C^\perp$, and hence the null space of H is $(C^\perp)^\perp = C$. Hence $x \in C$ if and only $Hx^T = 0$ or, equivalently, $xH^T = 0$. So a parity-check matrix completely determines the code. Thus we have proved the following result.

THEOREM 4.2.8 Let C be an $[n, k]$-code over F, and let H be a parity-check matrix of C. Then

$$C = \{x \in F^n \mid xH^T = 0 = Hx^T\}$$

The condition $Hx^T = 0$ gives a homogeneous system of $n - k$ linear equations for the components of x. These are referred to as *parity-check equations*.

The following theorem gives the mutual relation between a generator matrix and a parity-check matrix of a linear code.

THEOREM 4.2.9 Let C be an $[n, k]$-code. Let G and H be, respectively, a generator matrix and a parity-check matrix of C. Then

$$GH^T = 0 = HG^T$$

Conversely, suppose G is a $k \times n$ matrix of rank k, and H is an $(n - k) \times n$ matrix of rank $n - k$, such that $GH^T = 0$. Then H is a parity-check matrix of the code C if and only if G is a generator matrix of C.

Proof: By Theorem 4.2.8, $xH^T = 0$ for every $x \in C$. Hence, in particular, $G_iH^T = 0$ for every row G_i of G, and $GH^T = 0$. Taking the transpose, we get $HG^T = 0$. This proves the first part of the theorem.

To prove the second part, let G and H be matrices as stated, with $GH^T = 0$, and suppose H is a parity-check matrix of C. Then $G_iH^T = 0$ for each $i = 1, \ldots, k$.

Hence $G_1, \ldots, G_k \in C$. Since $rank\, G = k$, $G_1, \ldots, G_k$ are linearly independent and hence form a basis of C. This proves that G is a generator matrix of C. The converse is proved similarly. ∎

Example 4.2.10 Find a generator matrix and a parity-check matrix of the binary code $C = \{000, 111\}$.

Solution. The matrix $G = [1 \quad 1 \quad 1]$ is the unique generator matrix of C. To find a parity-check matrix H, we first find the dual code $C^\perp$. By checking the inner products of all vectors in $(\mathbb{F}_2)^3$ with vectors in C, we find that $C^\perp = \{000, 110, 011, 101\}$. Here, any two nonzero vectors in $C^\perp$ form a basis. Hence we can take $H = \begin{bmatrix} 1 & 1 & 0 \\ 0 & 1 & 1 \end{bmatrix}$.

Example 4.2.11 Find the binary linear code C with parity-check matrix

$$H = \begin{bmatrix} 1 & 1 & 1 & 1 & 1 \\ 1 & 0 & 1 & 1 & 0 \end{bmatrix}$$

and write a generator matrix G of C. Also find the dual code $C^\perp$.

Solution. Since H is a 2×5 matrix, we know that C is a $[5, 3]$-code. Moreover, C is the null space of the matrix H. So we solve the homogeneous linear system

$$x_1 + x_2 + x_3 + x_4 + x_5 = 0$$
$$x_1 + x_3 + x_4 = 0$$

We can choose x_1, x_2, and x_3 arbitrarily. Then $x_4 = x_1 + x_3$ and $x_5 = x_2$. Hence the general solution of the linear system is

$$x = [x_1 \quad x_2 \quad x_3 \quad x_1 + x_3 \quad x_2]$$
$$= x_1[1 \quad 0 \quad 0 \quad 1 \quad 0] + x_2[0 \quad 1 \quad 0 \quad 0 \quad 1] + x_3[0 \quad 0 \quad 1 \quad 1 \quad 0]$$
$$= [x_1 \quad x_2 \quad x_3]G$$

where x_1, x_2, x_3 can be assigned arbitrary values (0 or 1) and

$$G = \begin{bmatrix} 1 & 0 & 0 & 1 & 0 \\ 0 & 1 & 0 & 0 & 1 \\ 0 & 0 & 1 & 1 & 0 \end{bmatrix}$$

Thus G is a generator matrix of the code C. Assigning all possible values to x_1, x_2, x_3, we obtain

$$C = \{00000, 10010, 01001, 00110, 11011, 10100, 01111, 11101\}$$

Now H is a generator of the dual code $C^{\perp}$; hence

$$C^{\perp} = \{00000, 11111, 10110, 01001\}$$

We observed earlier that a generator matrix of a linear code C is not unique (except in the trivial case of a binary code of dimension 1). Let G be a generator matrix of C, and let G' be a matrix obtained as a result of performing elementary row operations on G. Then every row of G is a linear combination of the rows of G', and conversely. So G, G' have the same row space C. Hence G' is also a generating matrix of C. Conversely, if G, G' are both generating matrices of C, then each can be obtained by performing elementary row operations on the other. In particular, let C be an $[n, k]$-code, and suppose the first k columns of G are linearly independent. Then, by performing elementary row operations, we can transform G to a row-reduced echelon matrix of the form $G^* = [\, I_k : A \,]$, where I_k is the identity matrix of order k, and A is some $k \times (n - k)$ matrix. G^* is called the *canonical generator matrix* of C. Now let $H^* = [\, -A^T : I_{n-k} \,]$. Then H^* is an $(n - k) \times n$ matrix of rank $n - k$, and

$$G^*(H^*)^T = [\, I_k : A \,] \begin{bmatrix} -A \\ \cdots \\ I_{n-k} \end{bmatrix} = -A + A = 0$$

Hence, by Theorem 4.2.9, H^* is a parity-check matrix of C. We refer to H^* as the *canonical parity-check matrix* of C. Thus we have the following result:

THEOREM 4.2.12 Let C be an $[n, k]$-code. If C has a canonical generator matrix $G = [\, I_k : A \,]$, then $H = [\, -A^T : I_{n-k} \,]$ is the canonical parity-check matrix of C. Conversely, if $H = [\, B : I_{n-k} \,]$ is a parity-check matrix of C, then $G = [\, I_k : -B^T \,]$ is a generator matrix of C.

The generator matrix that we found for the code in Example 4.2.11 is in canonical form. We can obtain it more easily by using Theorem 4.2.12.

Example 4.2.13 Use Theorem 4.2.12 to find the canonical generator matrix of the code C in Example 4.2.11.

Solution. The given parity-check matrix

$$H = \begin{bmatrix} 1 & 1 & 1 & 1 & 1 \\ 1 & 0 & 1 & 1 & 0 \end{bmatrix}$$

is transformed (by interchanging rows and then adding the first row to the second) to canonical form

$$H^* = \begin{bmatrix} 1 & 0 & 1 & 1 & 0 \\ 0 & 1 & 0 & 0 & 1 \end{bmatrix} = [\, B : I_2 \,]$$

Hence the canonical generator matrix of C is

$$G^* = \left[\, I_3 \;\vdots\; -B^T \,\right] = \begin{bmatrix} 1 & 0 & 0 & 1 & 0 \\ 0 & 1 & 0 & 0 & 1 \\ 0 & 0 & 1 & 1 & 0 \end{bmatrix}$$

We mentioned earlier that a generator matrix G of an $[n, k]$-code C determines an encoding mapping $e : F^k \to F^n$ that maps a messsage word $u \in F^k$ to a codeword $x = uG \in C$. This encoding assumes a particularly simple form when G is canonical. If $G = [\, I_k \;\vdots\; A \,]$, then $x = uG = [\, u \;\vdots\; uA \,]$. So the first k components of x form the message word itself. The encoding simply adjoins to the message word $n - k$ *check symbols* given by uA.

If the first k columns of a generator matrix G of an $[n, k]$-code C are not linearly independent, then C does not have a canonical generator matrix. However, since G has rank k, there exist k linearly independent columns in G. By reordering the columns of G, we can obtain a matrix G' whose first k columns are linearly independent, but then G' is not a generator matrix of C. If C' is the linear code generated by G', the codes C, C' are said to be equivalent. The general definition of equivalent codes follows.

Definition 4.2.14 Let C and C' be $[n, k]$-codes over F. The codes C and C' are said to be *equivalent* if there exists a bijective mapping $f : C \to C'$ given by

$$f(x_1, \ldots, x_n) = (\alpha_1 x_{\sigma(1)}, \ldots, \alpha_n x_{\sigma(n)})$$

where $\alpha_1, \ldots, \alpha_n \in F$ are nonzero scalars, and σ is a permutation of the set $\{1, \ldots, n\}$.

The definition leads to the following result. The proof is left as an exercise.

THEOREM 4.2.15 Let C and C' be $[n, k]$-codes over F with generator matrices G and G', respectively. The codes C and C' are *equivalent* if and only if one matrix can be obtained from the other by performing operations of the following types:

(**a**)   Elementary row operations
(**b**)   Permutations of columns
(**c**)   Multiplication of any column by a nonzero scalar in F

A similar property holds for parity-check matrices of two equivalent codes.

4.2.2 Minimum Distance

We now consider the minimum distance of a linear code. Recall that the minimum distance of a code C is defined as $d(C) = \min\{d(x, y) \mid x, y \in C, x \neq y\}$, where $d(x, y)$ is the Hamming distance between x and y. For linear codes, we introduce a new concept.

Definition 4.2.16 The *weight* of a vector $x \in F^n$, written $w(x)$, is defined to be the number of nonzero components in x.

It follows immediately from the definition that for any vectors $x, y \in F^n$,

$$d(x, y) = w(x - y)$$

THEOREM 4.2.17 Let C be a linear code. Then the minimum distance of C is equal to the smallest weight of nonzero codewords in C; that is,

$$d(C) = \min\{w(x) \mid x \in C, x \neq 0\}$$

Proof: Let $d(C) = d$. Then there exist $c, c' \in C$ such that $d(c, c') = d$. Hence $c - c' \in C$ and $w(c - c') = d$. Let x be any nonzero vector in C. Then $w(x) = w(x - 0) = d(x, 0) \geq d$. This proves that d is the smallest weight of any nonzero codeword in C. ∎

We thus see that in order to find the minimum distance of a linear code with M codewords, we have to compare the weights of $M - 1$ nonzero codewords. But to find the minimum distance of a general code with M codewords, we have to compare the distances of $M(M - 1)/2$ pairs of distinct codewords.

Even with the simpler criterion, finding the minimum distance of a linear code with very large M can be difficult. Moreover, a linear code is often defined by specifying a parity-check matrix, and the codewords are not explicitly known. The following theorem, which gives the relation between the minimum distance and the parity-check matrix of a linear code, is therefore of fundamental importance.

THEOREM 4.2.18 Let H be a parity-check matrix of an $[n, k]$-code C over F. Then $d(C)$ is equal to the minimal number of linearly dependent columns in H.

Consequently, $d(C) \leq n - k + 1$.

Proof: Let $H^1, \ldots, H^n$ denote the columns of H. Let $x \in F^n$. Then, by Theorem 4.2.8, $x \in C$ if and only if

$$Hx^T = x_1 H^1 + \cdots + x_n H^n = 0$$

Let $d(C) = d$. Then $w(x) \geq d$ for all $x \in C$, and there exists $c \in C$ with $w(c) = d$. Let $c_{i_1}, \ldots, c_{i_d}$ be the nonzero components of c. Then

$$Hc^T = c_1 H^1 + \cdots + c_n H^n = c_{i_1} H^{i_1} + \cdots + c_{i_d} H^{i_d} = 0$$

Hence H has d linearly dependents columns—namely, $H^{i_1}, \ldots, H^{i_d}$. We claim that any list of fewer than d columns of H is linearly independent. Suppose the columns $H^{i_1}, \ldots, H^{i_c}$ are linearly dependent, where $c < d$. Then there exist scalars $\alpha_1, \ldots, \alpha_c$, not all zero, such that

$$\alpha_1 H^{i_1} + \cdots + \alpha_c H^{i_c} = 0$$

Let $x \in F^n$ such that the components $i_1, \ldots, i_c$ of x are equal to $\alpha_1, \ldots, \alpha_c$, respectively, and all other components are zero. Then $Hx^T = 0$; hence $x \in C$. But $w(x) < d$, a contradiction. This proves that d is the minimal number of linearly dependent columns in H.

Now H is an $(n-k) \times n$ matrix of rank $n-k$. Hence any $n-k+1$ columns of H are linearly dependent. Therefore $d \leq n-k+1$. ∎

Example 4.2.19 Find the minimum distance of the binary $[n, n-1]$-code C with parity-check matrix

$$H = [1 \quad 1 \quad \ldots \quad 1]$$

Solution. Clearly, any two columns of H are linearly dependent. Hence the minimal number of linearly dependent columns is 2. Therefore the minimum distance of the code C is 2.

This code is known as a *parity-check code* and can detect a single error. The components of every codeword x satisfy the parity-check equation $x_1 + \cdots + x_n = 0 \pmod 2$. Therefore every codeword contains an even number of 1's.

Example 4.2.20 Find the minimum distance of the $[10, 9]$-code C over $\mathbb{F}_{11}$ defined by the parity-check matrix

$$H = [1 \quad 2 \quad 3 \quad 4 \quad 5 \quad 6 \quad 7 \quad 8 \quad 9 \quad 10]$$

Solution. As in Example 4.2.19, the minimum distance of C is 2. The codewords satisfy the parity-check equation

$$xH^T = \sum_{i=1}^{10} i x_i = 0$$

In addition to detecting a single error in general, this code can detect a double error where two digits have been transposed. Suppose a codeword $x = x_1 \ldots x_{10}$ is transmitted, and the received vector $y = y_1 \ldots y_{10}$ differs from x only in that the jth and kth digits have been transposed; that is, $y_j = x_k$, $y_k = x_j$, and $y_i = x_i$ for all other i. Then

$$yH^T = xH^t + (y-x)H^t = (y-x)H^t$$

$$= \sum_{i=1}^{10} i(y_i - x_i) = (k-j)(x_j - x_k) \neq 0$$

Hence y is not a codeword, so the error is detected.

This code is of practical interest because it is used for the International Standard Book Numbers (ISBN) and is therefore called the *ISBN code*. All books published now are assigned an ISBN, which is a 10-digit codeword $x_1 \ldots x_{10}$. These ten digits are divided into four parts that are separated by hyphens or spaces. (The hyphens or spaces are inserted for the sake of convenience in reading and have no bearing on the code.) The first part is called the *group identifier* and identifies a country or a language area. For example, 0 represents the English language area. The second part is called the *publisher identifier* and represents a specific publisher in a specific group. The third part, called the *title identifier,* identifies a specific publication of a specific publisher. The lengths of these three parts are variable, but the total number of digits is 9. The fourth part is the *check-digit* x_{10}, which is determined by the parity-check equation written above. Since $10 = -1$ in $\mathbb{F}_{11}$, we have $x_{10} = \sum_{i=1}^{9} i x_i$. If x_{10} turns out to be 10, it is shown as X in the final position of the ISBN. The ISBN standard is currently under revision to enlarge it to a 13-digit number.

Example 4.2.21 Find the minimum distance of the [10, 8]-code C over $\mathbb{F}_{11}$ with parity-check matrix

$$H = \begin{bmatrix} 1 & 1 & 1 & 1 & 1 & 1 & 1 & 1 & 1 & 1 \\ 1 & 2 & 3 & 4 & 5 & 6 & 7 & 8 & 9 & 10 \end{bmatrix}$$

Solution. It is clear that no column in H is a scalar multiple of another. Hence any two columns are linearly independent. Therefore $d > 2$. On the other hand, by Theorem 4.2.18, $d \leq n - k + 1 = 3$. Hence $d = 3$. It follows that C is a single-error-correcting code.

4.2.3 Hamming Codes

We now describe an important class of linear codes known as Hamming codes. We first define the special case of a binary Hamming code and later consider the general q-ary Hamming codes.

Definition 4.2.22 Let r be a positive integer greater than 1. Let H be an $r \times (2^r - 1)$ matrix whose columns are the distinct nonzero vectors in $\mathbb{F}_2^r$. Then the code with H as its parity-check matrix is called a *binary Hamming code* and denoted by Ham$(r, 2)$.

Note that the definition does not specify the order in which the columns of H are written, but a permutation of the columns gives an equivalent code. Thus for a given r, there are $(2^r - 1)!$ equivalent binary Hamming codes. Ham$(r, 2)$ is a general notation for any of these equivalent codes.

Since H is an $r \times (2^r - 1)$ matrix, Ham$(r, 2)$ is a code of length $n = 2^r - 1$ and dimension $k = n - r = 2^r - 1 - r$. Hence Ham$(r, 2)$ is a $[2^r - 1, 2^r - 1 - r]$-code. The parameter $r = n - k$ represents the redundancy of the code.

For example, let $r = 2$. Then Ham$(2, 2)$ is a $[3, 2]$-code with a parity-check matrix $H = \begin{bmatrix} 1 & 1 & 0 \\ 1 & 0 & 1 \end{bmatrix}$. We have written the columns so that H is in canonical form. Hence, by Theorem 4.2.12, the generator matrix of the code is $G = [1 \ \ 1 \ \ 1]$. Thus Ham$(2, 2)$ is simply the binary repetition code $\{000, 111\}$.

On taking $r = 3$, we see that Ham$(3, 2)$ is a $[7, 4]$-code with a parity-check matrix

$$H = \begin{bmatrix} 1 & 0 & 1 & 0 & 1 & 0 & 1 \\ 0 & 1 & 1 & 0 & 0 & 1 & 1 \\ 0 & 0 & 0 & 1 & 1 & 1 & 1 \end{bmatrix}$$

(Here H is not in canonical form.) We observe that the third column in H is the sum of the first two columns. Hence the first three columns are linearly dependent. Moreover, no column is a scalar multiple of another, so 3 is the minimal number of linearly dependent columns in H. Hence, by Theorem 4.2.18, the minimum distance of Ham$(3, 2)$ is 3. The dimension of the code is 4. Hence the number of codewords in Ham$(3, 2)$ is $2^4 = 16$. So in the notation of Section 4.1, Ham$(3, 2)$ is a $(7, 16, 3)$-code. We showed in Example 4.1.7 that a $(7, 16, 3)$-code is perfect. This proves that Ham$(3, 2)$ is a perfect code with minimum distance 3.

The binary repetition code $\{111, 000\}$ is also a (trivial) perfect code with minimum distance 3. Thus we see that Ham$(2, 2)$ and Ham$(3, 2)$ are both perfect codes with minimum distance 3. We will presently show that this property holds for all Hamming codes, but first we give the definition of a general q-ary Hamming code, Ham(r, q).

Let $F = \mathbb{F}_q$ (where q is a power of some prime p) and r be a positive integer greater than 1. Then there are $q^r - 1$ nonzero vectors in F^r. Given a nonzero vector $v \in F^r$, we have a set of $q - 1$ nonzero scalar multiples of v—namely, $\{\alpha v \mid \alpha \in F, \alpha \neq 0\}$. So there are in all $n = (q^r - 1)/(q - 1)$ pairwise disjoint sets of this kind that form a partition of the set of nonzero vectors in F^r. By arbitrarily picking one vector from each of these sets, we obtain a set of n nonzero vectors such that no vector is a scalar multiple of another. In particular, we may take all those nonzero vectors in F^r in which the first nonzero entry is 1.

Definition 4.2.23 Let $F = \mathbb{F}_q$ and r be a positive integer greater than 1. Let $n = (q^r - 1)/(q - 1)$. Let H be an $r \times n$ matrix whose columns are nonzero vectors in F^r such that no column is a scalar multiple of another. Then the $[n, n-r]$-code with H as its parity-check matrix is called a *q-ary Hamming code* and is denoted by $\text{Ham}(r, q)$.

As in the case of binary Hamming codes, $\text{Ham}(r, q)$ is a general notation for a class of equivalent codes. For example, if $r = 2$ and $q = p$ (prime), then a parity-check matrix for $\text{Ham}(2, p)$ is

$$H = \begin{bmatrix} 0 & 1 & 1 & 1 & 1 & \cdots & 1 \\ 1 & 0 & 1 & 2 & 3 & \cdots & p-1 \end{bmatrix}$$

The following theorem states the most important property of Hamming codes.

THEOREM 4.2.24 $\text{Ham}(r, q)$ is a perfect code with minimum distance 3.

Proof: Let H a be parity-check matrix of $\text{Ham}(r, q)$. Then some three columns in H are scalar multiples of

$$\begin{bmatrix} 1 \\ 0 \\ 0 \\ \vdots \\ 0 \end{bmatrix}, \quad \begin{bmatrix} 0 \\ 1 \\ 0 \\ \vdots \\ 0 \end{bmatrix}, \quad \begin{bmatrix} 1 \\ 1 \\ 0 \\ \vdots \\ 0 \end{bmatrix}$$

and are therefore linearly dependent. So 3 is the minimal number of linearly dependent columns in H. Hence, by Theorem 4.2.18, $\text{Ham}(r, q)$ has minimum distance 3, which implies that it is a single-error-correcting code.

To show that the code is perfect, we invoke Theorem 4.1.6. The number M of codewords in $\text{Ham}(r, q)$ is q^{n-r}, where $n = (q^r - 1)/(q - 1)$. Further, $t = 1$. Hence

$$M \sum_{m=0}^{t} \binom{n}{m} (q-1)^m = q^{n-r}(1 + n(q-1))$$
$$= q^{n-r}(1 + q^r - 1)$$
$$= q^n$$

This proves that the code is perfect.   ∎

4.2.4 Decoding

We now consider the *decoding procedure* for a linear code. As we explained, the general principle of decoding is to pick the codeword nearest to the received vector. For this purpose, we prepare a look-up table that gives the nearest codeword for every possible received vector. The algebraic structure of a linear code as a subspace provides a convenient method for preparing such a table. Let C be a subspace of

F^n. Then C is a subgroup of the additive group F^n. Recall that for every $a \in F^n$, $a + C = \{a + c \mid c \in C\}$ is called a *coset* of C. Two elements x, $y \in F^n$ lie in the same coset if and only if $x - y \in C$. These cosets form a partition of the set F^n. Hence F^n is the disjoint union of distinct cosets of C.

Let y be any vector in F^n, and suppose $x \in C$ is the codeword nearest to y. Now y lies in the coset $y + C = \{y - c \mid c \in C\}$. For all $c \in C$, $d(y, x) \leq d(y, c)$; that is, $w(y - x) \leq w(y - c)$. Hence $y - x$ is the vector of least weight in the coset containing y. Writing $e = y - x$, we have $x = y - e$. Thus we have proved the following result.

THEOREM 4.2.25 Let $C \subset F^n$ be a linear code. Given a vector $y \in F^n$, the code-word x nearest to y is given by $x = y - e$, where e is the vector of least weight in the coset containing y.

If the coset containing y has more than one vector of least weight, then there are more than one codewords nearest to y.

Definition 4.2.26 Let C be a linear code in F^n. The *coset leader* of a given coset of C is defined to be the vector with the least weight in that coset.

If there are more than one vectors with the least weight in a coset, we choose any one of them to be the coset leader. Recall that if C is a t-error-correcting code, then a vector cannot be at a distance $\leq t$ from more than one codeword. Hence if e is a coset leader with weight $\leq t$, then e is the unique vector of least weight in its coset.

Let C be an $[n, k]$-code over $F = \mathbb{F}_q$. Since F^n has q^n elements and every coset of C has q^k elements, it follows that there are q^{n-k} distinct cosets of C. Let us denote their coset leaders by $e_1, e_2, \ldots, e_N$, where $N = q^{n-k}$. Further, let us suppose the coset leaders have been numbered in ascending order of weight; that is, $w(e_i) \leq w(e_{i+1})$ for all i. So $e_1 = 0$ is the coset leader of $C = 0 + C$. Let $C = \{c_1, c_2, \ldots, c_M\}$, where $M = q^k$ and $c_1 = 0$. We can arrange the q^n vectors in F^n in an $N \times M$ table, as shown below, in which the (i, j)-entry is the vector $e_i + c_j$. So the ith row contains the elements of the coset $e_i + C$, with the coset leader e_i as the first entry. The top row is C itself. This table is called a *standard array* for the code C.

$$
\begin{array}{ccccccc}
e_1 = 0 = c_1 & c_2 & \cdots & c_j & \cdots & c_M \\
e_2 & e_2 + c_2 & \cdots & e_2 + c_j & \cdots & e_2 + c_M \\
\vdots & \vdots & & \vdots & & \vdots \\
e_i & e_i + c_2 & \cdots & e_i + c_j & \cdots & e_i + c_M \\
\vdots & \vdots & & \vdots & & \vdots \\
e_N & e_N + c_2 & \cdots & e_N + c_j & \cdots & e_N + c_M
\end{array}
$$

The standard array is used for decoding as follows: Suppose a vector $y \in F^n$ is received. We find its position in the table. If y is the (i, j)-entry in the table, then $y = e_i + c_j$. Since e_i is the vector with the least weight in the coset, it follows by Theorem 4.2.25 that the codeword nearest to y is $x = y - e_i = c_j$. Thus a received vector y is decoded as the codeword at the top of the column in which y occurs.

If a codeword x is transmitted and a vector y is received, then $e = y - x$ is called the *error vector*. Thus a coset leader is the error vector for every vector y lying in that coset.

Example 4.2.27 Write a standard array for the binary code with the generator matrix

$$G = \begin{bmatrix} 1 & 0 & 1 & 0 & 1 \\ 0 & 1 & 0 & 1 & 1 \end{bmatrix}$$

and decode the received vector 01111.

Solution. The code C generated by G is a $[5, 2]$-code, with codewords given by

$$C = \{00000, 10101, 01011, 11110\}$$

Hence there are $2^3 = 8$ distinct cosets of C, so the standard array consists of 8 rows. The minimum distance of C is 3, and hence $t = 1$. Therefore the 5 vectors of weight 1 produce 5 distinct cosets. To get the remaining 2 rows of the array, we pick in each case a vector of weight 2 that has not already appeared in the preceding rows. Note that each of the last two rows has 2 vectors of least weight.

00000	10101	01011	11110
10000	00101	11011	01110
01000	11101	00011	10110
00100	10001	01111	11010
00010	10111	01001	11100
00001	10100	01010	11111
11000	01101	10011	00110
10010	00111	11001	01100

To decode the vector 01111, we find its position in the table. We note that 01111 occurs in the third column. The top entry in that column is 01011. Hence 01111 is decoded as the codeword 01011.

A standard array is useful for decoding when the code length n is small. For large n, it is not a convenient method. We now describe a more efficient decoding procedure.

Definition 4.2.28 Let C be an $[n, k]$-code over F with parity-check matrix H. For any vector $y \in F$, the *syndrome* of y, denoted by $S(y)$, is defined to be

$$S(y) = yH^T$$

Note that the syndrome is defined with respect to a specific parity-check matrix H. A different parity-check matrix will give a different syndrome. For an $[n, k]$-code, $S(y)$ is a vector of length $n - k$. It may be written as a row vector yH^T (as stated in the definition) or as a column vector Hy^T. In the latter case, $S(y) = Hy^T = y_1 H^1 + \cdots + y_n H^n$, where $H^1, \ldots, H^n$ are the columns of H.

By Theorem 4.2.8, $S(y) = 0$ if and only if $y \in C$. Let $y, y' \in F^n$. Then $S(y) = S(y')$ holds if and only if $(y - y')H^T = 0$—that is, $y - y' \in C$. Hence two vectors have the same syndrome if and only if they lie in the same coset of C. Thus there is a one-to-one correspondence between the cosets of C and the syndromes. A table with two columns showing the coset leaders e_i and the corresponding syndromes $S(e_i)$ is called the *syndrome table*. To decode a received vector y, we compute its syndrome $S(y)$ and then look at the table to find the coset leader e for which $S(e) = S(y)$. Then y is decoded as $x = y - e$. This procedure is known as *syndrome decoding*.

Example 4.2.29 Write the syndrome table for the code in Example 4.2.27. Decode the vector 11010.

Solution. The generator matrix of the code given in Example 4.2.27 is in canonical form. Hence, by Theorem 4.2.12, its parity-check matrix is

$$H = \begin{bmatrix} 1 & 0 & 1 & 0 & 0 \\ 0 & 1 & 0 & 1 & 0 \\ 1 & 1 & 0 & 0 & 1 \end{bmatrix}$$

Computing eH^T for every coset leader e, we get the following table:

Coset Leader	Syndrome
10000	101
01000	011
00100	100
00010	010
00001	001
11000	110
10010	111

The syndrome of the given vector $y = 11010$ is

$$yH^T = [1 \quad 1 \quad 0 \quad 1 \quad 0] \begin{bmatrix} 1 & 0 & 1 \\ 0 & 1 & 1 \\ 1 & 0 & 0 \\ 0 & 1 & 0 \\ 0 & 0 & 1 \end{bmatrix} = [1 \quad 0 \quad 0]$$

From the syndrome table we find that the coset leader with syndrome 100 is $e = 00100$. Hence we decode y as the codeword $x = y - e = 11110$.

EXERCISES 4.2

1. Show that a binary linear code having 20 codewords does not exist.

2. Show that the binary code $C = \{000, 011, 101, 110\}$ is linear. Find its generator matrix.

3. Show that the binary code $C = \{0000, 1010, 0101, 1111\}$ is linear and self-orthogonal.

4. Show that a binary linear code with generator matrix G is self-orthogonal if and only if the rows of G are of even weight and are orthogonal to one another.

5. Show that the binary linear code C with generator matrix

$$G = \begin{bmatrix} 1 & 0 & 0 & 0 & 1 & 1 & 1 \\ 0 & 1 & 0 & 1 & 0 & 1 & 1 \\ 0 & 0 & 1 & 1 & 1 & 0 & 1 \end{bmatrix}$$

is self-orthogonal. Find the dual code of C.

6. Find the dual of the binary repetition code of length n.

7. Show that if C_1, C_2 are linear codes in F^n, then $C_1 + C_2 = \{x_1 + x_2 \mid x_1 \in C_1, x_2 \in C_2\}$ and $C_1 \cap C_2$ are also linear codes. Further show that $(C_1 + C_2)^\perp = C_1^\perp \cap C_2^\perp$.

8. Let C be the set of all vectors of even weight in $\mathbb{F}_2^n$. Show that C is a linear code.

9. Show that, in a binary linear code, either all codewords have even weight or exactly half have even weight.

10. Find the canonical generator and parity-check matrices of the binary code with generator matrix

$$G = \begin{bmatrix} 1 & 1 & 1 & 0 & 1 & 1 & 0 \\ 1 & 0 & 1 & 1 & 1 & 0 & 1 \\ 1 & 1 & 0 & 0 & 1 & 0 & 1 \end{bmatrix}$$

11. Find the canonical generator matrix of the ternary [4, 2]-code with parity-check matrix

$$H = \begin{bmatrix} 0 & 1 & 1 & 1 \\ 1 & 0 & 1 & 2 \end{bmatrix}$$

and write all the codewords.

12. Find a generator matrix for the parity-check code (Example 4.2.19).

13. Find a generator matrix for the ISBN code (Example 4.2.20).

14. Find a generator matrix for the code in Example 4.2.21.

15. Check whether the following are valid ISBN codewords: 0256379418, 326173282X, 3753794961.

16. Replace ? with the correct digit in the following ISBN: 368576?969.

17. Show that G and G' are generator matrices of the same code if and only if $G' = PG$ for some invertible matrix P.

18. Write a parity-check matrix for a binary Hamming code Ham(4, 2), and find its canonical generator matrix.

19. Write a parity-check matrix for a ternary Hamming code Ham(2, 3), and find its canonical generator matrix.

20. Write a parity-check matrix for a ternary Hamming code Ham(3, 3), and find its canonical generator matrix .

21. Write the syndrome table for the binary Hamming code Ham(3, 2). Decode these received vectors: 1001011, 1100110, 1111001

4.3 CYCLIC CODES

We now discuss linear codes of a special kind, known as *cyclic codes*. These codes have interesting ring-theoretic properties and a richer algebraic structure than general linear codes. Many important codes belong to this category.

The mapping $\sigma : F^n \to F^n$ given by

$$\sigma(a_1, a_2, \ldots, a_n) = (a_n, a_1, \ldots, a_{n-1})$$

is called a *cyclic shift*. It is easily seen that σ is a linear mapping; that is, for all $a, b \in F^n$ and $\lambda \in F$,

$$\sigma(a + b) = \sigma(a) + \sigma(b) \quad \text{and} \quad \sigma(\lambda a) = \lambda \sigma(a)$$

Definition 4.3.1 A linear code $C \subset F^n$ is called a *cyclic code* if

$$\sigma(a) \in C \qquad \text{for all } a \in C$$

For example, the code $C = \{000, 110, 011, 101\}$ is a binary cyclic code.

THEOREM 4.3.2 Let G be a generator matrix of a linear $[n, k]$-code C. Then C is a cyclic code if and only if $\sigma(G_i) \in C$ for each row G_i of G.

Proof: If C is cyclic, then $\sigma(a) \in C$ for all $a \in C$. Hence, in particular, $\sigma(G_i) \in C$ for each $i = 1, \ldots, k$. Conversely, suppose $\sigma(G_i) \in C$ for each row G_i of G. Let $a \in C$. Then $a = \lambda_1 G_1 + \cdots + \lambda_k G_k$ for some scalars $\lambda_1, \ldots, \lambda_k$. Hence

$$\sigma(a) = \sigma(\lambda_1 G_1 + \cdots + \lambda_k G_k) = \lambda_1 \sigma(G_1) + \cdots + \lambda_k \sigma(G_k) \in C$$

Therefore C is cyclic.   ∎

Example 4.3.3 Show that the code C with generator matrix

$$G = \begin{bmatrix} 1 & 1 & 0 & 1 & 0 & 0 & 0 \\ 0 & 1 & 1 & 0 & 1 & 0 & 0 \\ 0 & 0 & 1 & 1 & 0 & 1 & 0 \\ 0 & 0 & 0 & 1 & 1 & 0 & 1 \end{bmatrix}$$

is cyclic.

Solution. Clearly, $\sigma(G_1) = G_2$, $\sigma(G_2) = G_3$, and $\sigma(G_3) = G_4$. Further,

$$\sigma(G_4) = 1000110 = G_1 + G_2 + G_3$$

Thus $\sigma(G_i) \in C$ for each $i = 1, 2, 3, 4$. Hence C is cyclic.

We will henceforth number the components of a vector in F^n as $0, 1, \ldots,$ $n - 1$; that is, $a \in F^n$ will be written $a = (a_0, a_1, \ldots, a_{n-1})$. Then corresponding to each vector a in F^n we have the polynomial $a(x) = a_0 + a_1 x + \cdots + a_{n-1} x^{n-1}$. Let $F[x]_n$ denote the set of all polynomials in x of degree less than n over the field F; that is,

$$F[x]_n = \{a_0 + a_1 x + \cdots + a_{n-1} x^{n-1} \mid a_0, a_1, \ldots, a_{n-1} \in F\}$$

Clearly, $F[x]_n$ is a vector space of dimension n over F. The mapping $a \mapsto a(x)$ is an isomorphism between the vector spaces F^n and $F[x]_n$. We will henceforth identify F^n with $F[x]_n$ and treat each vector $a \in F^n$ as the polynomial $a(x) \in F[x]_n$; that is,

$$(a_0, a_1, \ldots, a_{n-1}) = a_0 + a_1 x + \cdots + a_{n-1} x^{n-1}$$

We now impose on $F[x]_n$ the structure of a ring as follows: As usual, let $F[x]$ denote the ring of all polynomials in x over F. We showed in Chapter 0 that, given any polynomial $p(x) \in F$, we can construct the quotient ring $F[x]/(p(x))$, where $(p(x))$ denotes the principal ideal generated by $p(x)$. The elements of the ring are the cosets of the ideal $(p(x))$. The quotient ring is a field if and only if $p(x)$ is irreducible over F. Further, if $p(x)$ is of degree n, then

$$\frac{F[x]}{(p(x))} = \{a_0 + a_1 t + \cdots + a_{n-1} t^{n-1} \mid a_0, a_1, \ldots, a_{n-1} \in F\}$$

where t denotes the coset $x + (p(x))$, so $p(t) = 0$.

Let us take $p(x) = x^n - 1$. Then we have the quotient ring

$$\frac{F[x]}{(x^n - 1)} = \{a_0 + a_1 t + \cdots + a_{n-1} t^{n-1} \mid a_0, a_1, \ldots, a_{n-1} \in F\}$$

where t satisfies the relation $t^n - 1 = 0$. Note that this ring is not a field because $p(x)$ can be factored as $x^n - 1 = (x - 1)(x^{n-1} + \cdots + x + 1)$.

Let us now make a change in notation and write x in place of t. Then the ring $F[x]/(x^n - 1)$ becomes $F[x]_n$. Thus we have made $F[x]_n$ into a ring in which the relation $x^n - 1 = 0$ holds. So $F[x]_n$ is a ring as well as a vector space over F. In other words, $F[x]_n$ is an algebra over F.

The multiplication in the ring $F[x]_n$ is modulo $(x^n - 1)$. To avoid confusion between multiplications in the rings $F[x]$ and $F[x]_n$, we denote the latter by $*$. So, given $a(x), b(x) \in F[x]_n$, we write $a(x) * b(x)$ to denote their product in the ring $F[x]_n$, and $a(x)b(x)$ to denote their product in the ring $F[x]$. If $\deg a(x) + \deg b(x) < n$, then $a(x) * b(x) = a(x)b(x)$. Otherwise, $a(x) * b(x)$ is the remainder left on dividing $a(x)b(x)$ by $x^n - 1$. In other words, if $a(x) * b(x) = c(x)$, then $a(x)b(x) = c(x) + (x^n - 1)q(x)$ for some polynomial $q(x)$. In practice, to obtain $a(x) * b(x)$, we simply compute the ordinary product $a(x)b(x)$ and then put $x^n = 1$, $x^{n+1} = x$, and so on.

In particular, consider the product $x * a(x)$. In the ring $F[x]$,

$$x\,a(x) = x(a_0 + a_1 x + \cdots + a_{n-1} x^{n-1}) = a_0 x + a_1 x^2 + \cdots + a_{n-1} x^n$$

Hence, in the ring $F[x]_n$,

$$x * a(x) = a_{n-1} + a_0 x + a_1 x^2 + \cdots + a_{n-2} x^{n-1}$$

Thus we see that multiplication by x in the ring $F[x]_n$ corresponds to cyclic shift σ in F^n; that is, $x * a(x) = \sigma(a)(x)$.

Let $C \subset F^n$ be a linear code. As already agreed, we identify every vector a in F^n with the polynomial $a(x)$ in $F[x]_n$, so $C \subset F[x]_n$. The elements of the code C are now referred to as codewords or *code polynomials*. From the result proved above, we get the following theorem.

THEOREM 4.3.4 Let C be a linear code over F. Then C is cyclic if and only if $x * a(x) \in C$ for every $a(x) \in C$.

In the following theorem we prove the basic algebraic property of cyclic codes. Recall that an ideal in a commutative ring R is a nonempty subset A of R such that for all $a, b \in A$ and all $r \in R$, $a - b \in A$ and $ra \in A$.

THEOREM 4.3.5 A subset C of $F[x]_n$ is a cyclic code if and only if C is an ideal of the ring $F[x]_n$.

Proof: Suppose C is a cyclic code. Then C is a linear code over F. Hence for all $a(x), b(x) \in C$ and all $\lambda \in F$, $a(x) - b(x) \in C$ and $\lambda a(x) \in C$. Further, since C is cyclic, $x * a(x) \in C$ for all $a(x) \in C$. Hence $x^2 * a(x) = x * (x * a(x)) \in C$, and so on. Therefore, for every $r(x) = r_0 + r_1 x + \cdots + r_{n-1} x^{n-1} \in F[x]_n$,

$$r(x) * a(x) = r_0 a(x) + r_1 x * a(x) + \cdots + r_{n-1} x^{n-1} * a(x) \in C$$

This proves that C is an ideal in the ring $F[x]_n$.

Conversely, suppose C is an ideal. Let $a(x), b(x) \in C$ and $\lambda \in F$. Then $a(x) - b(x) \in C$ and, since $\lambda \in F[x]_n$, $\lambda a(x) \in C$. Hence C is a linear code. Further, $r(x) * a(x) \in C$ for all $r(x) \in F[x]_n$. Hence, in particular, $x * a(x) \in C$. This proves that C is a cyclic code. ∎

Given any polynomial $p(x) \in F[x]_n$, let $\langle p(x) \rangle$ denote the principal ideal generated by $p(x)$ in the ring $F[x]_n$; that is,

$$\langle p(x) \rangle = \{ f(x) * p(x) \mid f(x) \in F[x]_n \}$$

The following theorem shows that every ideal in $F[x]_n$ is a principal ideal.

THEOREM 4.3.6 Let C be a nonzero ideal in the ring $F[x]_n$. Then

(a)   There exists a unique monic polynomial $g(x)$ of least degree in C.
(b) $g(x)$ divides $x^n - 1$ in $F[x]$.
(c)   For all $a(x) \in C$, $g(x)$ divides $a(x)$ in $F[x]$.
(d) $C = \langle g(x) \rangle$.

Conversely, suppose C is the ideal generated by $p(x) \in F[x]_n$. Then $p(x)$ is a polynomial of least degree in C if and only if $p(x)$ divides $x^n - 1$ in $F[x]$.

Proof: **(a)** Suppose $g(x)$ and $h(x)$ are both monic polynomials of least degree k in C. Let $f(x) = g(x) - h(x)$. Then $f(x) \in C$ and $\deg f(x) < k$. If λ is the leading coefficient in $f(x)$, then $\lambda^{-1} f(x)$ is a monic polynomial of degree $< k$, a contradiction. Hence there is a unique monic polynomial $g(x)$ of least degree in C.

(b) By the division algorithm in the ring $F[x]$, there exist polynomials $q(x)$, $r(x)$ such that

$$x^n - 1 = q(x) g(x) + r(x)$$

and $\deg r(x) < \deg g(x)$ or $r(x) = 0$. Hence, in $F[x]_n$, $q(x) * g(x) + r(x) = 0$, so $r(x) = -q(x) * g(x) \in C$. Since $g(x)$ is of minimal degree in C, it follows that $r(x) = 0$. Hence $g(x)$ divides $x^n - 1$ in $F[x]$.

(c), (d) Since C is an ideal in $F[x]_n$ and $g(x) \in C$, for every $f(x) \in F[x]_n$, $f(x) * g(x) \in C$. Conversely, let $a(x) \in C$. Then, by the division algorithm in $F[x]$,

$$a(x) = q(x) g(x) + r(x)$$

where $\deg r(x) < \deg g(x)$ or $r(x) = 0$. Since $\deg a(x) < n$, $q(x) g(x) = q(x) * g(x)$. Hence $r(x) = a(x) - q(x) * g(x) \in C$. By the minimality of $\deg g(x)$,

$r(x) = 0$. Hence $a(x) = q(x) g(x)$. So $g(x)$ divides $a(x)$ in $F[x]$. Moreover, we have $C = \{f(x) * g(x) \mid f(x) \in F[x]_n\} = \langle g(x) \rangle$.

To prove the last part of the theorem, suppose $p(x)$ divides $x^n - 1$, and let $x^n - 1 = p(x)q(x)$. Let $a(x)$ be any nonzero polynomial in $C = \langle p(x) \rangle$. Then $a(x) = f(x) * p(x)$ for some $f(x) \in F[x]_n$. Hence

$$a(x) = f(x) p(x) + b(x) (x^n - 1) = \{f(x) + b(x)q(x)\}p(x)$$

for some $b(x)$. Therefore $\deg a(x) \geq \deg p(x)$, so $p(x)$ is of least degree in C. Conversely, if $p(x)$ is of least degree in C, then, by part (b), $p(x)$ divides $x^n - 1$. ∎

We see from Theorem 4.3.6 that the only ideals in the ring $F[x]_n$ are those generated by the factors of $x^n - 1$. Thus we can obtain all cyclic codes of length n over F if we find all factors of $x^n - 1$. In the case of trivial factors, we get trivial codes. When $g(x) = x^n - 1$, we get $\langle g(x) \rangle = (0)$. When $g(x) = 1$, we have $\langle g(x) \rangle = F[x]_n$.

Note that if $p(x)$ does not divide $x^n - 1$, then $p(x)$ cannot be of least degree in the ideal $\langle p(x) \rangle$. For example, consider $p(x) = x^{n-1} + 1 \in F[x]_n$. Then $x + 1 = x * (x^{n-1} + 1) \in \langle x^{n-1} + 1 \rangle$.

Example 4.3.7 Find all nontrivial ideals of $F[x]_3$, where $F = \mathbb{F}_2$. Hence obtain all binary cyclic codes of length 3.

Solution. The polynomial $x^3 - 1$ can be factored as $x^3 - 1 = (x - 1)(x^2 + x + 1)$. Now $x - 1$ and $x^2 + x + 1$ are both irreducible over $\mathbb{F}_2$, so the only nontrivial factors of $x^3 - 1$ are $x - 1$ and $x^2 + x + 1$. The ideals generated by them are

$$\langle x - 1 \rangle = \{0, 1 + x, x + x^2, 1 + x^2\}$$
$$\langle x^2 + x + 1 \rangle = \{0, 1 + x + x^2\}$$

Writing the polynomials as vectors, we get the cyclic codes $\{000, 110, 011, 101\}$ and $\{000, 111\}$.

Definition 4.3.8 Let C be a nonzero ideal in $F[x]_n$. Let $g(x)$ be the unique monic polynomial of least degree in C. Then $g(x)$ is called the *generator polynomial* of the cyclic code C.

Note that if $C = \langle p(x) \rangle$ is the ideal generated by $p(x)$, then $p(x)$ is the generator polynomial of C if and only if $p(x)$ is monic and divides $x^n - 1$.

In the following theorem, we show that binary Hamming codes are (equivalent to) cyclic codes. Recall that for every integer $r > 1$, Ham$(r, 2)$ stands for a class of equivalent codes defined by an $r \times (2^r - 1)$ parity-check matrix H whose columns are the nonzero vectors in $\mathbb{F}_2^r$. We show that by a suitable ordering of the columns of H we obtain a cyclic Hamming code Ham$(r, 2)$.

THEOREM 4.3.9 Let $p(x)$ be a primitive irreducible polynomial of degree r over $F = \mathbb{F}_2$. Let $n = 2^r - 1$. Then the cyclic code with the generator polynomial $p(x)$ in the ring $F[x]_n$ is Ham$(r, 2)$.

Proof: We show that $p(x)$ divides $x^n - 1$. The quotient ring $K = F[x]/\langle p(x) \rangle$ is a field of 2^r elements, given by

$$K = \{a_0 + a_1 t + \cdots + a_{r-1} t^{r-1} \mid a_0, a_1, \ldots, a_{r-1} \in F\}$$

where t denotes the coset $x + (p(x))$ and satisfies the relation $p(t) = 0$. The multiplicative group K^* is of order $2^r - 1 = n$, so $t^n = 1$. By the division algorithm,

$$x^n - 1 = q(x)p(x) + s(x)$$

where $s(x) = 0$ or $\deg s(x) < r$. Letting $x = t$, we get $s(t) = 0$, which implies $s(x) = 0$. This proves that $p(x)$ divides $x^n - 1$.

Let C be the cyclic code in $F[x]_n$ with generator polynomial $p(x)$. By Theorem 4.3.6, if $a(x) \in C$, then $a(x) = q(x)p(x)$ for some $q(x)$. Hence $a(t) = 0$. Conversely, suppose $a(x) \in F[x]_n$ such that $a(t) = 0$. Then, by using the same argument as for $x^n - 1$, we can show that $p(x)$ divides $a(x)$. Hence $a(x) \in C$. Thus

$$C = \{a(x) \in F[x]_n \mid a(t) = 0\}$$

Let us write each element $b_0 + b_1 t + \cdots + b_{r-1} t^{r-1} \in K$ as a column vector $[b_0 \ b_1 \ \ldots \ b_{r-1}]^T$. Let

$$H = \begin{bmatrix} 1 & t & \ldots & t^{n-1} \end{bmatrix}$$

Since $p(x)$ is a primitive polynomial, t is a primitive element of K, so $1, t, \ldots, t^{n-1}$ are distinct nonzero elements in K. Thus H is an $r \times n$ matrix whose columns are all the nonzero distinct vectors in $\mathbb{F}_2^r$.

Now, for any $a(x) = a_0 + a_1 x + \cdots + a_{n-1} x^{n-1} \in F[x]_n$,

$$a(t) = a_0 + a_1 t + \cdots + a_{n-1} t^{n-1}$$

$$= H \begin{bmatrix} a_0 \\ a_1 \\ \vdots \\ a_{n-1} \end{bmatrix}$$

Writing the polynomial $a(x)$ as vector $a = (a_0, a_1, \ldots, a_{n-1})$, we have

$$C = \{a \in F^n \mid Ha^T = 0\}$$

Hence H is a parity-check matrix for the code C. By definition, the code with parity-check matrix H is Ham$(r, 2)$. This proves that C is a Ham$(r, 2)$ code. ∎

Example 4.3.10 Find the generator polynomial of a cyclic Ham(3, 2) code.

Solution. Consider the polynomial $p(x) = x^3 + x + 1$ over $\mathbb{F}_2$. Clearly, neither 0 nor 1 is a root of $p(x)$, and hence $p(x)$ is irreducible. Moreover, the multiplicative group of the field $\mathbb{F}_2^3$ is of order 7 (which is prime). Hence every nonzero element in $\mathbb{F}_2^3$ is primitive. Therefore $p(x)$ is a primitive polynomial. By Theorem 4.3.9, the cyclic code with generator polynomial $x^3 + x + 1$ in the ring $\mathbb{F}_2[x]_7$ is Ham(3, 2).

We can similarly show that the code with generator polynomial $x^3 + x^2 + 1$ in the ring $\mathbb{F}_2[x]_7$ is Ham(3, 2).

The generator polynomial of a cyclic code C determines a generator matrix as well as a parity-check matrix for C. In the following theorem we obtain a generator matrix.

THEOREM 4.3.11 Let $C \subset F[x]_n$ be a cyclic code with generator polynomial

$$g(x) = g_0 + g_1 x + \cdots + g_r x^r$$

where $g_r = 1$. Then C is of dimension $n - r$.

Moreover, the $(n - r) \times n$ matrix

$$G = \begin{bmatrix} g_0 & g_1 & g_2 & \cdots & & \cdots & g_r & 0 & 0 & \cdots & 0 \\ 0 & g_0 & g_1 & \cdots & & \cdots & g_{r-1} & g_r & 0 & \cdots & 0 \\ \vdots & \vdots & \vdots & & & & & & & & \vdots \\ 0 & 0 & \cdots & 0 & g_0 & g_1 & & \cdots & & \cdots & g_r \end{bmatrix}$$

is a generator matrix of C.

Proof: By Theorem 4.3.6, $x^n - 1 = g(x)h(x)$ for some polynomial $h(x)$; hence $g_0 \neq 0$. Therefore G is a row echelon matrix, and the rows of G are linearly independent. Written as polynomials, the rows of G are $g(x), xg(x), \ldots, x^{n-r-1}g(x)$. Let $a(x) \in C$. By Theorem 4.3.6, $a(x) = q(x)g(x)$ for some $q(x)$. Since $\deg a(x) < n$, we have $\deg q(x) < n - r$. Thus $q(x)$ is of the form $q(x) = q_0 + q_1 x + \cdots + q_{n-r-1}x^{n-r-1}$. Hence

$$a(x) = q_0 g(x) + q_1 x g(x) + \cdots + q_{n-r-1}x^{n-r-1}g(x)$$

Thus $a(x)$ can be expressed as a linear combination of the rows of G. Hence G is a generator matrix of C. ∎

The generator polynomial of a cyclic code can be used for encoding. We saw in Section 4.2 that if C is a linear $[n, k]$-code with generator matrix G, then a message word $u \in F^k$ is encoded as uG. Now suppose C is a cyclic code with generator polynomial $g(x)$, and let G be the generator matrix obtained in Theorem 4.3.11. Then $r = \deg g(x) = n - k$, and the rows of G are $g(x), xg(x), \ldots, x^{k-1}g(x)$.

Let $u = (u_0, u_1, \ldots, u_{k-1})$, and $u(x) = u_0 + u_1 x + \cdots + u_{k-1} x^{k-1}$. Then

$$uG = u_0 g(x) + u_1 x g(x) + \cdots + u_{k-1} x^{k-1} g(x) = u(x) g(x)$$

Thus a message polynomial $u(x)$ is encoded as $u(x) g(x)$.

Definition 4.3.12 Let $g(x)$ be the generator polynomial of a cyclic code $C \subset F[x]_n$. Then the polynomial $h(x)$ such that $x^n - 1 = g(x) h(x)$ is called the *check polynomial* of C.

Since $g(x)$ is monic, $h(x)$ is also monic. The following theorem explains the nomenclature.

THEOREM 4.3.13 Let $C \subset F[x]_n$ be a cyclic code with check polynomial $h(x)$. Let $a(x) \in F[x]_n$. Then $a(x) \in C$ if and only if $a(x) * h(x) = 0$.

Proof: Let $g(x)$ be the generator polynomial of C. Then $g(x) h(x) = x^n - 1$; hence $g(x) * h(x) = 0$. Let $a(x) \in C$. Then, by Theorem 4.3.6, $a(x) = q(x) g(x)$ for some $q(x)$, and hence $a(x) * h(x) = q(x) g(x) * h(x) = 0$. Conversely, let $a(x) \in F[x]_n$ such that $a(x) * h(x) = 0$. Then $a(x) h(x) = f(x) (x^n - 1)$ for some $f(x)$, so $a(x) h(x) = f(x) g(x) h(x)$. Therefore $a(x) = f(x) g(x)$, so $a(x) \in C$.   ∎

Let C be a cyclic $[n, k]$-code. Then, by Theorem 4.3.11, its generator polynomial $g(x)$ is of degree $n - k$. Hence its check polynomial $h(x)$ is of degree k. Given any polynomial $f(x) = f_0 + f_1 x + \cdots + f_m x^m$ of degree m, the *reciprocal polynomial* of $f(x)$ is defined to be

$$\bar{f}(x) = f_m + f_{m-1} x + \cdots + f_0 x^m$$

The coefficients in $\bar{f}(x)$ are those of $f(x)$ taken in reverse order. Formally, we can write $\bar{f}(x) = x^m f(x^{-1})$. In particular, $\bar{g}(x) = x^{n-k} g(x^{-1})$ and $\bar{h}(x) = x^k h(x^{-1})$. Now $g(x) h(x) = x^n - 1$. Hence

$$\bar{g}(x) \bar{h}(x) = x^{n-k} g(x^{-1}) x^k h(x^{-1}) = x^n (x^{-n} - 1) = 1 - x^n$$

So $\bar{h}(x)$ divides $x^n - 1$.

In the following theorem, we find a parity-check matrix for C and show that the dual code $C^\perp$ is also cyclic.

THEOREM 4.3.14 Let C be a cyclic $[n, k]$-code with check polynomial

$$h(x) = h_0 + h_1 x + \cdots + h_k x^k$$

where $h_k = 1$. Then

(a) The $(n-k) \times n$ matrix

$$H = \begin{bmatrix} h_k & h_{k-1} & h_{k-2} & \cdots & & \cdots & h_0 & 0 & 0 & \cdots & 0 \\ 0 & h_k & h_{k-1} & \cdots & & \cdots & h_1 & h_0 & 0 & \cdots & 0 \\ \vdots & \vdots & \vdots & & & & & & & & \vdots \\ 0 & 0 & \cdots & 0 & h_k & h_{k-1} & & \cdots & & \cdots & h_0 \end{bmatrix}$$

is a parity-check matrix for C.

(b) The dual code $C^\perp$ is cyclic and generated by the polynomial

$$\bar{h}(x) = h_k + h_{k-1}x + \cdots + h_0 x^k$$

Proof: **(a)** Since H is in row echelon form, its rows are linearly independent. We claim that each row of H is orthogonal to every codeword in C and hence a vector in $C^\perp$. Let $a(x) = a_0 + a_1 x + \cdots + a_{n-1}x^{n-1} \in C$. Then, by Theorem 4.3.13, $a(x) *$ $h(x) = 0$. So the coefficient of x^i in the product on the left-hand side must be zero for each $i = 0, 1, \ldots, n-1$. In particular, on equating to zero the coefficient of x^i for $i = k, k+1, \ldots, n-1$, we get

$$a_{i-k}h_k + a_{i-k+1}h_{k-1} + \cdots + a_i h_0 = 0$$

for each $i = k, k+1, \ldots, n-1$. Hence

$$H \begin{bmatrix} a_0 \\ a_1 \\ \vdots \\ a_{n-1} \end{bmatrix} = 0$$

Thus the rows of H are $n-k$ linearly independent vectors in the dual space $C^\perp$. Hence H is a generator matrix of $C^\perp$ and therefore a parity-check matrix of C.

(b) Since $\bar{h}(x)$ divides $x^n - 1$, by Theorem 4.3.11, H is a generator matrix of the cyclic code $\langle \bar{h}(x) \rangle$. But H is a generator matrix of $C^\perp$, so $C^\perp = \langle \bar{h}(x) \rangle$. ∎

The following example illustrates the application of Theorems 4.3.11 and 4.3.14.

Example 4.3.15 Write a generator matrix and a parity-check matrix for a cyclic Hamming code Ham(3, 2). Obtain the complete code.

Solution. Recall that Ham(3, 2) is a linear [7, 4]-code. We showed in Example 4.3.10 that $g(x) = 1 + x + x^3$ is the generator polynomial of a cyclic Ham(3, 2) code C. Hence, by Theorem 4.3.11, a generator matrix of C is

$$G = \begin{bmatrix} 1 & 1 & 0 & 1 & 0 & 0 & 0 \\ 0 & 1 & 1 & 0 & 1 & 0 & 0 \\ 0 & 0 & 1 & 1 & 0 & 1 & 0 \\ 0 & 0 & 0 & 1 & 1 & 0 & 1 \end{bmatrix}$$

Taking all possible linear combinations of the rows of G, we obtain

$C = \{0000000, 1101000, 0110100, 0011010, 0001101, 1011100, 1110110,$
$\quad 1100101, 0101110, 0111001, 0010111, 1000110, 0100011, 1111111,$
$\quad 1010001, 1001011\}$

The check polynomial of C is

$$h(x) = (x^7 - 1)/(x^3 + x + 1) = x^4 + x^2 + x + 1$$

Hence, by Theorem 4.3.14,

$$H = \begin{bmatrix} 1 & 0 & 1 & 1 & 1 & 0 & 0 \\ 0 & 1 & 0 & 1 & 1 & 1 & 0 \\ 0 & 0 & 1 & 0 & 1 & 1 & 1 \end{bmatrix}$$

is a parity-check matrix of C.

The generator matrix and the parity-check matrix of a cyclic code C given by Theorems 4.3.11 and 4.3.14 are not in canonical form. In general, for a linear code, a generator matrix G is transformed to canonical form by applying elementary row operations. But here, in the case of a cyclic code, we can obtain the canonical form by using the generator polynomial and the division algorithm in $F[x]$.

Let $g(x)$ be the generator polynomial of a cyclic $[n, k]$-code C over F. For any polynomial $f(x) \in F[x]$, let $rem_{g(x)}(f(x))$ denote the remainder on dividing $f(x)$ by $g(x)$. For the sake of convenience, we simply write $r(f(x))$ for $rem_{g(x)}(f(x))$, so $f(x) = q(x)g(x) + r(f(x))$ for some $q(x) \in F(x)$. Hence $f(x) - r(f(x)) = q(x)g(x) \in C$. In particular, consider $r(x^j)$. Since $\deg g(x) = n - k$, we have $r(x^j) = x^j$ for $j < n - k$. Further, since $g(x)$ divides $x^n - 1$, $r(x^{n+j}) = r(x^j)$ for all $j \geq 0$. Thus we have to compute $r(x^j)$ for only $j = n - k, \ldots, n - 1$.

We define

$$G_i(x) = x^{i-1} - x^k r(x^{n-k+i-1})$$

for $i = 1, \ldots, k$. Clearly, $\deg G_i(x) < n$; hence $G_i(x) \in F[x]_n$. Moreover, $x^{n-k+i-1} - r(x^{n-k+i-1}) \in C$; therefore $G_i(x) = x^k * (x^{n-k+i-1} - r(x^{n-k+i-1})) \in C$. Let G be the $k \times n$ matrix whose ith row is $G_i(x)$, written as a row vector, $i = 1, \ldots, k$. Then

$$G = \begin{bmatrix} I_k & \vdots & -A \end{bmatrix}$$

where A is a $k \times (n - k)$ matrix whose ith row is $r(x^{n-k+i-1})$. The rows of G are elements of C and are linearly independent. Hence G is the canonical generator matrix of C. By Theorem 4.2.12, the canonical parity-check matrix of C is $H = [A^T \vdots I_{n-k}]$. Thus we have proved the following theorem:

THEOREM 4.3.16 Let $g(x)$ be the generator polynomial of a cyclic $[n, k]$-code C over F. Let A be a $k \times (n - k)$ matrix whose ith row is $rem_{g(x)}(x^{n-k+i-1})$,

$i = 1, \ldots, k$. Then the canonical generator and parity-check matrices of C are, respectively,

$$G = \begin{bmatrix} I_k & \vdots & -A \end{bmatrix} \quad \text{and} \quad H = \begin{bmatrix} A^T & \vdots & I_{n-k} \end{bmatrix}$$

Example 4.3.17 Find the canonical generator and parity-check matrices of the cyclic Ham(3, 2) code of Example 4.3.10.

Solution. Here $g(x) = 1 + x + x^3$, $k = 4$, $n = 7$. Dividing x^j by $g(x)$, $j = 3, 4, 5, 6$, we get the remainders

$$r(x^3) = 1 + x, \quad r(x^4) = x + x^2, \quad r(x^5) = 1 + x + x^2, \quad r(x^6) = 1 + x^2$$

Hence

$$A = \begin{bmatrix} 1 & 1 & 0 \\ 0 & 1 & 1 \\ 1 & 1 & 1 \\ 1 & 0 & 1 \end{bmatrix}$$

Therefore

$$G = \begin{bmatrix} 1 & 0 & 0 & 0 & 1 & 1 & 0 \\ 0 & 1 & 0 & 0 & 0 & 1 & 1 \\ 0 & 0 & 1 & 0 & 1 & 1 & 1 \\ 0 & 0 & 0 & 1 & 1 & 0 & 1 \end{bmatrix}, \quad H = \begin{bmatrix} 1 & 0 & 1 & 1 & 1 & 0 & 0 \\ 1 & 1 & 1 & 0 & 0 & 1 & 0 \\ 0 & 1 & 1 & 1 & 0 & 0 & 1 \end{bmatrix}$$

We now consider the syndrome function for a cyclic code. Recall that if C is a linear $[n, k]$-code over F with parity-check matrix H, then the syndrome of a vector $a \in F^n$ (with respect to H) is $S(a) = aH^T$. The following theorem gives the syndrome for a cyclic code with respect to the canonical parity-check matrix.

THEOREM 4.3.18 Let C be a cyclic $[n, k]$-code over F with generator polynomial $g(x)$. Let H be the canonical parity-check matrix of C. Then for any $a \in F^n$,

$$S(a) = rem_{g(x)}(x^{n-k}a(x))$$

Proof: By Theorem 4.3.16, $H^T = \begin{bmatrix} A \\ \cdots \\ I_{n-k} \end{bmatrix}$, where A is a $k \times (n-k)$ matrix whose ith row is $r(x^{n-k+i-1})$, $i = 1, \ldots, k$. The ith row of the identity matrix I_{n-k} is x^{i-1}, $i = 1, \ldots, n-k$. Hence, using the relation $r(x^{n+j}) = r(x^j)$, we see that the ith row of H^T is $r(x^{n-k+i-1})$, $i = 1, \ldots, n$.

Let $a = (a_0, a_1, \ldots, a_{n-1}) \in F$,   so $a(x) = a_0 + a_1 x + \cdots + a_{n-1} x^{n-1} \in$
$F[x]_n$. Then

$$S(a) = [\, a_0 \quad a_1 \quad \ldots \quad a_{n-1} \,] H^T$$

$$= \sum_{i=1}^{n} a_{i-1} r(x^{n-k+i-1})$$

$$= r \left(\sum_{i=1}^{n} a_{i-1} x^{n-k+i-1} \right)$$

$$= r(x^{n-k} a(x)) \qquad \blacksquare$$

We see from Theorem 4.3.18 that for a cyclic code we need not know the canonical parity-check matrix in order to find the syndrome with respect to it. This suggests that we may choose a simpler form for the syndrome. We showed in Section 4.2 that for any linear code C, two vectors have the same syndrome (with respect to any parity-check matrix) if and only if they lie in the same coset of C. If C is a cyclic code with generator polynomial $g(x)$, then vectors a, b lie in the same coset if and only if $g(x)$ divides $a(x) - b(x)$—that is, $r(a(x)) = r(b(x))$. Therefore we may define the syndrome of a vector a as

$$S(a) = rem_{g(x)}(a(x))$$

This syndrome is not with respect to the canonical parity-check matrix of C obtained above. But if we take H to be the matrix whose jth column is $r(x^{j-1})$, $j = 1, \ldots, n$, then the syndrome with respect to H is given by $S(a) = r(a(x))$. It can be verified that the matrix H thus defined is indeed a parity-check matrix.

From the various results proved above, we observe that generator and parity-check matrices G and H do not have a direct important role in cyclic codes. The uniquely determined generator polynomial $g(x)$ of a cyclic code C enables us to carry out all the tasks for which G and H are used in the case of general linear codes. The elements of the code are of the form $a(x)g(x)$. Encoding is done by the mapping $a(x) \mapsto a(x)g(x)$. The syndrome is given by $rem_{g(x)}(a(x))$.

We have already shown that binary Hamming codes are cyclic. We now describe two other important cyclic codes, known as *Golay codes*.

The polynomial $x^{23} - 1$ can be factored over $\mathbb{F}_2$ as

$$x^{23} - 1 = (x - 1)g(x)g_1(x)$$

where

$$g(x) = x^{11} + x^9 + x^7 + x^6 + x^5 + x + 1$$
$$g_1(x) = x^{11} + x^{10} + x^6 + x^5 + x^4 + x^2 + 1$$

These two polynomials are reciprocals of each other and so generate equivalent codes. By Theorem 4.3.11, these codes are of dimension 12. The cyclic [23, 12]-code with generator polynomial $g(x)$ in $\mathbb{F}_2[x]_{23}$, or any code equivalent to it, is called the *binary Golay code* and denoted by G_{23}. Now $g(x)$ is itself an element of the code, and the number of nonzero terms in $g(x)$ is 7. So, written as a vector, $g(x)$ is a codeword of weight 7. Hence, by Theorem 4.2.17, the minimum distance of the code is at most 7.

The polynomial $x^{11} - 1$ can be factored over $\mathbb{F}_3$ as

$$x^{11} - 1 = (x - 1)g(x)g_1(x)$$

where

$$g(x) = x^5 + x^4 - x^3 + x^2 - 1$$
$$g_1(x) = x^5 - x^3 + x^2 - x + 1$$

These two polynomials generate equivalent codes of dimension 6. The cyclic [11, 6]-code with generator polynomial $g(x)$ in $\mathbb{F}_3[x]_{11}$, or any code equivalent to it, is called the *ternary Golay code* and denoted by G_{11}. Since $g(x)$ has 5 nonzero terms, the minimum distance of the code is at most 5.

We state the following theorem without proof.

THEOREM 4.3.19 The binary [23, 12] Golay code has minimum distance 7. The ternary [11, 6] Golay code has minimum distance 5.

With this result, we can now easily prove that both the Golay codes are perfect.

THEOREM 4.3.20 The binary [23, 12, 7] Golay code and the ternary [11, 6, 5] Golay code are both perfect.

Proof: By Theorem 4.1.6, a q-ary (n, M, d)-code with $d = 2t + 1$ is perfect if and only if

$$M \sum_{m=0}^{t} \binom{n}{m} (q - 1)^m = q^n$$

We show that this condition holds for each Golay code. For the binary Golay code, $q = 2, n = 23, t = 3, M = 2^{12}$. It is easily verified by direct computation that

$$2^{12} \left\{ 1 + 23 + \binom{23}{2} + \binom{23}{3} \right\} = 2^{23}$$

For the ternary Golay code, $q = 3, n = 11, t = 2, M = 3^6$. Again, it can be verified that

$$3^6 \left\{ 1 + 11 \cdot 2 + \binom{11}{2} \cdot 2^2 \right\} = 3^{11}$$

This proves that both the Golay codes are perfect. ∎

In Section 4.2 we showed that all Hamming codes are perfect. Essentially, the Hamming codes and the two Golay codes are the only perfect codes. We state the following theorem without proof.

THEOREM 4.3.21 Every nontrivial single-error-correcting perfect code has the parameters of a Hamming code.

Every nontrivial multiple-error-correcting perfect code is equivalent to either the binary [23, 12, 7] Golay code or the ternary [11, 6, 5] Golay code.

EXERCISES 4.3

1. Let $u(x) = 1 + x + \cdots + x^{n-1} \in F[x]_n$. Show that $x * u(x) = u(x)$. Hence deduce that for any $f(x) \in F[x]_n$, $f(x) * u(x) = ku(x)$ for some scalar $k \in F$.

2. Find all binary cyclic codes of length 4.

3. Find all binary cyclic codes of length 5.

4. Find all binary cyclic codes of length 6.

5. Find all binary cyclic codes of length 7. Write the generator matrix for each.

6. Find all ternary cyclic codes of length 4.

7. Write a parity-check matrix for every ternary cyclic code of length 4.

8. Show that Ham(2, 3) is not cyclic.

9. Show that the binary code {0000, 1001, 0110, 1111} is equivalent to a cyclic code.

10. Find the number of binary cyclic codes of length 10.

4.4 BCH CODES

In this section we describe a special class of cyclic codes, known as *BCH codes* (named after Bose, Chaudhuri, and Hocquenghem). The concept of a BCH code is a generalization of the method we used in Section 4.3 to construct a cyclic binary Hamming code. We showed that if t is a primitive element in the field $\mathbb{F}_2^r$ and $p(x) \in \mathbb{F}_2^r[x]$ is an irreducible polynomial such that $p(t) = 0$, then the cyclic code with generator polynomial $p(x)$ in $\mathbb{F}_2[x]_n$, where $n = 2^r - 1$, is Ham(r, 2). A BCH code is constructed by a generalization of this technique.

Before defining a BCH code, we give a summary of some important properties of finite fields and irreducible polynomials. The order of every finite field is a power of some prime p. For every prime p and positive integer r, there is a unique field (up to isomorphism) of order p^r, denoted by $\mathbb{F}_{p^r}$ or $GF(pr)$. If $r = 1$, then $\mathbb{F}_p = \mathbb{Z}_p$. Let $F = \mathbb{F}_q$, where $q = p^r$. Then the set F^* of nonzero elements in F is a cyclic group of order $q - 1$ under multiplication. Hence $a^{q-1} = 1$; that is, $a^q = a$ for every $a \in F$. For any divisor n of $q - 1$, there is an element a of order $o(a) = n$ in the group F^*. Such an element is called a *primitive nth root of unity* in the field F. If

$o(a) = q - 1$, then a is a generator of the cyclic group F^* and is called a *primitive element* in F. The characteristic of the field F is p. Hence for all $a \in F$, $pa = 0$ and therefore also $qa = 0$.

Given a finite field F and any positive integer m, there exists an irreducible polynomial $p(x) \in F[x]$ of degree m. The field $\mathbb{F}_{p^r}$ is obtained by constructing the quotient ring $\mathbb{Z}_p/(p(x))$, where $p(x)$ is an irreducible polynomial of degree r in $\mathbb{Z}_p[x]$. More generally, starting from $F = \mathbb{F}_q$, we construct the field $\mathbb{F}_{q^m}$ as the quotient ring $\mathbb{F}_q[x]/(p(x))$, where $p(x)$ is an irreducible polynomial of degree m in $\mathbb{F}_q[x]$. If we write t to denote the coset $x + (p(x))$, then $p(t) = 0$ and

$$\mathbb{F}_{q^m} = \{a_0 + a_1 t + \cdots + a_{m-1} t^{m-1} \mid a_0, a_1, \ldots, a_{m-1} \in \mathbb{F}_q\}$$

The field $\mathbb{F}_{q^m}$ is called an *extension of* $\mathbb{F}_q$ *of degree m.*

On using the fact that $a^q = a$ for all $a \in \mathbb{F}_q$ and $q\beta = 0$ for all $\beta \in \mathbb{F}_{q^m}$, we immediately get the following result:

THEOREM 4.4.1 Let $a_1, \ldots, a_n \in \mathbb{F}_q$ and $\beta_1, \ldots, \beta_n \in \mathbb{F}_{q^m}$. Then

$$(a_1 \beta_1 + \cdots + a_n \beta_n)^q = a_1 \beta_1^q + \cdots + a_n \beta_n^q$$

Let $\alpha \in \mathbb{F}_{q^m}$. Then there exists a monic polynomial $g(x) \in \mathbb{F}_q[x]$ of least degree such that $g(\alpha) = 0$. The polynomial $g(x)$ is irreducible over $\mathbb{F}_q$ and is called the *minimal polynomial* of α over $\mathbb{F}_q$. If $f(x) \in \mathbb{F}_q[x]$ is any polynomial such that $f(\alpha) = 0$, then, by using the division algorithm in $\mathbb{F}_q[x]$, we can show that $g(x)$ divides $f(x)$. Further, $\deg g(x)$ divides m. If α is a primitive element in $\mathbb{F}_{q^m}$, then $\deg g(x) = m$.

By Theorem 4.4.1, it follows that if $\alpha \in \mathbb{F}_{q^m}$ is a root of a polynomial $f(x) \in \mathbb{F}_q[x]$, then α^q is also a root of $f(x)$. In particular, we have the following result.

THEOREM 4.4.2 Let $\alpha \in \mathbb{F}_{q^m}$. Then $\alpha, \alpha^q, \alpha^{q^2}, \ldots$ have the same minimal polynomial over $\mathbb{F}_q$.

A BCH code is defined as follows: Let c, d, q, n be positive integers such that $2 \le d \le n$, q is a prime power, and n is relatively prime to q. Let m be the least positive integer such that $q^m \equiv 1 \pmod{n}$. [By Euler's theorem, $q^{\varphi(n)} \equiv 1 \pmod{n}$, so m divides $\varphi(n)$.] Thus n divides $q^m - 1$. Let ζ be a primitive nth root of unity in $\mathbb{F}_{q^m}$. Let $m_i(x) \in \mathbb{F}_q[x]$ denote the minimal polynomial of ζ^i. Let $g(x)$ be the product of distinct polynomials among $m_i(x)$, $i = c, c + 1, \ldots, c + d - 2$; that is,

$$g(x) = \text{lcm}\{m_i(x) \mid i = c, c + 1, \ldots, c + d - 2\}$$

Since $m_i(x)$ divides $x^n - 1$ for each i, it follows that $g(x)$ divides $x^n - 1$. Let C be the cyclic code with generator polynomial $g(x)$ in the ring $\mathbb{F}_q[x]_n$. Then C is called a *BCH code* of length n over $\mathbb{F}_q$ with *designed distance d.*

If $n = q^m - 1$ in the foregoing definition, then the BCH code C is called *primitive*. If $c = 1$, then C is called a *narrow sense BCH code.*

THEOREM 4.4.3 Let C be a *BCH code* of length n over $\mathbb{F}_q$ with *designed distance* d. Then, with the notation used above,

$$C = \{v(x) \in \mathbb{F}_q[x]_n \mid v(\zeta^i) = 0 \text{ for all } i = c, c+1, \ldots, c+d-2\}$$

Equivalently, C is the null space of the matrix

$$H = \begin{bmatrix} 1 & \zeta^c & \zeta^{2c} & \cdots & \zeta^{(n-1)c} \\ 1 & \zeta^{c+1} & \zeta^{2(c+1)} & \cdots & \zeta^{(n-1)(c+1)} \\ \vdots & \vdots & \vdots & \ddots & \vdots \\ 1 & \zeta^{c+d-2} & \zeta^{2(c+d-2)} & \cdots & \zeta^{(n-1)(c+d-2)} \end{bmatrix}$$

Proof: Let $v(x) \in C$. Then, by Theorem 4.3.11, $v(x) = q(x)g(x)$ for some $q(x)$, where $g(x)$ is the generator polynomial of C. Hence $v(\zeta^i) = 0$ for all $i = c$, $c+1, \ldots, c+d-2$. Conversely, let $v(x) \in \mathbb{F}_q[x]_n$ such that $v(\zeta^i) = 0$ for all $i = c, c+1, \ldots, c+d-2$. Then $m_i(x)$ divides $v(x)$ for all $i = c, c+1, \ldots$, $c+d-2$. Hence $g(x)$ divides $v(x)$, so $v(x) \in C$. This proves the first part of the theorem.

To prove the second part, let $v(x) = v_0 + v_1x + \cdots + v_{n-1}x^{n-1} \in \mathbb{F}_q[x]_n$. Then $v(\zeta^i) = 0$ holds for all $i = c, c+1, \ldots, c+d-2$ if and only if $Hv^T = 0$, where $v = (v_0, v_1, \ldots, v_{n-1}) \in \mathbb{F}_q^n$. This proves that C is the null space of H.   ∎

We note that H is a $(d-1) \times n$ matrix over $\mathbb{F}_{q^m}$. Every element in $\mathbb{F}_{q^m}$ is of the form $a_0 + a_1t + \cdots + a_{m-1}t^{m-1}$, where $a_0, a_1, \ldots, a_{m-1} \in \mathbb{F}_q$, and hence can be written as a column vector of length m. Therefore H can be written as an $m(d-1) \times n$ matrix over $\mathbb{F}_q$. The rows of H are not necessarily linearly independent. Hence H is not a parity-check matrix of C in the strict sense of the term. Let us call H a *quasi parity-check matrix*. Since *rank* $(H) \leq m(d-1)$, it follows that dim $C \geq n - m(d-1)$.

In the following theorem we show that the minimum distance of a BCH code is at least equal to its designed distance. We showed in Theorem 4.2.18 that the minimum distance $d(C)$ of a linear code C is equal to the minimal number of linearly dependent columns in a parity-check matrix H of C. The proof of the theorem shows that this property does not depend on whether or not the rows of H are linearly independent. Hence the theorem also holds when H is a quasi parity-check matrix.

THEOREM 4.4.4 Let C be a BCH code of designed distance d. Then $d(C) \geq d$.

Proof: Let H be the quasi parity-check matrix of C given in Theorem 4.4.3. We show that any $d-1$ columns of H are linearly independent. Let K be the $(d-1) \times (d-1)$ matrix formed by the columns with first entries $\zeta^{i_1c}, \zeta^{i_2c}, \ldots, \zeta^{i_{d-1}c}$, respectively. Then

$$\det K = \zeta^{(i_1+i_2+\cdots+i_{d-1})c} \begin{vmatrix} 1 & 1 & \cdots & 1 \\ \zeta^{i_1} & \zeta^{i_2} & \cdots & \zeta^{i_{d-1}} \\ \vdots & \vdots & \ddots & \vdots \\ \zeta^{i_1(d-2)} & \zeta^{i_2(d-2)} & \cdots & \zeta^{i_{d-1}(d-2)} \end{vmatrix}$$

The determinant on the right-hand side is a Vandermonde determinant. Now ζ is an element of order n in the multiplicative group of $\mathbb{F}_{q^m}$; therefore $\zeta^{i_1}, \zeta^{i_2}, \ldots, \zeta^{i_{d-1}}$ are all distinct and hence $\det K \neq 0$. Therefore the columns of K are linearly independent. It follows that the minimal number of linearly dependent columns in H is greater than $d - 1$. Hence $d(C) \geq d$. ∎

We now present some examples of BCH codes. The simplest example is a *Reed–Solomon code,* which is defined as follows: Let q be any prime power, and let $n = q - 1$. Then $m = 1$. Let ζ be a primitive element in $\mathbb{F}_q$. Then the minimal polynomial of ζ over $\mathbb{F}_q$ is $x - \zeta$. Take $c = 1$ and let d be any positive integer, $2 \leq d \leq n$. Then the cyclic code C with generator polynomial

$$g(x) = (x - \zeta)(x - \zeta^2) \cdots (x - \zeta^{d-1})$$

is a primitive narrow sense BCH code of designed distance d. It is called a Reed–Solomon code. Since $\deg g(x) = d - 1$, the code C has dimension $k = n - d + 1$. By Theorem 4.4.4, $d(C) \geq d$. On the other hand, by Theorem 4.2.18, $d(C) \leq n - k + 1 = d$, so $d(C) = d$. Hence C is a $[q - 1, q - d, d]$-code.

We next show that the binary Hamming codes and the two Golay codes are all BCH codes.

To obtain $\text{Ham}(r, 2)$ as a BCH code, we take $q = 2$ and $n = 2^r - 1$. Then $m = r$, so $\mathbb{F}_{q^m} = \mathbb{F}_{2^r}$. Let ζ be a primitive nth root of unity in $\mathbb{F}_{2^r}$. Then ζ is in fact a primitive element in $\mathbb{F}_{2^r}$. Let $g(x) \in \mathbb{F}_2[x]$ be the minimal polynomial of ζ. Then $g(x)$ is a primitive polynomial of degree r. Now ζ and ζ^2 have the same minimal polynomial, so $m_1(x) = m_2(x) = g(x)$. Thus

$$g(x) = \text{lcm}\{m_i(x) \mid i = 1, 2\}$$

Hence, by definition, the cyclic code C with generator polynomial $g(x)$ is a narrow sense primitive BCH code of designed distance 3. On the other hand, by Theorem 4.3.9, C is $\text{Ham}(r, 2)$. This proves that $\text{Ham}(r, 2)$ is a BCH code. We have shown in Theorem 4.2.24 that the minimum distance of $\text{Ham}(r, 2)$ is 3. Thus, in this case, $d(C) = d$.

For the binary Golay code, we take $q = 2$ and $n = 23$. Then $m = 11$, so $\mathbb{F}_{q^m} = \mathbb{F}_{2^{11}}$. Let ζ be a primitive 23rd root of unity in $\mathbb{F}_{2^{11}}$, and let $g(x)$ be the minimal polynomial of ζ. Now $\zeta, \zeta^2, \zeta^{2^2}, \zeta^{2^3}, \ldots$ all have the same minimal polynomial. Using the relation $\zeta^{23} = 1$, we have $\zeta^{2^8} = \zeta^{256} = \zeta^3$. Thus $\zeta, \zeta^2, \zeta^3, \zeta^4$ have the same minimal polynomial $g(x)$. Thus

$$g(x) = \text{lcm}\{m_i(x) \mid i = 1, 2, 3, 4\}$$

The cyclic code C with generator polynomial $g(x)$ is a narrow sense BCH code of designed distance 5 over $\mathbb{F}_2$. Now $g(x)$ divides $x^{23} - 1$ and $\deg g(x)$ divides $m = 11$, so $\deg g(x) = 11$. Hence C is the binary $[23, 12, 7]$ Golay code described in Section 4.3. In this case, $d(C) = 7 > d$.

The ternary Golay code is also a BCH code. Let $q = 3$ and $n = 11$. Then $m = 5$, so $\mathbb{F}_{q^m} = \mathbb{F}_{3^5}$. Let ζ be a primitive 11th root of unity in $\mathbb{F}_{3^5}$, and let $g(x)$ be the minimal polynomial of ζ. Then $\zeta, \zeta^3, \zeta^{3^2}, \ldots$ have the same minimum polynomial. Since $\zeta^{11} = 1$, we have $\zeta^{27} = \zeta^5$ and $\zeta^{81} = \zeta^4$. Thus $\zeta^3, \zeta^4, \zeta^5$ have the same minimum polynomial $g(x)$. Thus

$$g(x) = \text{lcm}\{m_i(x) \mid i = 3, 4, 5\}$$

The cyclic code C with generator polynomial $g(x)$ is a BCH code of designed distance 4. Now $g(x)$ is an irreducible polynomial of degree 5 and divides $x^{11} - 1$. Hence C is the ternary $[11, 6, 5]$ Golay code. Here again $d(C) > d$.

The following example illustrates how we construct a BCH code of a given length and designed distance. The first step is to have a primitive element ζ in the field $\mathbb{F}_{q^m}$ for which we need a primitive polynomial over $\mathbb{F}_q$ of degree m. For reference, here are some primitive polynomials over $\mathbb{F}_2$ of degrees 4 to 8:

$$x^4 + x + 1$$
$$x^5 + x^2 + 1$$
$$x^6 + x + 1$$
$$x^7 + x + 1$$
$$x^8 + x^4 + x^3 + x^2 + 1$$

Example 4.4.5 Construct a binary narrow sense BCH code of length 15 and designed distance 7. Show that its minimum distance is 7.

Solution. Here $q = 2$, $n = 15$, so $m = 4$ and $2^4 - 1 = 15$. The polynomial

$$p(x) = x^4 + x + 1$$

is a primitive irreducible polynomial over $\mathbb{F}_2$. So we can represent the field $\mathbb{F}_{2^4}$ as

$$\mathbb{F}_{2^4} = \{a_0 + a_1\zeta + a_2\zeta^2 + a_3\zeta^3 \mid a_0, a_1, a_2, a_3 \in \mathbb{F}_2\}$$

where ζ satisfies the relation $\zeta^4 + \zeta + 1 = 0$. Using this relation, we obtain the following table for the powers of ζ:

$$\zeta^4 = 1 + \zeta \qquad \zeta^8 = 1 + \zeta^2 \qquad \zeta^{12} = 1 + \zeta + \zeta^2 + \zeta^3$$
$$\zeta^5 = \zeta + \zeta^2 \qquad \zeta^9 = \zeta + \zeta^3 \qquad \zeta^{13} = 1 + \zeta^2 + \zeta^3$$
$$\zeta^6 = \zeta^2 + \zeta^3 \qquad \zeta^{10} = 1 + \zeta + \zeta^2 \qquad \zeta^{14} = 1 + \zeta^3$$
$$\zeta^7 = 1 + \zeta + \zeta^3 \qquad \zeta^{11} = \zeta + \zeta^2 + \zeta^3 \qquad \zeta^{15} = 1$$

Now ζ is a primitive 15th root of unity in $\mathbb{F}_{2^4}$, and $p(x)$ is the minimal polynomial of ζ. To obtain a BCH code of designed distance $d = 7$, we need the minimal polynomials of ζ^i for $i = 1, \ldots, 6$. By Theorem 4.4.2, ζ, ζ^2, ζ^4 have the same minimal polynomial $p(x)$. Let $q(x)$ be the minimal polynomial of ζ^3. Then $\zeta^3, \zeta^6, \zeta^{12}, \zeta^{24}, \ldots$ all have the same minimal polynomial $q(x)$. Using the relation $\zeta^{15} = 1$, we see that the roots of $q(x)$ are $\zeta^3, \zeta^6, \zeta^9, \zeta^{12}$. Hence

$$q(x) = (x - \zeta^3)(x - \zeta^6)(x - \zeta^9)(x - \zeta^{12})$$
$$= x^4 - (\zeta^3 + \zeta^6 + \zeta^9 + \zeta^{12})x^3 + (\zeta^3 + \zeta^6 + \zeta^9 + \zeta^{12})x^2$$
$$- (\zeta^3 + \zeta^6 + \zeta^9 + \zeta^{12})x + 1$$
$$= x^4 + x^3 + x^2 + x + 1$$

Similarly the minimal polynomial $h(x)$ of ζ^5 has roots ζ^5, ζ^{10}, so

$$h(x) = (x - \zeta^5)(x - \zeta^{10})$$
$$= x^2 + x + 1$$

Hence the generator polynomial of the desired BCH code is

$$g(x) = \text{lcm}\{m_i(x) \mid i = 1, 2, 3, 4, 5, 6\}$$
$$= p(x)q(x)h(x)$$
$$= x^{10} + x^8 + x^5 + x^4 + x^2 + x + 1$$

By Theorem 4.3.11, the cyclic code C generated by $g(x)$ in $\mathbb{F}_2[x]_{15}$ has dimension 5, so C is a $[15, 5]$ primitive narrow sense BCH code of designed distance 7. Hence $d(C) \geq 7$. Now $g(x)$ is itself a code polynomial and has 7 nonzero terms, so it is a codeword of weight 7. Hence, by Theorem 4.2.17, $d(C) \leq 7$. This proves that the minimum distance of C is 7, so C is a $[15, 5, 7]$-code.

Example 4.4.6 Suppose C is a narrow sense BCH code of length 31 and designed distance 5 over $\mathbb{F}_2$. Find its dimension.

Solution. With $n = 31$, $q = 2$, we have $m = 5$ and $2^5 - 1 = 31$. Let ζ be a primitive element in $\mathbb{F}_{2^5}$, so ζ is a 31st primitive root of unity. Let $p(x)$ be the minimal polynomial of ζ. Then ζ, ζ^2, ζ^4 have the same minimal polynomial $p(x)$. Let $q(x)$ be the minimal polynomial of ζ^3. We write $g(x) = p(x)q(x)$.

Then $g(x) = \text{lcm}\{m_i(x) \mid i = 1, 2, 3, 4\}$. Hence the cyclic code C with generator polynomial $g(x)$ in $\mathbb{F}_2[x]_{31}$ is of length 31 and designed distance 5. Now $p(x)$ and $q(x)$ are both monic polynomials of degree 5; hence $\deg g(x) = 10$. Therefore the dimension of the code is 21, so C is a $[31, 21]$-code.

We now proceed to discuss decoding with BCH codes. We explained in Section 4.2 that for a linear code in general we prepare a syndrome table for the purpose of decoding. When a vector y is received, we compute its syndrome $S(y) = yH^T$ and then look up the syndrome table to find the error vector e such that $S(y) = S(e)$. Then y is decoded as the transmitted vector $x = y - e$. For a BCH code, however, we have an algebraic method for finding the error vector e from the syndrome vector $S(y)$. We present below a brief description of this method.

Let C be a BCH code over $F = \mathbb{F}_q$ of length n and designed distance d. Let H be the $(d-1) \times n$ matrix over $\mathbb{F}_{q^m}$ given in Theorem 4.4.3. (As pointed out earlier, H is a quasi parity-check matrix for C.) We use this matrix to define the syndrome of a vector $a \in F^n$ as $S(a) = aH^T$. Writing $a = (a_0, a_1, \ldots, a_{n-1})$ and $a(x) = a_0 + a_1x + \cdots + a_{n-1}x^{n-1}$, we have

$$S(a) = \begin{bmatrix} a_0 & a_1 & \cdots & a_{n-1} \end{bmatrix} \begin{bmatrix} 1 & \zeta^c & \cdots & \zeta^{(n-1)c} \\ 1 & \zeta^{c+1} & \cdots & \zeta^{(n-1)(c+1)} \\ \vdots & \vdots & \ddots & \vdots \\ 1 & \zeta^{c+d-2} & \cdots & \zeta^{(n-1)(c+d-2)} \end{bmatrix}^T$$

$$= \begin{bmatrix} S_c & S_{c+1} & \cdots & S_{c+d-2} \end{bmatrix}$$

where

$$S_j = a_0 + a_1\zeta^j + \cdots + a_{n-1}\zeta^{(n-1)j} = a(\zeta^j)$$

for $j = c, c+1, \ldots, c+d-2$.

Now suppose a codeword $v \in C$ is transmitted and the vector received is $a = v + e$, where e is the error vector. Then $S(e) = S(a)$. Let $e = (e_0, e_1, \ldots, e_{n-1})$ and $e(x) = e_0 + e_1x + \cdots + e_{n-1}x^{n-1}$. Let $i_1, \ldots, i_r$ be the positions where an error has occurred. Then $e_i \neq 0$ if and only if $i \in I = \{i_1, \ldots, i_r\}$. Hence $e(x) = \sum_{i \in I} e_i x^i$. The code C can correct up to t errors, where $t = \lfloor (d-1)/2 \rfloor$. So we assume $r \leq t$—that is, $2r < d$. Since $S(e) = S(a)$, we have $e(\zeta^j) = S_j$ for all $j = c, c+1, \ldots, c+d-2$. Thus the $2r$ unknowns $i_1, \ldots, i_r$ and $e_{i_1}, \ldots, e_{i_r}$ satisfy the following system of $d-1$ linear equations in $e_{i_1}, \ldots, e_{i_r}$:

$$\sum_{i \in I} e_i \zeta^{ji} = S_j \qquad j = c, c+1, \ldots, c+d-2 \tag{1}$$

We first obtain a solution for the error positions $i_1, \ldots, i_r$. We define the *error locator polynomial* $f(x)$ as

$$f(x) = \prod_{i \in I}(x - \zeta^i) = f_0 + f_1x + \cdots + f_{r-1}x^{r-1} + x^r$$

Since $f(\zeta^i) = 0$ for each $i \in I$, we have

$$f_0 + f_1\zeta^i + \cdots + f_{r-1}\zeta^{i(r-1)} + \zeta^{ir} = 0$$

for each $i \in I$. On multiplying this equation by $e_i\zeta^{ji}$, we get

$$f_0 e_i \zeta^{ji} + f_1 e_i \zeta^{(j+1)i} + \cdots + f_{r-1} e_i \zeta^{(j+r-1)i} + e_i \zeta^{(j+r)i} = 0$$

for each $i \in I$. Summing these r equations for $i = i_1, \ldots, i_r$ and using the relations (1), we have

$$f_0 S_j + f_1 S_{j+1} + \cdots + f_{r-1} S_{j+r-1} + S_{j+r} = 0$$

for each $j = c, c + 1, \ldots, c + r - 1$. Thus the r unknowns $f_0, f_1, \ldots, f_{r-1}$ satisfy the following $r \times r$ system of linear equations:

$$
\begin{bmatrix}
S_c & S_{c+1} & \cdots & S_{c+r-1} \\
S_{c+1} & S_{c+2} & \cdots & S_{c+r} \\
\vdots & \vdots & \ddots & \vdots \\
S_{c+r-1} & S_{c+r} & \cdots & S_{c+2r-2}
\end{bmatrix}
\begin{bmatrix}
f_0 \\
f_1 \\
\vdots \\
f_{r-1}
\end{bmatrix}
=
\begin{bmatrix}
-S_{c+r} \\
-S_{c+r+1} \\
\vdots \\
-S_{c+2r-1}
\end{bmatrix}
\tag{2}
$$

Let S denote the coefficient matrix in the above linear system. It can be verified by direct computation that $S = VDV^T$, where

$$
V =
\begin{bmatrix}
1 & 1 & \cdots & 1 \\
\zeta^{i_1} & \zeta^{i_2} & \cdots & \zeta^{i_r} \\
\vdots & \vdots & \ddots & \vdots \\
\zeta^{i_1(r-1)} & \zeta^{i_2(r-1)} & \cdots & \zeta^{i_r(r-1)}
\end{bmatrix},
\quad
D =
\begin{bmatrix}
e_{i_1}\zeta^{i_1 c} & 0 & \cdots & 0 \\
0 & e_{i_2}\zeta^{i_2 c} & \cdots & 0 \\
\vdots & \vdots & \ddots & \vdots \\
0 & 0 & \cdots & e_{i_r}\zeta^{i_r c}
\end{bmatrix}
$$

Now V is a Vandermonde matrix. Since ζ is a primitive nth root of unity in $\mathbb{F}_{q^m}$ and $i_1, \ldots, i_r$ are distinct integers in $\{0, \ldots, n-1\}$, we have $\zeta^{i_1}, \ldots, \zeta^{i_r}$ are all distinct. Hence $\det V \neq 0$. Further, $e_{i_1}, \ldots, e_{i_r}$ are all nonzero and hence $\det D \neq 0$. Therefore $\det S \neq 0$, and linear system (2) has a unique solution.

In the foregoing discussion we assumed that the number of positions where an error has occurred is $r \leq t$. If the actual number of error positions is less than r, then for any choice of distinct positions $i_1, \ldots, i_r$, the coefficients $e_{i_1}, \ldots, e_{i_r}$ cannot be all nonzero. So $\det D = 0$. Hence r is the greatest positive integer $\leq t$ such that system (2) has a unique solution. Therefore we find the value of r by taking successively $r = t, t - 1, \ldots$ in system (2) until we have a value for which system (2) has a unique solution.

The unique solution of (2) thus obtained gives us the error locator polynomial $f(x) = f_0 + f_1 x + \cdots + f_{r-1}x^{r-1} + x^r$. We now find the roots of $f(x)$ by trying $x = \zeta^i$, $i = 0, 1, \ldots$. By the definition of $f(x)$, these roots are $\zeta^{i_1}, \ldots, \zeta^{i_r}$. Thus we obtain a unique solution for the unknowns $i_1, \ldots, i_r$. If the code C is binary, then $e_{i_1}, \ldots, e_{i_r}$ are all equal to 1, so the error polynomial is $e(x) = x^{i_1} + \cdots + x^{i_r}$. In the general nonbinary case, we find $e_{i_1}, \ldots, e_{i_r}$ by solving the system of equations (1). Having thus found the error vector e, we decode the received vector a as the codeword $v = a - e$.

Note that the matrix H is formally determined by the numbers c, n, d. To compute the syndrome we need not know the generator polynomial $g(x)$ of the BCH code. Further, by Theorem 4.4.1, if the code is over $\mathbb{F}_q$, then $S_q = (S_1)^q$. In particular, for a binary code, $S_2 = (S_1)^2$, $S_4 = (S_2)^2$, $S_6 = (S_3)^2$, and so on. We can compute the syndrome more easily by using the division algorithm. If $p(x)$ is the minimal polynomial of ζ, then $S_1 = a(\zeta)$ can be obtained by finding the remainder on dividing $a(x)$ by $p(x)$ and then putting $x = \zeta$ in it. In general, to find S_j, we divide $a(x^j)$ by $p(x)$ and find the remainder.

Further note that if only one error has occurred, say in the ith position, then $S(a)$ is equal to the ith column of H. Therefore, if $S(a)$ is nonzero and not equal to any column of H, then at least two errors have occurred.

The following example illustrates the decoding procedure.

Example 4.4.7 Suppose the $[15, 5, 7]$ BCH code C of Example 4.4.5 is used and the vector $a = 110001001101000 \in \mathbb{F}_2^{15}$ is received. Find the error vector and determine the corrected codeword. Then find the message word.

Solution. The polynomial $a(x)$ for the received vector $a = 110001001101000$ is

$$a(x) = 1 + x + x^5 + x^8 + x^9 + x^{11} \in \mathbb{F}_2[x]_{15}$$

Hence the components S_i of the syndrome vector $S(a)$ are given by

$$S_i = a(\zeta^i) = 1 + \zeta^i + \zeta^{5i} + \zeta^{8i} + \zeta^{9i} + \zeta^{11i}$$

for each $i = 1, \ldots, 6 = d - 1$. Using the relations given in the table for powers of ζ in Example 4.4.5, we obtain

$$
\begin{aligned}
S_1 &= 1 + \zeta + \zeta^5 + \zeta^8 + \zeta^9 + \zeta^{11} = \zeta^2 \\
S_2 &= (S_1)^2 = 1 + \zeta = \zeta^4 \\
S_3 &= 1 + \zeta^3 + \zeta^{15} + \zeta^{24} + \zeta^{27} + \zeta^{33} = 1 + \zeta^2 = \zeta^8 \\
S_4 &= (S_2)^2 = 1 + \zeta^2 = \zeta^8 \\
S_5 &= 1 + \zeta^5 + \zeta^{25} + \zeta^{40} + \zeta^{45} + \zeta^{55} = 1 \\
S_6 &= (S_3)^2 = \zeta
\end{aligned}
$$

The code C corrects up to $t = (d - 1)/2 = 3$ errors, so we assume that the number of errors that have occurred in the received vector a is $r \le 3$. Now $S(a)$ is not equal to any column of H. Therefore at least two errors have occurred, so $r = 2$ or 3. We first try $r = 3$. Let

$$f(x) = f_0 + f_1 x + f_2 x^2 + x^3$$

be the error locator polynomial. Then we have the following system of equations for the unknown coefficients f_0, f_1, f_2:

$$\begin{bmatrix} S_1 & S_2 & S_3 \\ S_2 & S_3 & S_4 \\ S_3 & S_4 & S_5 \end{bmatrix} \begin{bmatrix} f_0 \\ f_1 \\ f_2 \end{bmatrix} = - \begin{bmatrix} S_4 \\ S_5 \\ S_6 \end{bmatrix}$$

The system has a unique solution if and only if the coefficient matrix S is nonsingular. Now

$$\det S = \begin{vmatrix} \zeta^2 & \zeta^4 & \zeta^8 \\ \zeta^4 & \zeta^8 & \zeta^8 \\ \zeta^8 & \zeta^8 & 1 \end{vmatrix}$$

$$= \zeta^{14} \begin{vmatrix} 1 & 1 & 1 \\ \zeta^2 & \zeta^4 & 1 \\ \zeta^6 & \zeta^4 & \zeta^7 \end{vmatrix}$$

$$= \zeta^{18}(1 + \zeta^5 + \zeta^6 + \zeta^7)$$

$$= 0$$

Hence r cannot be equal to 3, so $r = 2$. Then the error locator polynomial is $f(x) = f_0 + f_1 x + x^2$, and the linear system for f_0, f_1 is

$$\begin{bmatrix} S_1 & S_2 \\ S_2 & S_3 \end{bmatrix} \begin{bmatrix} f_0 \\ f_1 \end{bmatrix} = - \begin{bmatrix} S_3 \\ S_4 \end{bmatrix}$$

Now

$$\det S = \begin{vmatrix} \zeta^2 & \zeta^4 \\ \zeta^4 & \zeta^8 \end{vmatrix} = \zeta^{10} + \zeta^8 = \zeta$$

Hence we have a unique solution given by

$$f_0 = \frac{S_3^2 - S_2 S_4}{\det S} = \frac{\zeta^{16} + \zeta^{12}}{\zeta} = \zeta^{12}$$

$$f_1 = \frac{S_2 S_3 - S_1 S_4}{\det S} = \frac{\zeta^{12} + \zeta^{10}}{\zeta} = \zeta^2$$

So the error locator polynomial is $f(x) = \zeta^{12} + \zeta^2 x + x^2$. By trial and error we find that one root is ζ^{13}, so the other root is $f_0/\zeta^{13} = \zeta^{14}$. Thus the error positions are 13 and 14, and the error polynomial is $e(x) = x^{13} + x^{14}$.

The corrected code polynomial is

$$v(x) = a(x) - e(x) = 1 + x + x^5 + x^8 + x^9 + x^{11} + x^{13} + x^{14}$$

so $v = 110001001101011$. To obtain the message word, recall that for a cyclic code with generator polynomial $g(x)$, a message polynomial $u(x)$ is encoded as code polynomial $v(x) = u(x)g(x)$. So, in the present case,

$$u(x) = \frac{v(x)}{g(x)} = \frac{x^{14} + x^{13} + x^{11} + x^9 + x^8 + x^5 + x + 1}{x^{10} + x^8 + x^5 + x^4 + x^2 + x + 1} = x^4 + x^3 + x^2 + 1$$

Hence the message word is 10111.

We mentioned that for a BCH code we need not know the generator polynomial in order to compute the syndrome of a received vector and decode it. The following example illustrates this.

Example 4.4.8 Let C be the binary $[31, 21]$ BCH code of Example 4.4.6, and suppose the minimal polynomial of ζ is $p(x) = x^5 + x^2 + 1$. Decode the received vector

$$a = 1011000100100111000110111001011$$

Solution. Using the relation $\zeta^5 + \zeta^2 + 1 = 0$, we compute the powers of ζ and obtain the following table:

$\zeta^5 = 1 + \zeta^2$	$\zeta^{14} = 1 + \zeta^2 + \zeta^3 + \zeta^4$	$\zeta^{23} = 1 + \zeta + \zeta^2 + \zeta^3$
$\zeta^6 = \zeta + \zeta^3$	$\zeta^{15} = 1 + \zeta + \zeta^2 + \zeta^3 + \zeta^4$	$\zeta^{24} = \zeta + \zeta^2 + \zeta^3 + \zeta^4$
$\zeta^7 = \zeta^2 + \zeta^4$	$\zeta^{16} = 1 + \zeta + \zeta^3 + \zeta^4$	$\zeta^{25} = 1 + \zeta^3 + \zeta^4$
$\zeta^8 = 1 + \zeta^2 + \zeta^3$	$\zeta^{17} = 1 + \zeta + \zeta^4$	$\zeta^{26} = 1 + \zeta + \zeta^2 + \zeta^4$
$\zeta^9 = \zeta + \zeta^3 + \zeta^4$	$\zeta^{18} = 1 + \zeta$	$\zeta^{27} = 1 + \zeta + \zeta^3$
$\zeta^{10} = 1 + \zeta^4$	$\zeta^{19} = \zeta + \zeta^2$	$\zeta^{28} = \zeta + \zeta^2 + \zeta^4$
$\zeta^{11} = 1 + \zeta + \zeta^2$	$\zeta^{20} = \zeta^2 + \zeta^3$	$\zeta^{29} = 1 + \zeta^3$
$\zeta^{12} = \zeta + \zeta^2 + \zeta^3$	$\zeta^{21} = \zeta^3 + \zeta^4$	$\zeta^{30} = \zeta + \zeta^4$
$\zeta^{13} = \zeta^2 + \zeta^3 + \zeta^4$	$\zeta^{22} = 1 + \zeta^2 + \zeta^4$	$\zeta^{31} = 1$

Written as a polynomial, the received vector is

$$a(x) = 1 + x^2 + x^3 + x^7 + x^{10} + x^{13} + x^{14} + x^{15} + x^{19} + x^{20}$$
$$+ x^{22} + x^{23} + x^{24} + x^{27} + x^{29} + x^{30}$$

Since the code C is of designed distance 5, the syndrome of a is

$$S(a) = \begin{bmatrix} S_1 & S_2 & S_3 & S_4 \end{bmatrix}$$

where $S_j = a(\zeta^j)$, $j = 1, 2, 3, 4$. Dividing $a(x)$ by $p(x)$, we get the remainder x^3; hence $S_1 = \zeta^3$. Similarly $S_3 = \zeta + \zeta^2 + \zeta^3$. Using the above table, we have

$$S_1 = a(\zeta) = \zeta^3$$
$$S_2 = S_1^2 = \zeta^6$$
$$S_3 = a(\zeta^3) = \zeta + \zeta^2 + \zeta^3 = \zeta^{12}$$
$$S_4 = S_2^2 = \zeta^{12}$$

Now $t = 2$; hence we assume the number of errors to be $r \leq 2$. But the syndrome vector is not equal to any column of H, so at least two errors have occurred. Thus $r = 2$. Let the error locator polynomial be $f(x) = f_0 +$

$f_1 x + x^2$. The coefficients f_0, f_1 satisfy the equations

$$S_1 f_0 + S_2 f_1 = S_3$$
$$S_2 f_0 + S_3 f_1 = S_4$$

Solving these equations, we get

$$f_0 = \frac{\zeta^{18} + \zeta^{24}}{\zeta^{15} + \zeta^{12}} = \zeta^4$$

$$f_1 = \frac{\zeta^{18} + \zeta^{15}}{\zeta^{15} + \zeta^{12}} = \zeta^3$$

Hence the error locator polynomial is $f(x) = \zeta^4 + \zeta^3 x + x^2$. Trying successively $x = 1, \zeta, \zeta^2, \ldots$, we find that ζ^{10} is a root. So the other root is $\zeta^4 / \zeta^{10} = \zeta^{25}$. Hence the error polynomial is $e(x) = x^{10} + x^{25}$. Therefore we decode $a(x)$ as

$$v(x) = a(x) - e(x)$$
$$= 1 + x^2 + x^3 + x^7 + x^{13} + x^{14} + x^{15} + x^{19} + x^{20}$$
$$+ x^{22} + x^{23} + x^{24} + x^{25} + x^{27} + x^{29} + x^{30}$$

The corrected codeword is $v = 1011000100000111000110111101011$.

EXERCISES 4.4

1. Construct a binary BCH code of length 15 and designed distance 5. Show that its minimum distance is 5.

2. Construct a binary BCH code of length 31 and designed distance 5.

3. Construct a binary BCH code of length 31 and designed distance 7.

4. Construct a binary BCH code of length 31 and designed distance 11.

5. Construct a binary BCH code of length 21 and designed distance 5.

6. Construct a binary BCH code of length 63 and designed distance 7.

7. Find the dimension of the 3-error-correcting binary BCH code of length 127.

8. Find the dimension of the 3-error-correcting binary BCH code of length 255.

9. Construct a ternary BCH code of length 8 and designed distance 5.

10. Find the dimension of the 5-error-correcting ternary BCH code of length 80.

11. If C is a binary BCH code of length n and designed distance d, where d divides n, show that $d(C) = d$.

12. Suppose the [15, 5] BCH code C of Example 4.4.5 is used and the vector $a = 110101110110011 \in \mathbb{F}_2^{15}$ is received. Find the error vector and determine the corrected codeword. Then find the message word.

13. Suppose the BCH code of length 31 and designed distance 5, with $\zeta^5 + \zeta^2 + 1 = 0$, is used and the vector $1000010010101010011101101110101$ is received. Decode it.

CHAPTER 5

SYMMETRY GROUPS AND COLOR PATTERNS

5.1 PERMUTATION GROUPS

Let X be any nonempty set. Then the set of all permutations of X is a group under composition of mappings. S_X is called the *symmetric group* on the set X. The identity e of the group S_X is the identity mapping $e : X \to X$ given by $e(x) = x$ for all $x \in X$. The composite of permutations $\alpha, \beta \in S_X$ is called their *product* and written $\alpha\beta$. Any subgroup of the group S_X is called a *permutation group*.

If the set X has more than two elements, the group S_X is nonabelian. Let a, b, c be three distinct elements in X, and let $\alpha, \beta \in S_X$ be defined as:

$$\alpha(a) = b, \quad \alpha(b) = a, \quad \alpha(x) = x \text{ for all other } x \in X$$
$$\beta(a) = c, \quad \beta(c) = a, \quad \beta(x) = x \text{ for all other } x \in X$$

Then $\alpha\beta(a) = c$, but $\beta\alpha(a) = b$. Hence $\alpha\beta \neq \beta\alpha$.

If X is an infinite set, then clearly S_X is an infinite group. If X has n elements, then S_X is of order $n!$. Let $X = \{x_1, \ldots, x_n\}$ and consider the number of ways in which an arbitrary permutation σ of X can be constructed. We can take $\sigma(x_1)$ to be any one of the n elements $x_1, \ldots, x_n$. Having chosen $\sigma(x_1)$, we can take $\sigma(x_2)$ to be any one of the remaining $n - 1$ elements in the set $\{x_1, \ldots, x_n\}$. Continuing in this manner, we can construct σ in $n(n - 1) \cdots 1 = n!$ ways. This proves that there are exactly $n!$ permutations of X.

If $X = \{1, \ldots, n\}$, the group S_X is written S_n and called the *symmetric group of degree n*. It is easily seen that for any set X of n elements, the group S_X is isomorphic with S_n. Therefore the symbol S_n may be used to denote the abstract group of permutations of a set of n elements.

A permutation σ of a finite set $X = \{x_1, \ldots, x_n\}$ can be exhibited as

$$\sigma = \begin{pmatrix} x_1 & \cdots & x_n \\ y_1 & \cdots & y_n \end{pmatrix}$$

where $y_i = \sigma(x_i), i = 1, \ldots, n$. This notation for a permutation is referred to as *two-row* notation. Presently we will introduce a simpler one-row notation.

Given a set X, let $a \in X$ and $\sigma \in S_X$. If $\sigma(a) \neq a$, we say σ *moves* a. If $\sigma(a) = a$, we say σ *fixes* a. Because σ is injective, $\sigma(a) \neq a$ implies $\sigma(\sigma(a)) \neq \sigma(a)$. Therefore, if σ moves a, then σ also moves $\sigma(a)$.

Definition 5.1.1 Two permutations α and β of a set X are said to be *disjoint* if there is no element in X that is moved by both α and β.

The most important property of disjoint permutations is that their product is commutative, as shown by the following theorem.

THEOREM 5.1.2 Let $\alpha, \beta \in S_X$ be disjoint permutations. Then $\alpha\beta = \beta\alpha$.

Proof: Let $x \in X$. Because α and β cannot both move x, we have the following cases:

(a) *Neither α nor β moves x.* Then $\alpha\beta(x) = x = \beta\alpha(x)$.
(b) *Only α moves x.* Then α also moves $\alpha(x)$, so β fixes both x and $\alpha(x)$. Hence $\alpha\beta(x) = \alpha(x) = \beta\alpha(x)$.
(c) *Only β moves x.* Then $\alpha\beta(x) = \beta(x) = \beta\alpha(x)$.

Thus we see that $\alpha\beta(x) = \beta\alpha(x)$ for every $x \in X$. This proves that $\alpha\beta = \beta\alpha$. ∎

Every list of distinct elements in X determines a permutation of a special type called a *cycle*.

Definition 5.1.3 Let $\sigma \in S_X$. If there exist distinct elements $a_1, \ldots, a_r$ in X such that

$$\sigma(a_i) = a_{i+1} \qquad \text{for each } i = 1, \ldots, r-1$$
$$\sigma(a_r) = a_1$$
$$\sigma(x) = x \qquad \text{for all other elements in } X$$

then σ is called a *cycle of length r* (or an *r-cycle*) and written $\sigma = (a_1\, a_2 \ldots a_r)$.
A cycle of length 2 is called a *transposition*.

For example, let

$$\sigma = \begin{pmatrix} 1 & 2 & 3 & 4 & 5 \\ 4 & 2 & 1 & 3 & 5 \end{pmatrix}$$

Then $\sigma = (1\,4\,3)$, so σ is a cycle of length 3.

It should be noted that in this one-row notation for a cycle, there in no indication of the underlying set X, which could be any set containing the elements in the cycle. So one has to make out X from context. Moreover, a cycle of length $r > 1$, say $\sigma = (a_1 a_2 \ldots a_r)$, can be written in r different ways—namely,

$$\sigma = (a_i \ldots a_r a_1 \ldots a_{i-1}), \qquad i = 1, \ldots, r$$

For example, $(1\,4\,3)$, $(4\,3\,1)$, and $(3\,1\,4)$ all represent the same cycle.

It is clear that a cycle of length 1 does not move any element. So for each $a \in X$, the cycle (a) is the identity permutation of X. Because a cycle $(a_1 \ldots a_r)$ of length greater than 1 moves only the elements $a_1, \ldots, a_r$, it follows that two cycles $(a_1 \ldots a_r)$ and $(b_1 \ldots b_s)$, both of lengths greater than 1, are disjoint permutations if and only if the sets $\{a_1, \ldots, a_r\}$ and $\{b_1, \ldots, b_s\}$ are disjoint.

Let $\sigma \in S_X$ and $a \in X$. The *orbit* of a under σ is defined to be the set

$$Orb_\sigma(a) = \{\sigma^i(a) \mid i \in \mathbb{Z}\}$$

The following theorem shows that any permutation can be expressed uniquely as a product of disjoint cycles. Given a permutation $\sigma \in S_X$, let us define the *rank* of σ to be the number of elements in X moved by σ. The identity permutation e is obviously of rank zero. If σ moves an element a, then σ also moves $\sigma(a)$. Hence there is no permutation of rank 1.

THEOREM 5.1.4 Let X be a finite set. Let $\sigma \in S_X$, $\sigma \neq e$. Then σ can be expressed as a product of pairwise disjoint cycles of lengths greater than 1. This decomposition is unique, except for the order in which the cycles are written.

Proof: We prove the theorem by induction on the rank k of σ. The statement holds vacuously for $k = 1$. Suppose σ is of rank k greater than 1, and assume that the theorem holds for all permutations of rank less than k. Let a be an element moved by σ. Since X is finite, there exists a smallest positive integer m such that $\sigma^m(a) = a$. Since $\sigma(a) \neq a$, we have $m > 1$. Therefore the orbit of a is given by

$$A = Orb_\sigma(a) = \{a, \sigma(a), \ldots, \sigma^{m-1}(a)\}$$

Let $\gamma_1 \in S_X$ be the cycle

$$\gamma_1 = (a\ \sigma(a) \ldots \sigma^{m-1}(a))$$

Then for each $x \in A$, we have $\sigma(x) = \gamma_1(x)$. Let $\sigma_1 = \gamma_1^{-1}\sigma$, so $\sigma = \gamma_1\sigma_1$. Now for each $x \in A$, $\sigma_1(x) = \gamma_1^{-1}(\sigma(x)) = \gamma_1^{-1}(\gamma_1(x)) = x$. Thus σ_1, γ_1 are disjoint permutations. Hence, for each $x \notin A$, $\sigma(x) = \gamma_1\sigma_1(x) = \sigma_1\gamma_1(x) = \sigma_1(x)$. Thus

$$\sigma_1(x) = \begin{cases} x & \text{if } x \in A \\ \sigma(x) & \text{if } x \notin A \end{cases}$$

It follows that the number of elements moved by σ_1 is $k - m$. If $k - m = 0$, then $\sigma_1 = e$ and hence $\sigma = \gamma_1$. Otherwise, by the induction hypothsis, σ_1 is a product of pairwise disjoint cycles of lengths greater than 1, say $\sigma_1 = \gamma_2 \cdots \gamma_r$, so $\sigma = \gamma_1 \gamma_2 \cdots \gamma_r$. Since γ_1, σ_1 are disjoint permutations, it follows that $\gamma_1, \gamma_2, \ldots, \gamma_r$ are all pairwise disjoint cycles. This proves the first part of the theorem.

To prove the uniqueness of the decomposition, suppose σ can be decomposed in two ways, say

$$\sigma = \gamma_1 \cdots \gamma_r \tag{1}$$

$$\sigma = \delta_1 \cdots \delta_s \tag{2}$$

Let a be an element that occurs in some cycle γ_i in decomposition (1). Then a is moved by σ, and hence a must occur in some cycle δ_j in decomposition (2). So γ_i and δ_j are both cycles generated by a under σ; hence $\gamma_i = \delta_j = (a\,\sigma(a)\ldots\sigma^{m-1}(a))$, where m is the least positive integer such that $\sigma^m(a) = a$. Thus, every cycle in (1) is equal to some cycle in (2). By the same argument, every cycle in (2) is equal to some cycle in (1). Hence the two decompositions are identical, except for the order in which the cycles are written. ∎

The decomposition of σ as a product of pairwise disjoint cycles is called the *cycle decomposition* of σ. Since the product of disjoint cycles is commutative, the order in which the cycles are written is immaterial. Moreover, since a cycle of length 1 is the identity permutation e, we can include in the decomposition the cycle (a) for every a in X that is fixed by σ. When this is done, every element in X occurs in some cycle.

To find the cycle decomposition of a given permutation $\sigma \in S_X$, we start with any element a in X, find its orbit, and write the cycle $(a\,\sigma(a)\ldots\sigma^{m-1}(a))$ determined by the orbit. Then we pick any element b not in this cycle, and write the cycle determined by the orbit of b. We continue in this fashion until each element in X has appeared in some cycle.

Example 5.1.5 Find the cycle decomposition of the permutation

$$\sigma = \begin{pmatrix} 1 & 2 & 3 & 4 & 5 & 6 & 7 & 8 & 9 & 10 & 11 & 12 & 13 & 14 & 15 & 16 \\ 4 & 3 & 8 & 7 & 1 & 6 & 5 & 2 & 16 & 9 & 10 & 15 & 11 & 14 & 12 & 13 \end{pmatrix}$$

Solution. Let us start by finding the orbit of 1 under σ. σ takes 1 to 4, 4 to 7, 7 to 5, 5 to 1. Hence the cycle determined by 1 is (1 4 7 5). Similarly, the cycle determined by the orbit of 2 is (2 3 8). Continuing in this manner, we obtain the decomposition

$$\sigma = (1\,4\,7\,5)(2\,3\,8)(6)(9\,16\,13\,11\,10)(12\,15)(14)$$

If we omit cycles of length 1, we have

$$\sigma = (1\,4\,7\,5)(2\,3\,8)(9\,16\,13\,11\,10)(12\,15)$$

Let $\sigma \in S_X$, where X is a set of n elements. Suppose

$$\sigma = \gamma_1 \cdots \gamma_r$$

is a cycle decomposition of σ that includes all cycles of length 1. Further suppose that the cycles $\gamma_1, \ldots, \gamma_r$ are written in ascending order of length, and let $n_1, \ldots, n_r$ be their lengths, respectively. Then $n_1 \leq \cdots \leq n_r$. Further, each element in X occurs in one and only one of the cycles. Hence $n_1 + \cdots + n_r = n$. The list $(n_1, \ldots, n_r)$ is called the *cycle structure* of σ. For instance, the cycle structure of the permutation σ in Example 5.1.5 is $(1, 1, 2, 3, 4, 5)$.

Given a positive integer n, a list $n_1, \ldots, n_r$ of positive integers is called a *partition* of n if $n_1 \leq \cdots \leq n_r$ and $n_1 + \cdots + n_r = n$. For example, there are five partitions of the integer 4: $(1, 1, 1, 1)$, $(1, 1, 2)$, $(1, 3)$, $(2, 2)$, and (4).

Thus we see that the cycle structure of any permutation of a set X of n elements is a partition of n. Conversely, given any partition $n_1, \ldots, n_r$ of n, it is clear that we can write a permutation of X with cycle structure $n_1, \ldots, n_r$. Thus there is a one-to-one correspondence between the set of all partitions of n and the set of all possible cycle structures of permutations of a set of n elements.

The permutations in S_X can be classified by their cycle structure. The following table gives the cycle structure classification for S_4. It shows a typical permutation (expressed as a cycle decomposition) for each cycle structure and the number of permutations that have this cycle structure.

Cycle Structure	Typical Permutation	Number of Permutations
$(1, 1, 1, 1)$	$(1) = e$	1
$(1, 1, 2)$	$(1\,2)$	6
$(1, 3)$	$(1\,2\,3)$	8
$(2, 2)$	$(1\,2)(3\,4)$	3
(4)	$(1\,2\,3\,4)$	6

Such a classification makes it easier to write all the permutations. For example, the group S_6 is of order 720. It would be tedious to write directly all the 720 permutations, but by classifying them according to cycle structure, the task becomes easy. In certain applications, in fact, one need know only the number of permutations of a given cycle structure.

The following theorem gives an important group-theoretic property of permutations of X that have the same cycle structure.

Definition 5.1.6 Let G be a group, and let $a, b \in G$. If there exists some element x in G such that $b = xax^{-1}$, then a is said to be *conjugate* to b, written $a \sim b$.

It is easily seen that $\sim$ is an equivalence relation in G. (See Exercise 4 in Section 5.1.) The equivalence classes under $\sim$ are called *conjugate classes* in G.

THEOREM 5.1.7 Let α, β be permutations of a finite set X. Then α, β have the same cycle structure if and only if α, β are conjugate elements in the group S_X.

Proof: Suppose α, β have the same cycle structure $(n_1, \ldots, n_r)$. Then they have cycle decompositions of the form

$$\alpha = (a_1 \ldots a_{n_1})(a_{n_1+1} \ldots a_{n_1+n_2}) \cdots (a_{n_1+\cdots+n_{r-1}+1} \ldots a_n)$$
$$\beta = (b_1 \ldots b_{n_1})(b_{n_1+1} \ldots b_{n_1+n_2}) \cdots (b_{n_1+\cdots+n_{r-1}+1} \ldots b_n)$$

where $n = n_1 + \cdots + n_r$ is the number of elements in X. Let $\sigma \in S_X$ be the permutation given by $\sigma(a_i) = b_i$ for all $i = 1, \ldots, n$. Then $\sigma\alpha\sigma^{-1}(b_i) = \sigma\alpha(a_i) = \beta(b_i)$ for all $i = 1, \ldots, n$. Hence $\sigma\alpha\sigma^{-1} = \beta$, so α, β are conjugate elements in the group S_X.

Conversely, suppose α, β are conjugate elements in S_X. Then $\beta = \sigma\alpha\sigma^{-1}$ for some $\sigma \in S_X$. Given any $x \in X$, if $\alpha(x) = y$, then $\beta(\sigma(x)) = \sigma\alpha\sigma^{-1}\sigma(x) = \sigma(y)$. It follows that if $(x_1 \ldots x_k)$ is a cycle in the decomposition of α, then $(\sigma(x_1) \ldots \sigma(x_k))$ is a cycle in the decomposition of β. Therefore the cycle decomposition of β can be obtained by replacing each x in the decomposition of α by $\sigma(x)$. Hence α, β have the same cycle structure. ∎

It follows from the theorem that there is a one-to-one correspondence between the conjugate classes in the group S_n and the partitions of n.

5.1.1 Even and Odd Permutations

The permutations of a finite set can be classified into two categories, called *even* and *odd*. We have seen that every permutation can be decomposed as a product of disjoint cycles of length greater than 1. Further, every cycle of length $r > 1$ can be expressed as a product of $r - 1$ transpositions as follows:

$$(a_1 \ldots a_r) = (a_r\, a_{r-1}) \cdots (a_r\, a_2)(a_r\, a_1)$$

Hence it follows that every permutation on a finite set can be expressed as a product of transpositions. However, this factorization is not unique. For example, the cycle $(a\,b\,c\,d)$ can be factored as $(a\,d)(a\,c)(a\,b)$ and also as $(c\,d)(b\,d)(b\,c)(a\,d)(b\,c)$.

But it can be proved that if a given permutation σ is a product of r transpositions and also a product of s transpositions, then the integers r and s are either both even or both odd. (The proof may be found in any standard book on abstract algebra.) This justifies the following definition.

Definition 5.1.8 A permutation σ of a finite set is called *even* or *odd* according to whether σ is a product of an even or an odd number of transpositions.

The classification of permutations into even and odd kinds is called *parity*. Two permutations are said to have the same parity if they are both even or both odd.

The set of even permutations in S_n is written A_n. It can be easily shown that A_n is a subgroup of S_n; A_n is called the *alternating group of degree n*.

EXERCISES 5.1

1. Find the number of distinct cycles of length r in S_n.

2. Write the cycle structure classification table for S_5.

3. Write the cycle structure classification table for S_6.

4. Prove that the relation $\sim$ (of conjugate elements) in a group is an equivalence relation.

5. Show that the permutations

$$\alpha = \begin{pmatrix} 1 & 2 & 3 & 4 & 5 & 6 & 7 \\ 7 & 5 & 1 & 6 & 4 & 2 & 3 \end{pmatrix}$$

$$\beta = \begin{pmatrix} 1 & 2 & 3 & 4 & 5 & 6 & 7 \\ 4 & 1 & 6 & 2 & 3 & 7 & 5 \end{pmatrix}$$

   have the same cycle structure. Find σ such that $\beta = \sigma\alpha\sigma^{-1}$.

6. Let k_n denote the number of permutations in S_n that do not fix any element. Prove that

$$k_n = n!\left(1 - \frac{1}{1!} + \frac{1}{2!} - \frac{1}{3!} + \cdots + \frac{(-1)^{n-1}}{n!}\right)$$

   Hence show that k_n is equal to the integer closest to $\dfrac{n!}{e}$.

7. Let X be an infinite set. Let N denote the set of those permutations of X that move a finite number of elements in X. Prove that N is a normal subgroup of S_X.

8. Show that for every $n > 1$, A_n is a normal subgroup of S_n and $[S_n : A_n] = 2$.

9. Show that for every $n > 2$, the center of the group S_n is trivial.

5.2 GROUPS OF SYMMETRIES

In this section we consider the application of group theory to study the symmetries of a geometrical figure (in a plane or three-dimensional space).

Let $d(x, y)$ denote the distance between the points x and y.

Definition 5.2.1 Let X be a set of points. A bijective mapping $\sigma : X \to X$ is called a *symmetry* of X if

$$d(\sigma(x), \sigma(y)) = d(x, y) \qquad \text{for all } x, y \in X$$

In other words, a symmetry of a set of points X is a permutation of X that preserves the distance between every two points in X.

We write $Sym(X)$ to denote the set of all symmetries of X. Obviously, $Sym(X)$ is a subset of S_X.

THEOREM 5.2.2 Let X be a set of points. Then $Sym(X)$ is a subgroup of the symmetric group S_X.

Proof: The identity permutation $e \in S_X$ is obviously a symmetry; hence $e \in Sym(X)$. Let $\alpha, \beta \in Sym(X)$. Then, for all $x, y \in X$,

$$d(\alpha^{-1}\beta(x), \alpha^{-1}\beta(y)) = d(\alpha(\alpha^{-1}\beta(x)), \alpha(\alpha^{-1}\beta(y)))$$
$$= d(\beta(x), \beta(y))$$
$$= d(x, y)$$

Hence $\alpha^{-1}\beta \in Sym(X)$. This proves that $Sym(X)$ is a subgroup of S_X. ∎

The group $Sym(X)$ is called the *group of symmetries* (or *symmetry group*) of X. Note that the symmetric group S_X is defined for any set X, but $Sym(X)$ is defined for only a set X of points in which the distance between every two points is given.

Consider the symmetries of a polygon P. (By P we mean the set of points constituting the polygon.) Let V be the set of vertices of the polygon. It is clear from geometrical consideration that any symmetry σ of the polygon must map a vertex to a vertex. Thus σ determines a symmetry $\bar{\sigma}$ of the set V. Conversely, given any symmetry $\bar{\sigma}$ of V, it determines uniquely a symmetry σ of the polygon that coincides with $\bar{\sigma}$ on the vertices. Hence we can identify the symmetries of the polygon with the symmetries of the set of its vertices. In other words, speaking more formally, the group of symmetries of P is isomorphic with the group of symmetries of V.

Let us now consider a regular polygon of n sides ($n \geq 3$). Let us label the vertices in counterclockwise order as $1, 2, \ldots, n$. Consider any symmetry σ of the set of vertices. Suppose σ maps vertex 1 to vertex i. Then σ must take vertex 2 to a vertex adjacent to i—that is, either $i + 1$ or $i - 1$. Once $\sigma(1)$ and $\sigma(2)$ are fixed, the mapping σ is completely determined by the fact that it preserves the distance between every two points. So if σ maps 2 to $i + 1$, then it must map $3, 4, \ldots$ to $i + 2, i + 3, \ldots$, respectively. On the other hand, if σ maps 2 to $i - 1$, then it must map $3, 4, \ldots$ to $i - 2, i - 3, \ldots$, respectively. Thus there are exactly two symmetries, σ_i and τ_i, that take vertex 1 to i. These are given by

$$\sigma_i = \begin{pmatrix} 1 & 2 & 3 & \cdots & n \\ i & i+1 & i+2 & \cdots & i-1 \end{pmatrix}$$

$$\tau_i = \begin{pmatrix} 1 & 2 & 3 & \cdots & n \\ i & i-1 & i-2 & \cdots & i+1 \end{pmatrix}$$

Thus we see that a regular polygon of n sides has in all $2n$ symmetries, σ_i, τ_i, $i = 1, \ldots, n$.

Note that the vertex after n is 1, so $i + 1 = 1$ when $i = n$. Therefore, in the above representation for σ_i, τ_i, $+$ is to be understood as addition modulo n.

The mapping σ_i preserves the cyclic order of the vertices, but τ_i reverses the cyclic order. Geometrically, σ_i represents a rotation of the polygon about its center through an angle $2\pi(i - 1)/n$, and τ_i represents a reflection in the diameter lying midway between vertices 1 and i. (By a rotation, we mean a rotation of the polygon in its own plane. A reflection in a diameter is equivalent to a rotation about the diameter through an angle π, but this rotation takes place in the third dimension and not in the plane of the polygon.) It is obvious that σ_1 is the identity permutation, and τ_1 represents reflection in the diameter through vertex 1. The identity permutation is equivalent to a rotation through an angle 2π.

The $2n$ symmetries σ_i, τ_i, $i = 1, \ldots, n$, can be expressed in terms of two basic symmetries. We write $\alpha = \sigma_2$, $\beta = \tau_1$, so

$$\alpha = \begin{pmatrix} 1 & 2 & \cdots & n \\ 2 & 3 & \cdots & 1 \end{pmatrix}, \quad \beta = \begin{pmatrix} 1 & 2 & \cdots & n \\ 1 & n & \cdots & 2 \end{pmatrix}$$

Geometrically, α represents a rotation through angle $2\pi/n$ and moves each vertex i to $i + 1$. For any integer $m = 1, \ldots, n$, α^{m-1} represents a rotation through angle $2\pi(m - 1)/n$; hence $\alpha^{m-1} = \sigma_m$. Further $\alpha^{m-1}\beta(1) = \alpha^{m-1}(1) = m$ and $\alpha^{m-1}\beta(2) = \alpha^{m-1}(n) = m - 1$. Since a symmetry is determined uniquely by its effect on vertices 1 and 2, it follows that $\alpha^{m-1}\beta = \tau_m$. Thus the $2n$ symmetries are given by α^{m-1}, $\alpha^{m-1}\beta$, $m = 1, \ldots, n$.

It is clear that $\alpha^m = e$ and $\beta^2 = e$. Further, consider $\beta\alpha$: $\beta\alpha(1) = \beta(2) = n$ and $\beta\alpha(2) = \beta(3) = n - 1$. Hence $\beta\alpha = \tau_n = \alpha^{n-1}\beta$. Thus we have proved the following result.

THEOREM 5.2.3 The group G of symmetries of a regular polygon of n sides is given by

$$G = \{e, \alpha, \ldots, \alpha^{n-1}, \beta, \alpha\beta, \ldots, \alpha^{n-1}\beta\}$$

where α represents a rotation through an angle $2\pi/n$, and β represents reflection in a diameter through a vertex. Moreover, the following relations hold in the group G:

$$\alpha^n = e, \quad \beta^2 = e, \quad \beta\alpha = \alpha^{n-1}\beta$$

Any group of $2n$ elements that has the same structure as the group G above is called a *dihedral group* of degree n and denoted by D_n. That is, we have the following definition.

Definition 5.2.4 A *dihedral group* of degree n, written D_n, is a group of order $2n$ given by

$$D_n = \{e, a, \ldots, a^{n-1}, b, ab, \ldots, a^{n-1}b\}$$

with the following defining relations:

$$a^n = e, \quad b^2 = e, \quad ba = a^{n-1}b$$

We have thus shown that the group of symmetries of a regular polygon of n sides ($n \geq 3$) is a dihedral group of degree n. We shall shortly explain the geometrical interpretation of the dihedral groups D_1 and D_2 as groups of symmetries. For the present, let us observe that $D_1 = \{e, b\}$, with $b^2 = e$, is a cyclic group of order 2. Further, $D_2 = \{e, a, b, ab\}$ has the defining relations $a^2 = e, b^2 = e$, and $ba = ab$. Hence D_2 is identical with a Klein's 4-group.

If we interpret the elements of the dihedral group D_n, $n > 2$, as permutations of the set $\{1, \ldots, n\}$ of vertices of a regular polygon, then D_n is a subgroup of the symmetric group S_n. In particular, D_3 has six elements and hence $D_3 = S_3$. For $n > 3$, D_n is a proper subgroup of S_n.

An equilateral triangle is a regular polygon of three sides. Hence its group of symmetries is D_3. In this case, we can also arrive at this result directly as follows: The distance between every pair of vertices of an equilateral triangle is the same, and hence every permutation of the vertices is a symmetry. Therefore the group of symmetries of an equilateral triangle is S_3, which, as noted above, is the same as D_3.

Consider now an isosceles (but not equilateral) triangle. It has only one symmetry β in addition to the identity permutation—namely, the one given by reflection in the median bisecting the angle between the two equal sides. So the group G of symmetries of an isosceles triangle is given by $G = \{e, \beta\}$, with $\beta^2 = e$. As noted above, G is a dihedral group of degree 1.

Consider next the symmetries of a rectangle (other than a square). It is easily seen that there are only three symmetries α, β, γ in addition to e. Geometrically, these represent a rotation through an angle π and reflections in the lines through the center and parallel to the sides of the rectangle. Labeling the vertices as 1, 2, 3, 4 in order, we can write these symmetries as permutations of the vertices as follows:

$$\alpha = (1\,3)(2\,4), \quad \beta = (1\,2)(3\,4), \quad \gamma = (1\,4)(2\,3)$$

It is easily verified that $\alpha\beta = \gamma = \beta\alpha$. Hence the group of symmetries of a rectangle is given by $G = \{e, \alpha, \beta, \alpha\beta\}$, with the defining relations $\alpha^2 = e, \beta^2 = e, \beta\alpha = \alpha\beta$. So G is a dihedral group of degree 2.

We can summarize the results proved above as follows: The group of symmetries of an isosceles triangle is D_1. The group of symmetries of a rectangle is D_2. For any $n > 2$, the group of symmetries of a regular polygon of n sides is D_n.

The dihedral group D_n has a subgroup $C_n = \{e, a, \ldots, a^{n-1}\}$ that, geometrically, consists of all rotational symmetries of the polygon.

The symmetry group of a geometric figure may be infinite. For example, a circle has infinitely many symmetries. It can be shown that if the symmetry group G of a plane figure is finite, then G is either D_n or C_n for some n. For example, the symmetry group of this figure is C_4.

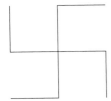

Let us now consider the symmetries of three-dimensional geometric objects. Consider first a regular tetrahedron. Let the vertices be labeled as 1, 2, 3, 4. As in the case of an equilateral triangle, the distance between every two vertices of a regular tetrahedron is the same. Hence every permutation of the vertices is a symmetry. Therefore the symmetry group of a regular tetrahedron is S_4. How many of the 24 permutations in S_4 are rotations? It is clear that by a suitable rotation we can take vertex 1 to any vertex $i = 1, 2, 3, 4$. Having done that, we can rotate the tetrahedron about an axis through the new position of vertex 1 through angles 0, $2\pi/3$, and $4\pi/3$ to obtain three symmetries. Thus there are in all $4 \times 3 = 12$ rotational symmetries. They form a subgroup of the group of all symmetries of the tetradedron.

The following table gives the 12 rotational symmetries of a regular tetrahedron as permutations of the vertices and their geometric description as rotations. The edge $i - j$ denotes the edge joining vertices i and j.

Permutation	Axis and Angle of Rotation
$(1) = e$	any axis, rotation through angle 2π
$(2\,3\,4)$, $(2\,4\,3)$	axis through vertex 1, angles $2\pi/3$ and $4\pi/3$
$(1\,3\,4)$, $(1\,4\,3)$	axis through vertex 2, angles $2\pi/3$ and $4\pi/3$
$(1\,2\,4)$, $(1\,4\,2)$	axis through vertex 3, angles $2\pi/3$ and $4\pi/3$
$(1\,2\,3)$, $(1\,3\,2)$	axis through vertex 4, angles $2\pi/3$ and $4\pi/3$
$(1\,2)(3\,4)$	axis through middle points of edges 1-2 and 3-4, angle π
$(1\,3)(2\,4)$	axis through middle points of edges 1-3 and 2-4, angle π
$(1\,4)(2\,3)$	axis through middle points of edges 1-4 and 2-3, angle π

Let us now consider the symmetries of a cube. Let the vertices of the cube be labeled $1, 2, \ldots, 8$ such that vertices 2, 3, and 4 are adjacent to vertex 1. Let $\sigma \in S_8$ be a symmetry. Suppose σ takes 1 to i. Then σ must take 2, 3, and 4 to the three vertices adjacent to i, which can be done in 6 ways. Hence there are $8 \times 6 = 48$ symmetries in all. Of these, 24 are rotational symmetries. The vertex 1 can be taken to any vertex i by a rotation, and then we can rotate the cube around the diameter through the new position of vertex 1 to obtain three symmetries.

We can arrive at the same result by considering the symmetries of the cube as permutations of its six faces. By a rotation, face 1 can be taken to face i ($i = 1, \ldots 6$). Having done that, we can rotate the cube around the diameter perpendicular to the new position of the face 1 through angles $\pi/2, \pi$, and $3\pi/2$ and obtain three symmetries. Hence there are $6 \times 4 = 24$ rotational symmetries.

The following table gives a geometric description of the various types of rotational symmetries of a cube and their numbers.

Axis and Angle of Rotation	Number
any axis, angle 2π	1
axis through opposite vertices, angles $2\pi/3$ and $4\pi/3$	$4 \times 2 = 8$
axis through centers of opposite faces, angles $\pi/2, \pi$, and $\frac{3\pi}{2}$	$3 \times 3 = 9$
axis through middle points of opposite edges, angle π	6

EXERCISES 5.2

1. Find the symmetry groups of the following figures.

2. Find the symmetry groups of the letters of the alphabet: A, B, ..., Z.

3. Find the symmetry groups of the conic sections ellipse, parabola, and hyperbola.

4. Find the symmetry groups of the following curves:
   (a) $y^2 = x(1 - x^2)$
   (b) $y^2 = x^2(1 - x^2)$
   (c) $r = 1 + \cos\theta$
   (d) $r = \sin 2\theta$
   (e) $r = \sin 3\theta$

5. Write the rotational symmetries of a cube as permutations of the vertices.

6. Write the rotational symmetries of a cube as permutations of the edges.

7. Show that the rotational symmetry group of a regular tetrahedron is isomorphic with A_4.

8. Show that the rotational symmetry group of a cube is isomorphic with S_4.

9. Show that the center of the group D_n is of order 1 or 2 according to whether n is odd or even.

5.3 COLORINGS AND COLOR PATTERNS

In this chapter we consider problems of the following type: Suppose we color each vertex of an equilateral triangle white or black. Then there are $2 \times 2 \times 2 = 8$ ways in which the three vertices can be colored. Let us refer to them as *color assignments* or *colorings*. We say that two color assignments are *equivalent* (or have the same *pattern*) if one of them can be obtained from the other by rotating the triangle through an appropriate angle or flipping it over. The second operation—namely, flipping over—is equivalent to reflection in some mirror line. We then find that the eight color assignments fall into four distinct patterns, as shown here.

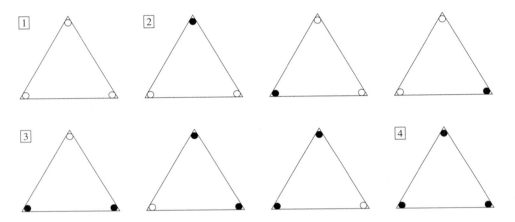

If, instead of an equilateral triangle, we consider an isosceles triangle, then we find that the eight color assignments fall into six distinct patterns.

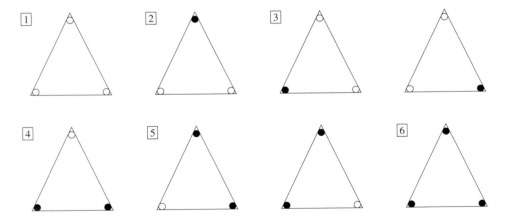

Finally, if we consider a triangle whose sides are all of unequal lengths, then no two colorings are equivalent, and hence all eight colorings are distinct patterns.

Now suppose we make the criterion for equivalent colorings stricter by stipulating that two colorings are equivalent if one can be obtained from the other by a rotation only (and not reflection). Then we find that no two colorings of an isosceles triangle are equivalent, so we have eight different patterns. In the case of an equilateral triangle, however, this makes no difference, and we still have only four patterns. But now suppose we use three colors, instead of only two. Then the two colorings of an equilateral triangle shown below are equivalent if we allow reflection, but they are not equivalent if we use the stricter criterion of allowing only rotation.

We now formulate the general problem that we intend to consider. Let S be a finite set whose elements are specified points or parts of some given geometric figure. Suppose we assign to each element in S a color out of some given set of m colors. The total number of ways in which such color assignments can be made is $m \times m \times \cdots \times m = m^n$, where n is the number of elements in S. The problem is to find the number of distinct patterns in which these m^n color assignments fall. For recognizing distinct patterns, we may use either the weaker criterion of allowing both rotations and reflections or the stricter criterion of allowing only rotations. It is obvious that the number of distinct patterns under the weaker criterion is less than or equal to the number under the stricter criterion.

It is clear from the examples discussed above that the number of distinct patterns into which the m^n color assignments fall depends not only on the numbers m and n but also on the symmetry properties of the underlying geometric figure. The greater the symmetry possessed by the figure, the larger the number of equivalent pairs of colorings and therefore the smaller the number of distinct patterns.

5.4 ACTION OF A GROUP ON A SET

Let X be a nonempty set, and let G be a permutation group on X; that is, G is a subgroup of the symmetric group S_X. So each element of G is a permutation of the set X. Hence, for all $g \in G$, $x \in X$, $g(x)$ is again an element of X. Moreover, since the group operation in G is composition of mappings, we have the following two properties for all $x \in X$:

1. $e(x) = x$, where e is the identity in G.
2. $(gh)(x) = g(h(x))$ for all $g, h \in G$.

These two properties motivate the general concept of action of a group on a set.

Definition 5.4.1 Let G be a group, and let X be a nonempty set. A mapping $* : G \times X \to X$, with $*(g, x)$ written $g * x$, is called an *action* of G on X if the following conditions hold for all $x \in X$:

(a) $e * x = x$, where e is the identity in G.
(b) $(gh) * x = g * (h * x)$ for all $g, h \in G$.

If there is an action of a group G on a set X, we say that G *acts* on X and call X a G-*set*.

EXAMPLES

1. Given any nonempty set X, let G be a subgroup of the symmetric group S_X. For any $g \in G$ and $x \in X$, we define $g * x = g(x)$. Then it follows from conditions (a) and (b) in Definition 5.4.1 that $*$ is an action of G on X. We say in this case that G acts *naturally* on X. In particular, if G is the group of symmetries of a set X of points in space, then G acts naturally on X.
2. Given any group G, let $X = G$. We define $g * x = gx$ (the product of g and x in the group G). Then $*$ is an action of G on X. We say in this case that G acts on itself by *left translation*. Further, if H is a subgroup of G, then G acts on the quotient set G/H by left translation on taking $g * (aH) = (ga)H$.
3. Given any group G, again take $X = G$. We define $g * x = gxg^{-1}$. Then $e * x = exe^{-1} = x$ and $(gh) * x = ghx(gh)^{-1} = g(hxh^{-1})g^{-1} = g * (h * x)$ for all $x \in X$ and $g, h \in G$. Hence G acts on X. We say in this case that G acts on itself by *conjugation*. Further, if N is a normal subgroup of G, then G acts on the quotient set G/N by conjugation on taking $g * (aH) = g(aH)g^{-1}$.

We saw in Example 1 that if G is a subgroup of S_X, then G acts naturally on X. Conversely, we can show that if a group G acts on a set X, then G determines a subgroup of S_X that is a homomorphic image of G.

Let G be a group acting on a set X. Given $g \in G$, we define the mapping $\sigma_g : X \to X$ by the rule $\sigma_g(x) = g * x$. Then σ_g is a permutation of X. Let $x, y \in X$. Then $\sigma_g(x) = \sigma_g(y) \Rightarrow g * x = g * y \Rightarrow g^{-1} * (g * x) = g^{-1} * (g * y) \Rightarrow (g^{-1}g) * x = (g^{-1}g) * y \Rightarrow e * x = e * y \Rightarrow x = y$. Hence σ_g is injective. Further, given $y \in X$, let $x = g^{-1} * y$. Then $\sigma_g(x) = g * x = g * (g^{-1} * y) = (g^{-1}g) * y = e * y = y$. Hence σ_g is surjective. This proves that σ_g is a permutation of X.

Consider now the mapping $\phi : G \to S_X$ given by $g \mapsto \sigma_g$. Let $g, h \in G$. Then for all $x \in X$, $\sigma_{gh}(x) = (gh) * x = g * (h * x) = \sigma_g(\sigma_h(x)) = (\sigma_g \sigma_h)(x)$. Hence $\phi(gh) = \phi(g)\phi(h)$. This proves that ϕ is a homomorphism. Hence Im ϕ is a subgroup of S_X and a homomorphic image of G. We write G_X to denote Im ϕ and refer to it as the permutation group on X *induced* by G.

Moreover, if the identity e in G is the only element such that $e * x = x$ for all $x \in X$, then ker $\phi = \{e\}$, and so ϕ is an isomorphism and G is isomorphic with G_X. In this case we can identify each g in G with σ_g, and consider G as a subgroup of S_X.

For the sake of convenience, we shall henceforth write simply gx instead of $g * x$.

Let G be a group acting on a set X. For any $a \in X$, the *orbit* of a under G is defined to be the set

$$Orb(a) = \{ga \mid g \in G\}$$

THEOREM 5.4.2 Let G be a group acting on a set X. Then the orbits in X under G form a partition of X.

Proof: For any $x, y \in X$, let $x \sim y$ mean that $x = gy$ for some $g \in G$. We claim that $\sim$ is an equivalence relation. For any $x \in X$, $x = ex$ and hence $x \sim x$. If $x = gy$ for some $g \in G$, then $g^{-1}x = g^{-1}(gy) = (g^{-1}g)y = ey = y$. Hence $x \sim y \Rightarrow y \sim x$. If $x = gy$ and $y = hz$ for some $g, h \in G$, then $x = (gh)z$ and hence $x \sim y, y \sim z \Rightarrow x \sim z$. This proves that $\sim$ is an equivalence relation in X. For any $a \in X$, the equivalence class of a under $\sim$ is

$$Cl_\sim(a) = \{x \in X \mid x \sim a\} = \{ga \mid g \in G\} = Orb(a)$$

Hence the orbits in X under G are the equivalence classes under $\sim$ and therefore form a partition of X. ∎

The partition of X formed by the set of orbits under G is denoted by X/G and called the *orbit decomposition* of X under G.

Again, let G be a group acting on a set X. Let $g \in G$ and $x \in X$. If $gx = x$, we say g *fixes* x. Given $x \in X$, the set of all elements in G that fix x is called the *stabilizer* of x and written $Stab(x)$; that is,

$$Stab(x) = \{g \in G \mid gx = x\}$$

THEOREM 5.4.3 Let G be a group acting on a set X, and let $x \in X$. Then

(a) $Stab(x)$ is a subgroup of G.
(b) The index of the subgroup $Stab(x)$ in G is

$$(G : Stab(x)) = |Orb(x)|$$

Proof: **(a)** We write $S = Stab(x)$. Since $ex = x$, we have $e \in S$. If $g, h \in S$, then $(g^{-1}h)x = g^{-1}(hx) = g^{-1}x = g^{-1}(gx) = (g^{-1}g)x = ex = x$ and hence $g^{-1}h \in S$. This proves that S is a subgroup of G.

(b) As usual, let G/S denote the set of left cosets of S in G. We define

$$\phi : G/S \rightarrow Orb(x)$$

by the rule

$$\phi(gS) = gx$$

We claim that ϕ is a bijective mapping. If $gS = hS$, then $g^{-1}h \in S$; hence $g^{-1}hx = x$ and therefore $gx = g(g^{-1}hx) = hx$. This shows that ϕ is well defined. If $\phi(gS) = \phi(hS)$, then $gx = hx$ and hence $g^{-1}hx = x$, which implies that $g^{-1}h \in S$; therefore $gS = hS$. This proves that ϕ is injective. If $y \in Orb(x)$, then $y = gx = \phi(gS)$ for some $g \in G$. Hence ϕ is surjective and therefore bijective. Hence $(G : Stab(x)) = |G/S| = |Orb(x)|$. ∎

Now we prove the main theorem that enables us to count the number of color patterns. Given $g \in G$, the set of all elements in X that are fixed by g is called the *fixture* of g and written $Fix(g)$; that is,

$$Fix(g) = \{x \in X \mid gx = x\}$$

THEOREM 5.4.4 (Burnside theorem) Let G be a finite group acting on a finite set X. Then the number k of orbits in X under G is given by

$$k = \frac{1}{|G|} \sum_{g \in G} F(g)$$

where $F(g) = |Fix(g)|$ is the number of elements in X that are fixed by g.

Proof: We count in two ways the number of ordered pairs (g, x) in $G \times X$ such that g fixes x. We write

$$P = \{(g, x) \in G \times X \mid gx = x\}$$

If $gx = x$, then $g \in Stab(x)$ and $x \in Fix(g)$. Hence, given $x \in X$, the number of elements in G that fix x is equal to $|Stab(x)|$. On the other hand, given $g \in G$, the number of elements in X that are fixed by g is equal to $|Fix(g)|$. Hence

$$\sum_{x \in X} |Stab(x)| = |P| = \sum_{g \in G} |Fix(g)| \tag{1}$$

By Theorem 5.4.3,

$$|Orb(x)| = (G : Stab(x)) = \frac{|G|}{|Stab(x)|}$$

Hence

$$\sum_{x \in X} |Stab(x)| = |G| \sum_{x \in X} \frac{1}{|Orb(x)|} \tag{2}$$

Now for any orbit $T \in X/G$,

$$\sum_{x \in T} \frac{1}{|Orb(x)|} = \sum_{x \in T} \frac{1}{|T|} = |T| \frac{1}{|T|} = 1$$

Therefore, since the orbits in X form a partition of X,

$$\sum_{x \in X} \frac{1}{|Orb(x)|} = \sum_{T \in X/G} \sum_{x \in T} \frac{1}{|Orb(x)|} = \left| \frac{X}{G} \right| = k$$

Hence, using (1) and (2), we obtain

$$k = \sum_{x \in X} \frac{1}{|Orb(x)|} = \frac{1}{|G|} \sum_{x \in X} |Stab(x)| = \frac{1}{|G|} \sum_{g \in G} |Fix(g)|$$

This completes the proof. ∎

Burnside theorem can be paraphrased as follows: If a group G acts on a set X, then the number of orbits in X under G is equal to the average number of elements in X fixed by an element in G.

5.5   BURNSIDE THEOREM AND COLOR PATTERNS

We now take up the original problem that we posed earlier in this chapter: finding the number of patterns in the colorings of a given set of points. We shall see how Burnside theorem is used to obtain the solution.

Using our earlier notation, we let S be a set of n elements representing some specified points (or parts) of a given geometric figure. To each element in S we assign some color out of a given set of m colors. This can be done in m^n ways, which we refer to as *m-colorings* of S.

Let X denote the set of all *m-colorings* of S. Let G be a group of symmetries of the set S. The group G acts naturally on the set S. Therefore G also acts on the set X. Given $g \in G$ and $x \in X$, gx represents the color assignment obtained by performing on the coloring x the symmetry operation (rotation or reflection) represented by g. Two color assignments x and y are equivalent if and only if $y = gx$ for some g in G. Hence all color assignments that are equivalent to x lie in the orbit of x under G. Each orbit represents a color pattern. Thus the number of distinct patterns is equal to the number of orbits in the set X under the action of the group G. This number is given by Burnside theorem.

Suppose the points in the set S are coplanar. Then the group of symmetries of S is either some dihedral group

$$D_q = \{e, \alpha, \dots, \alpha^{q-1}, \beta, \alpha\beta, \dots, \alpha^{q-1}\beta\}$$

(where α represents a rotation through angle $2\pi/q$, and β is a reflection) or its cyclic subgroup $C_q = \{e, \alpha, \dots, \alpha^{q-1}\}$, consisting of all rotations in D_q. If $G = D_q$, then the number of orbits in X under G gives the number of color patterns under rotations and reflections. If we wish to find the number of patterns under the stricter criterion of allowing rotations only, then we apply Burnside theorem to find the number of orbits under the group $H = C_q$ instead of under G.

In the application of Burnside theorem, we have to compute the numbers $F(g)$. For each $g \in G$, $F(g)$ is the number of color assignments that remain invariant under the action of the symmetry operation (rotation or reflection) represented by g. The identity element e in G fixes every $x \in X$, and hence $F(e) = m^n$. The following examples illustrate how we find these numbers $F(g)$ for other elements in G. The first example is the one with which we started our discussion in this chapter—namely, coloring the vertices of an equilateral triangle—but now we consider the general case of m colors.

Example 5.5.1 Each vertex of an equilateral triangle is colored by one of m given colors. Find the number of distinct patterns among all possible colorings.

Solution. Since each vertex can be colored in m ways, the total number of color assignments is m^3. The group G of symmetries of an equilateral triangle is the dihedral group of degree 3; that is,

$$G = D_3 = \{e, \alpha, \alpha^2, \beta, \alpha\beta, \alpha^2\beta\}$$

where α represents a rotation through angle $2\pi/3$, and β is a reflection in a diameter.

As mentioned above, every color assignment is invariant under the identity e; hence $F(e) = m^3$. To find the number of color assignments invariant under the other elements of G, let us number the vertices 1, 2, and 3. Then α takes vertex 1 to 2, 2 to 3, and 3 to 1. If a color assignment is invariant under α, then all three vertices must have the same color. This common color can be any one of the m given colors. Hence there are m color assignments that are invariant under α, so $F(\alpha) = m$. The same reasoning applies to α^2, so $F(\alpha^2) = m$.

Now suppose β is the reflection in the diameter passing through vertex 1. Then β takes vertex 2 to 3 and 3 to 2. If a color assignment is invariant under β, then vertices 2 and 3 must have the same color, so vertices 1 and 2 can have arbitrary colors. Hence the number of color assignments invariant under β is m^2. The same argument holds for the other two reflections $\alpha\beta$ and $\alpha^2\beta$. Hence $F(\beta) = F(\alpha\beta) = F(\alpha^2\beta) = m^2$.

By Burnside theorem, the number of patterns (that is, the number of orbits under G) is

$$k = \frac{1}{|G|}\{F(e) + F(\alpha) + F(\alpha^2) + F(\beta) + F(\alpha\beta) + F(\alpha^2\beta)\}$$

$$= \frac{1}{6}\{m^3 + 2m + 3m^2\}$$

To find the number of patterns under the stricter criterion of rotations only, we take the group of rotations $H = \{e, \alpha, \alpha^2\}$. By Burnside theorem, the number of orbits under the group H is

$$k' = \frac{1}{|H|}\{F(e) + F(\alpha) + F(\alpha^2)\} = \frac{1}{3}(m^3 + 2m)$$

In the particular case of only two colors, on putting $m = 2$ in the above results, we obtain

$$k = k' = 4$$

In the case $m = 3$, we have $k = 10$, $k' = 11$.

Example 5.5.2 A rectangular dining table seats six persons, two along each longer side and one on each shorter side. A colored napkin, having one of m given colors, is placed for each person. Find the number of distinct patterns among all possible color assignments.

Solution. The group of symmetries of the rectangle is

$$G = D_2 = \{e, \alpha, \beta, \alpha\beta\}$$

where α is a rotation through angle π, and β is a reflection. Let us take β to be the reflection in the line through the center parallel to the longer side of the rectangle. Then $\alpha\beta$ represents the reflection in the line parallel to the shorter side.

Let us number the six napkins as shown in the diagram above. Then α takes napkin 1 to 4, 2 to 5, 3 to 6, and vice versa. If a color assignment is invariant under the rotation α, then the napkins 1 and 4 must have the same color, 2 and 5 must have the same color, and 3 and 6 must have the same color. So we can assign arbitrary colors to napkins 1, 2, and 3. Hence the number of color assignments invariant under α is m^3.

Now β keeps napkins 3 and 6 fixed, takes 1 to 5, 2 to 4, and vice versa. If a color assignment is invariant under β, then napkins 1 and 5 must have the same color, and 2 and 4 must have the same color. So we can assign arbitrary colors to napkins 1, 2, 3, and 6. Hence the number of color assignments invariant under β is m^4. By a similar reasoning, we find that the number of color assignments invariant under $\alpha\beta$ is m^3. Therefore, by Burnside theorem, the number of patterns is

$$k = \frac{1}{|G|}\{F(e) + F(\alpha) + F(\beta) + F(\alpha\beta)\}$$

$$= \frac{1}{4}(m^6 + m^4 + 2m^3)$$

The number of patterns under the stricter criterion of rotations only is

$$k' = \frac{1}{2}\{F(e) + F(\alpha)\} = \frac{1}{2}(m^6 + m^4)$$

In the particular case of two colors, we have $k = 24$, $k' = 36$.

Example 5.5.3 (Polya's neckties) A straight necktie in the form of a long rectangular strip is divided into n bands of equal width parallel to the shorter side. Each band is colored by one of m given colors. Find the number of ties with distinct patterns.

Solution. The group of symmetries of a rectangle is the dihedral group D_2. But in the present case the reflection in the line parallel to the longer side doesn't play any role. The relevant group here is

$$G = D_1 = \{e, \alpha\}$$

where α may represent a rotation through angle π, or a reflection in the line through the center parallel to the shorter side of the rectangle. (The two operations are equivalent in this case.)

If a color assignment is invariant under α, then the bands 1 and n must have the same color, the bands 2 and $n - 1$ must have the same color, and so on. In general, the bands i and $n + 1 - i$ must have the same color. If n is even, we can assign arbitrary colors to bands $1, \ldots, \frac{n}{2}$; hence $F(\alpha) = m^{n/2}$.

But if n is odd, the bands $1, \ldots, \frac{n+1}{2}$ can be assigned arbitrary colors. (The $\frac{n+1}{2}$th band is the band in the middle.) Hence $F(\alpha) = m^{(n+1)/2}$.

Therefore, by Burnside theorem, the number of patterns is

$$k = \frac{1}{|G|}\{F(e) + F(\alpha)\}$$

$$= \begin{cases} \dfrac{1}{2}(m^n + m^{n/2}) & \text{if } n \text{ is even} \\[2mm] \dfrac{1}{2}(m^n + m^{(n+1)/2}) & \text{if } n \text{ is odd} \end{cases}$$

For example, if $m = 2$ and $n = 8$, then $k = \frac{1}{2}(2^8 + 2^4) = 136$. If $m = 2$ and $n = 9$, then $k = \frac{1}{2}(2^9 + 2^5) = 272$.

Example 5.5.4 Suppose each vertex of a regular hexagon is colored by one of m given colors. Find the number of distinct patterns among all colorings.

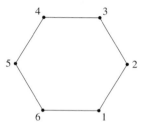

Solution. The group of symmetries of a regular hexagon is

$$G = D_6 = \{e, \alpha, \alpha^2, \alpha^3, \alpha^4, \alpha^5, \beta, \alpha\beta, \alpha^2\beta, \alpha^3\beta, \alpha^4\beta, \alpha^5\beta\}$$

where α represents a rotation through angle $\pi/3$, and β is a reflection in a diameter. Let us number the vertices, taken in order, as 1, 2, 3, 4, 5, and 6. If a color assignment is invariant under the rotation α, then each vertex must have the same color; hence $F(\alpha) = m$. If a color assignment is invariant under α^2, then vertices 1, 3, and 5 have the same color, and vertices 2, 4, and 6 have the same color. Hence $F(\alpha^2) = m^2$. If a color assignment is invariant under α^3, then vertices 1 and 4 have the same color, 2 and 5 have the same color, and 3 and 6 have the same color. Hence $F(\alpha^3) = m^3$. Similarly, we find $F(\alpha^4) = m^2$ and $F(\alpha^5) = m$.

Suppose β represents reflection in the diameter through vertex 1. If a color assignment is invariant under the reflection β, then the vertices 2 and 6 have the same color and 3 and 5 have the same color. So we can assign arbitrary colors to vertices 1, 2, 3 and 4. Hence $F(\beta) = m^4$.

Now $\alpha\beta$ is a reflection in the diameter passing through the middle point between vertices 1 and 2. If a color assignment is invariant under $\alpha\beta$, then vertices 1 and 2 must have the same color, 3 and 6 have the same color, and 4 and 5 have the same color. Hence $F(\alpha\beta) = m^3$.

By similar arguments, we obtain $F(\alpha^2\beta) = F(\alpha^4\beta) = m^4$ and $F(\alpha^3\beta) = F(\alpha^5\beta) = m^3$. Hence, by Burnside theorem, the number of patterns is

$$k = \frac{1}{|G|} \left\{ \begin{array}{l} F(e) + F(\alpha) + F(\alpha^2) + F(\alpha^3) + F(\alpha^4) + F(\alpha^5) \\ + F(\beta) + F(\alpha\beta) + F(\alpha^2\beta) + F(\alpha^3\beta) + F(\alpha^4\beta) + F(\alpha^5\beta) \end{array} \right\}$$

$$= \frac{1}{12}(m^6 + 3m^4 + 4m^3 + 2m^2 + 2m)$$

Under the stricter criterion of rotations only, the number of patterns is

$$k' = \frac{1}{6}\{F(e) + F(\alpha) + F(\alpha^2) + F(\alpha^3) + F(\alpha^4) + F(\alpha^5)\}$$

$$= \frac{1}{6}(m^6 + m^3 + 2m^2 + 2m)$$

In the particular case $m = 2$, we have $k = 13$, $k' = 14$.

Example 5.5.4 has an important application in chemistry. From the carbon ring consisting of six carbon atoms (figure a) one can obtain several chemically different molecules by attaching to each carbon atom either a hydrogen atom H or the group CH_3. For example, if a hydrogen atom is attached to each carbon atom, the result is a molecule of benzene (figure b). The question is: How many chemically different molecules can be obtained in this manner?

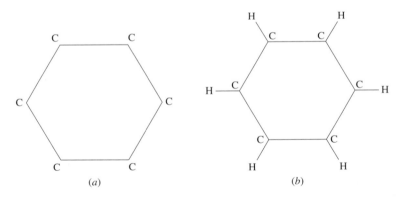

(a) (b)

It is obvious that the problem is mathematically the same as finding the number of patterns in coloring the vertices of a regular hexagon with two colors. This problem was solved in Example 5.5.4. There are 13 chemically different molecules that can be obtained from the carbon ring.

From the examples above, we see that the main work involved in the application of Burnside theorem is the computation of the numbers $F(g)$. Now we obtain a modified version of Burnside theorem that makes this task more systematic and somewhat easier.

As before, let S denote the set of points to be colored, and let C denote the set of colors. Then any coloring of the points in S with colors from the set C is in fact a mapping from S to C, so the set X of all color assignments is the set of all mappings from S to C, which we write as $X = C^S$. Let G be a group of symmetries of the set S. Then G acts naturally on the set S. Geometrically, it is evident that G also acts on the set X, a fact we have used in the examples above. But let us now give a general algebraic proof of this property.

Let G be a group acting on a set S, let C be an arbitrary nonempty set, and let $X = C^S$ be the set of all mappings from S to C. The action of G on S induces an action of G on X as follows: For any $g \in G$, $f \in X$, we define $g * f \in X$ as the mapping $g * f : S \to C$ with

$$(g * f)(s) = f(g^{-1}s)$$

for all $s \in S$. Then $e * f(s) = f(e^{-1}s) = f(s)$ for all $s \in S$; hence $e * f = f$. Further, for all $g, h \in G$ and $f \in X$,

$$((gh) * f)(s) = f((gh)^{-1}s) = f(h^{-1}g^{-1}s) = (h * f)(g^{-1}s) = (g * (h * f))(s)$$

holds for all $s \in S$. Hence $(gh) * f = (g * (h * f))$. This proves that $*$ is an action of G on X.

Recall that each element $g \in G$ induces a permutation σ_g of the set S given by the rule $\sigma_g(s) = gs$ for all $s \in S$.

THEOREM 5.5.5 Let G be a finite group acting on a finite set S, let C be a finite set of m elements, and let $X = C^S$ be the set of all mappings from S to C. Then the number of elements in X fixed by $g \in G$ is

$$F(g) = m^{\lambda(g)}$$

where $\lambda(g)$ is the number of disjoint cycles (including cycles of length 1) in the cycle decomposition of the permutation σ_g of S induced by g.

Consequently, the number k of orbits in X under the action of G is given by

$$k = \frac{1}{|G|} \sum_{g \in G} m^{\lambda(g)}$$

Proof: Let $g \in G$ and $f \in X$. If $g * f = f$, then $f(s) = (g * f)(s) = f(g^{-1}s)$ for all $s \in S$. Hence $f(gs) = f(g^{-1}gs) = f(s)$ for all $s \in S$. Conversely, if $f(gs) = f(s)$ for all $s \in S$, then $(g * f)(s) = f(g^{-1}s) = f(gg^{-1}s) = f(s)$ for all $s \in S$; hence $g * f = f$. Thus $f \in Fix(g)$ if and only if $f(gs) = f(s)$ for all $s \in S$.

Let σ_g be the permutation of S determined by g; that is, $\sigma_g(s) = gs$ for all $s \in S$. Let $\sigma_g = \gamma_1 \cdots \gamma_\lambda$ be the decomposition of σ_g into disjoint cycles (including cycles of length 1). Any cycle in this decomposition is of the form

$$\gamma = (a \; ga \; g^2a \ldots g^{r-1}a)$$

If $f \in Fix(g)$, then $f(a) = f(ga) = \cdots = f(g^{r-1}a)$; hence f is constant on the elements in the cycle γ. This holds for every cycle γ_i in the decomposition of σ_g.

Conversely, if f is constant on every cycle γ_i, then $f(gs) = f(s)$ for all $s \in S$. Hence $f \in Fix(g)$ if and only if f is constant on each cycle in the decomposition of σ_g.

Let $f \in Fix(g)$, and let $f_1, \ldots, f_\lambda \in C$ be the values of f on the cycles $\gamma_1, \ldots, \gamma_\lambda$, respectively. Then $f_1, \ldots, f_\lambda$ can be each chosen in m ways. Hence there are exactly m^λ elements in $Fix(g)$. This proves the first part of the theorem. Hence, by Burnside theorem,

$$k = \frac{1}{|G|} \sum_{g \in G} F(g) = \frac{1}{|G|} \sum_{g \in G} m^{\lambda(g)} \qquad \blacksquare$$

To illustrate the use of Theorem 5.5.5, we work out again the problem of Example 5.5.2 by the new method.

Example 5.5.6 Do the problem of Example 5.5.2 by using Theorem 5.5.5.

Solution. With the notation of Example 5.5.2, we identify each element in the group G with the permutation induced by it on the set $\{1, 2, 3, 4, 5, 6\}$ and find its cycle decomposition:

$$e = \begin{pmatrix} 1 & 2 & 3 & 4 & 5 & 6 \\ 1 & 2 & 3 & 4 & 5 & 6 \end{pmatrix} = (1)(2)(3)(4)(5)(6)$$

$$\alpha = \begin{pmatrix} 1 & 2 & 3 & 4 & 5 & 6 \\ 4 & 5 & 6 & 1 & 2 & 3 \end{pmatrix} = (1\,4)(2\,5)(3\,6)$$

$$\beta = \begin{pmatrix} 1 & 2 & 3 & 4 & 5 & 6 \\ 5 & 4 & 3 & 2 & 1 & 6 \end{pmatrix} = (1\,5)(2\,4)(3)(6)$$

$$\alpha\beta = \begin{pmatrix} 1 & 2 & 3 & 4 & 5 & 6 \\ 2 & 1 & 6 & 5 & 4 & 3 \end{pmatrix} = (1\,2)(3\,6)(4\,5)$$

Hence $\lambda(e) = 6$, $\lambda(\alpha) = 3$, $\lambda(\beta) = 4$, $\lambda(\alpha\beta) = 3$. Therefore, by Theorem 5.5.5,

$$k = \frac{1}{4}(m^6 + m^4 + 2m^3)$$

In the following example we solve the general problem of which Examples 5.5.1 and 5.5.4 are special cases.

Example 5.5.7 Suppose each vertex of a regular polygon of n sides is colored by one of m given colors. Find the number of distinct patterns among all colorings.

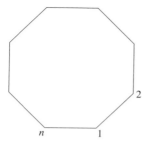

Solution. The group G of symmetries of a regular polygon of n sides is the dihedral group of degree n,

$$G = D_n = \{e, \alpha, \dots, \alpha^{n-1}, \beta, \alpha\beta, \dots, \alpha^{n-1}\beta\}$$

where α is a rotation through angle $2\pi/n$ and β is reflection in a diameter, say through the vertex 1. Interpreting the elements of G as permutations of the set $S = \{1, \dots, n\}$ of the vertices, we have

$$\alpha = \begin{pmatrix} 1 & 2 & 3 & \dots & n \\ 2 & 3 & 4 & \dots & 1 \end{pmatrix}, \quad \beta = \begin{pmatrix} 1 & 2 & 3 & \dots & n \\ 1 & n & n-1 & \dots & 2 \end{pmatrix}$$

Consider first the set $H = \{e, \alpha, \dots, \alpha^{n-1}\}$ of rotations in G. H is a cyclic subgroup of G of order n. Let g be an element of order r in the group H. We show that the number of cycles in the decomposition of the permutation g is $\lambda(g) = n/r$. Since $g^r = e$, we have $g^r(i) = i$ for each $i = 1, \dots, n$. We claim that given any i, r is the least positive integer such that $g^r(i) = i$. Suppose there exists a positive integer $s < r$ such that $g^s(j) = j$ for some j. Now g^s is a rotation of the polygon. If it takes j to j, it must take each i to i, so $g^s(i) = i$ for all $i = 1, \dots, n$. This means $g^s = e$, which contradicts the fact that g is of order r. Therefore given any $i \in \{1, \dots, n\}$, i generates the cycle $(i \ g(i) \dots g^{r-1}(i))$ of length r. Hence each cycle in the cyclic decomposition of the permutation g is of length r. It follows that the number of cycles is $\lambda(g) = n/r$.

Since H is a cyclic group of order n, the order of every element in H is a divisor of n. Moreover, given any divisor r of n, the number of elements in H of order r is $\varphi(r)$, where $\varphi(r)$ denotes the number of positive integers less than r and relatively prime to r. Hence the contribution to the summation $\sum m^{\lambda(g)}$ (in Theorem 5.5.5) from the rotations in the group G is

$$\sum_{g \in H} m^{\lambda(g)} = \sum_{r \mid n} \varphi(r) m^{n/r}$$

Now consider the set K of elements other than rotations in G; that is, $K = G - H = \{\beta, \alpha\beta, \dots, \alpha^{n-1}\beta\}$. Here two cases arise.

1. If n is odd, then each element g in K represents a reflection in a diameter through some vertex j. So g fixes the vertex j and interchanges vertices $j+i$ and $j-i$ for $i = 1, \ldots, \dfrac{n-1}{2}$. Thus the cyclic decomposition of g consists of one cycle (j) of length 1 and $\dfrac{n-1}{2}$ cycles $(j-1 \; j+1)$ of length 2. Hence $\lambda(g) = \dfrac{n+1}{2}$.

2. Suppose n is even. Then exactly half of the elements in K represent reflections in a diameter passing through two opposite vertices, and the remaining $\dfrac{n}{2}$ elements represent reflections in a diameter passing through the middle points of two opposite sides. It is easily seen that if g is a reflection of the first type, then $\lambda(g) = \dfrac{n+2}{2}$. If g is of the second type, then $\lambda(g) = \dfrac{n}{2}$.

Hence the contribution to the summation $\sum m^{\lambda(g)}$ from the reflections in the group G is

$$\sum_{g \in K} m^{\lambda(g)} = \begin{cases} nm^{(n+1)/2} & \text{if } n \text{ is odd} \\ \dfrac{n}{2}m^{(n+2)/2} + \dfrac{n}{2}m^{n/2} & \text{if } n \text{ is even} \end{cases}$$

If we combine the contributions from H and K, the number k of patterns is given by

$$k = \frac{1}{|G|} \sum_{g \in G} m^{\lambda(g)}$$

$$= \begin{cases} \dfrac{1}{2n}\left(\displaystyle\sum_{r|n} \varphi(r) m^{n/r} + nm^{(n+1)/2} \right) & \text{if } n \text{ is odd} \\ \dfrac{1}{2n}\left(\displaystyle\sum_{r|n} \varphi(r) m^{n/r} + \tfrac{n}{2}(m^{(n+2)/2} + m^{n/2}) \right) & \text{if } n \text{ is even} \end{cases}$$

Under the stricter criterion of rotations only, the number of patterns is

$$k' = \frac{1}{|H|} \sum_{g \in H} m^{\lambda(g)} = \frac{1}{n} \sum_{r|n} \varphi(r) m^{n/r}$$

EXERCISES 5.5

1. Find (without using the result of Example 5.5.7) the number of patterns obtained on coloring the vertices of a square with m colors.

2. Repeat Exercise 1 for a regular pentagon.

3. Repeat Exercise 1 for a rectangle.

4. Find the number of distinct necklaces with p beads (p prime), where each bead can have any one of n colors.

5. Find the number of distinct bracelets of six beads, where each bead is red, blue, or white.

6. A rectangular design consists of 11 parallel stripes of equal width. If each stripe can be painted red, blue, or green, find the number of possible patterns.

7. Each side of an equilateral triangle is divided into two equal parts, and each part is colored red or green. Find the number of patterns.

8. Each side of a square is divided into three equal parts, and each part is colored red, yellow, or green. Find the number of patterns.

9. Each side of a regular polygon of n sides is divided into q equal parts, and each part is painted with one of m colors. Find the number of patterns.

10. Each vertex of an equilateral triangle is colored with one of four colors such that at least two vertices have different colors. Find the number of patterns.

11. Repeat Exercise 10 for a square.

12. The interior of an equilateral triangle is divided into six parts by the medians. Each part is painted with one of m colors. Find the number of patterns.

13. The sides of a rectangle are 3 feet and 4 feet long. The rectangle is divided into 12 equal squares, and each square is painted with one of m colors. Find the number of patterns.

14. Repeat Exercise 13 for a rectangle with sides of lengths 4 feet and 6 feet, divided into 24 squares.

15. Repeat Exercise 13 for a rectangle with sides of lengths 5 feet and 7 feet, divided into 35 squares.

16. Repeat Exercise 13 for a rectangle with sides of lengths p feet and q feet, divided into pq squares.

17. Find the number of ways in which the faces of a regular tetrahedron can be painted with m colors.

18. Find the number of ways in which the faces of a cube can be painted with m colors.

5.6 POLYA'S THEOREM AND PATTERN INVENTORY

We now consider the problem of finding the number of color patterns in which the colors occur with preassigned frequencies. For instance, we found in Example 5.5.2 the number of distinct patterns in m-colorings of six napkins arranged on a rectangular table. We may now ask the question: What is the number of patterns in which there are exactly one yellow, two red, and three green napkins? It is questions of this type that can be answered by using Polya's theorem, which we are about to prove now.

5.6.1 Polya's Theorem

Let G be a group acting on a set X. Let $R = \mathbb{Q}[t_1, \ldots, t_q]$ be the set of all polynomials in some given indeterminates $t_1, \ldots, t_q$ with rational coefficients. A mapping $w : X \to R$ is called a *weight* function on X under G if $w(gx) = w(x)$ for all $g \in G, x \in X$. If this condition holds, then every element in the orbit $T = Orb(x)$ has the same weight $w(x)$. The common weight of all elements in an orbit T is called the *weight* of T and written $w(T)$.

The following theorem, known as *weighted* Burnside theorem, is a generalization of Theorem 5.4.4.

THEOREM 5.6.1 Let G be a finite group acting on a finite set X. Let $w : X \to R$ be a weight function on X under G. Then the sum of the weights of the orbits in X under G is

$$\sum_{T \in X/G} w(T) = \frac{1}{|G|} \sum_{g \in G} \sum_{x \in Fix(g)} w(x)$$

Proof: As in the proof of Burnside theorem, let

$$P = \{(g, x) \in G \times X \mid gx = x\}$$

We compute the sum

$$S = \sum_{(g,x) \in P} w(gx)$$

in two ways. On the one hand,

$$S = \sum_{g \in G} \sum_{x \in Fix(g)} w(gx) = \sum_{g \in G} \sum_{x \in Fix(g)} w(x) \tag{1}$$

On the other hand,

$$S = \sum_{x \in X} \sum_{g \in Stab(x)} w(gx) = \sum_{x \in X} |Stab(x)| w(x) \tag{2}$$

By Theorem 5.4.3,

$$|Stab(x)| = \frac{|G|}{|Orb(x)|}$$

Moreover, as shown in the proof of Theorem 5.4.4,

$$\sum_{x \in T} \frac{1}{|Orb(x)|} = 1$$

for any orbit $T \in X/G$. Hence

$$\sum_{x \in T} |Stab(x)| w(x) = \sum_{x \in T} \frac{|G|}{|Orb(x)|} w(x)$$

$$= |G|w(T) \sum_{x \in T} \frac{1}{|Orb(x)|}$$

$$= |G|w(T)$$

Therefore, since the orbits form a partition of X, we have

$$\sum_{x \in X} |Stab(x)|w(x) = \sum_{T \in X/G} \sum_{x \in T} |Stab(x)|w(x)$$

$$= |G| \sum_{T \in X/G} w(T)$$

Hence, equating the two expressions for S given in (1) and (2), we obtain

$$\sum_{T \in X/G} w(T) = \frac{1}{|G|} \sum_{g \in G} \sum_{x \in Fix(g)} w(x)$$

This completes the proof. ∎

If we take $w(x) = 1$ for every $x \in X$ in Theorem 5.6.1, we recover the original Burnside theorem. Then $w(T) = 1$ for every orbit T, and so the left-hand side gives the number k of orbits.

Now suppose G acts on a finite set S, and $X = C^S$ is the set of all mappings from S to C. Then we have seen that G has an induced action on X given by

$$(g * f)(s) = f(g^{-1}s)$$

for all $g \in G$, $f \in X$, $s \in S$.

Given any mapping $u : C \to R$, u induces a weight function w on X under G as follows: Let us define $w : X \to R$ by the rule

$$w(f) = \prod_{s \in S} u(f(s))$$

for all $f \in X$. Let $g \in G$. Then

$$w(g * f) = \prod_{s \in S} u((g * f)(s)) = \prod_{s \in S} u(f(g^{-1}s))$$

Now $s \mapsto g^{-1}s$ represents a permutation of S. Hence, as s ranges over S, $g^{-1}s$ also ranges over S. Therefore, since multiplication of polynomials in R is commutative,

$$\prod_{s \in S} u(f(g^{-1}s)) = \prod_{s \in S} u(f(s))$$

Hence $w(g * f) = w(f)$. This proves that w is a weight function on X under G.

THEOREM 5.6.2 (Polya's theorem) Let G be a finite group acting on a finite set S having n elements, let C be a finite nonempty set, and let $X = C^S$ be the set of all mappings from S to C. Let $u : C \to R$, and let $w : X \to R$ be the weight function on X under G induced by u. Then the sum of the weights of the orbits in X under G is

$$\sum_{T \in X/G} w(T) = \frac{1}{|G|} \sum_{g \in G} \prod_{i=1}^{n} \left(\sum_{c \in C} (u(c))^i \right)^{\lambda_i(g)}$$

where $\lambda_i(g)$ is the number of cycles of length i $(i = 1, \ldots, n)$ in the cycle decomposition of the permutation σ_g of S induced by g.

Proof: Let $g \in G$, and let $\sigma_g = \gamma_1 \cdots \gamma_r$ be the cycle decomposition of σ_g (including cycles of length 1). Let $f \in X$. As shown in the proof of Theorem 5.5.5, $f \in Fix(g)$ if and only if f is constant on each cycle γ_j. Let f_j be the value of f on the cycle γ_j, $j = 1, \ldots, r$. Then

$$w(f) = \prod_{s \in S} u(f(s)) = \prod_{j=1}^{r} (u(f_j))^{|\gamma_j|}$$

where $|\gamma_j|$ denotes the length of the cycle γ_j. Hence

$$\sum_{f \in Fix(g)} w(f) = \sum_{f_1 \in C} \cdots \sum_{f_r \in C} \prod_{j=1}^{r} (u(f_j))^{|\gamma_j|}$$

Changing the order of summation and multiplication on the right-hand side, we have

$$\sum_{f \in Fix(g)} w(f) = \prod_{j=1}^{r} \sum_{c \in C} (u(c))^{|\gamma_j|}$$

$$= \prod_{i=1}^{n} \left(\sum_{c \in C} (u(c))^i \right)^{\lambda_i(g)}$$

where $\lambda_i(g)$ is the number of cycles of length i in the cycle decomposition of σ_g $(i = 1, \ldots, n)$. Hence, by the weighted Burnside theorem, the sum of the weights of the orbits in X is

$$\sum_{T \in X/G} w(T) = \frac{1}{|G|} \sum_{g \in G} \sum_{f \in Fix(g)} w(f)$$

$$= \frac{1}{|G|} \sum_{g \in G} \prod_{i=1}^{n} \left(\sum_{c \in C} (u(c))^i \right)^{\lambda_i(g)} \qquad \blacksquare$$

If in Theorem 5.6.2 we take $u(c) = 1$ for each $c \in C$, then we recover Theorem 5.5.5. Then $w(T) = 1$ for every orbit T and, assuming C has m elements, we have

$$\prod_{i=1}^{n} \left(\sum_{c \in C} (u(c))^i \right)^{\lambda_i(g)} = \prod_{i=1}^{n} (m)^{\lambda_i(g)} = m^{\lambda(g)}$$

where $\lambda(g) = \sum_{i=1}^{n} \lambda_i(g)$ is the total number of cycles in the decomposition of σ_g.

Now let $C = \{c_1, \ldots, c_m\}$ be the set of colors and $R = \mathbb{Q}[t_1, \ldots, t_m]$. Let w be the weight function on X induced by the mapping $u : C \to R$ with

$$u(c_i) = t_i, \qquad i = 1, \ldots, m$$

Then Polya's theorem gives

$$\sum_{T \in X/G} w(T) = \frac{1}{|G|} \sum_{g \in G} \prod_{i=1}^{n} \left(t_1^i + \cdots + t_m^i \right)^{\lambda_i(g)} \tag{3}$$

5.6.2 Pattern Inventory

Consider a color assignment $f : S \to C$ in which the colors $c_1, \ldots, c_m$ occur with frequencies $\beta_1, \ldots, \beta_m$, respectively, where $\beta_1, \ldots, \beta_m$ are nonnegative integers such that $\beta_1 + \cdots + \beta_m = n$. So β_i $(i = 1, \ldots, m)$ is the number of elements s in S such that $f(s) = c_i$. Therefore the weight of f and hence also the weight of the orbit T containing f are

$$w(T) = w(f) = \prod_{s \in S} u(f(s)) = t_1^{\beta_1} \cdots t_m^{\beta_m}$$

Hence the sum of the weights of the orbits in X is equal to

$$\sum_{T \in X/G} w(T) = \sum p(\beta_1, \ldots, \beta_m) t_1^{\beta_1} \cdots t_m^{\beta_m} \tag{4}$$

where $p(\beta_1, \ldots, \beta_m)$ denotes the number of orbits having the same weight $t_1^{\beta_1} \cdots t_m^{\beta_m}$, and the summation on the right-hand side is over all m-tuples $(\beta_1, \ldots, \beta_m)$ of nonnegative integers such that $\beta_1 + \cdots + \beta_m = n$. Equivalently, in terms of colorings, $p(\beta_1, \ldots, \beta_m)$ is the number of patterns in which the colors $c_1, \ldots, c_m$ occur with frequencies $\beta_1, \ldots, \beta_m$, respectively. The polynomial on the right-hand side of (4) is a homogeneous polynomial of degree n in m indeterminates. We denote it by $P_{X/G}(t_1, \ldots, t_m)$ and refer to it as the *pattern inventory* of orbits in X under G. Thus the number of patterns with given color frequencies $\beta_1, \ldots, \beta_m$ is the coefficient of the monomial $t_1^{\beta_1} \cdots t_m^{\beta_m}$ in the pattern inventory polynomial $P_{X/G}(t_1, \ldots, t_m)$.

Equating the two expressions for $\sum w(T)$ in (3) and (4), we obtain

$$P_{X/G}(t_1, \ldots, t_m) = \frac{1}{|G|} \sum_{g \in G} \prod_{i=1}^{n} (t_1^i + \cdots + t_m^i)^{\lambda_i(g)} \tag{5}$$

We mentioned that $P_{X/G}(t_1, \ldots, t_m)$ is a homogeneous polynomial of degree n in $t_1, \ldots, t_m$. This fact is also clear from the right-hand side in (5). Since σ_g is a permutation of a set of n elements, the sum of the lengths of the cycles in the decomposition of σ_g is equal to n, so $\sum i\lambda_i(g) = n$. Hence every term in the summation in (5) is of degree n. Further, we see that $P_{X/G}(t_1, \ldots, t_m)$ is symmetric in the indeterminates $t_1, \ldots, t_m$.

5.6.3 Cycle Index Polynomial

The result of Polya's theorem and the formula for the pattern inventory can be expressed in a more compact and elegant form by introducing the concept of the cycle index polynomial of a permutation group.

Given a permutation σ of a set of n elements, the *cycle index* of σ is defined to be the n-tuple $(\lambda_1, \ldots, \lambda_n)$, where λ_i denotes the number of cycles of length i in the cycle decomposition of σ.

Definition 5.6.3 Given a permutation group G of degree n, the cycle *index polynomial* of G is defined to be the polynomial Z_G in n indeterminates $y_1, \ldots, y_n$ given by

$$Z_G(y_1, \ldots, y_n) = \frac{1}{|G|} \sum_{g \in G} \prod_{i=1}^{n} y_i^{\lambda_i(g)}$$

where $\lambda_i(g)$, $i = 1, \ldots, n$, denotes the number of cycles of length i in the cycle decomposition of the permutation g.

It is worth noting that the sum of the coefficients in the polynomial Z_G is 1. This is a consequence of the fact that the number of terms in the summation on the right-hand side is equal to $|G|$.

Example 5.6.4 Find the cycle index polynomial of S_3.

Solution. Writing the elements of S_3 as cycles, we have

$$S_3 = \{(1), (1\,2), (1\,3), (2\,3), (1\,2\,3), (1\,3\,2)\}$$

The identity permutation $(1) = (1)(2)(3)$ has the cycle index $(3, 0, 0)$. The three permutations $(1\,2)$, $(1\,3)$, and $(2\,3)$ all have the index $(1, 1, 0)$. The remaining two permutations both have the index $(0, 0, 1)$. Hence the cycle index polynomial of S_3 is

$$Z_{S_3} = \frac{1}{6}(y_1^3 + 3y_1 y_2 + 2y_3)$$

More generally, let G be a finite group acting on a finite set S of n elements. Let G_S be the permutation group induced by the action of G on S. Then the cycle index polynomial of G_S is called the *cycle index polynomial* of G acting on S and is denoted by Z_{G_S} or (if there is no confusion) simply Z_G.

The result of Polya's theorem can now be expressed as

$$\sum_{T \in X/G} w(T) = Z_G \left(\sum_{c \in C} u(c), \sum_{c \in C} (u(c))^2, \ldots, \sum_{c \in C} (u(c))^n \right)$$

where $Z_G(y_1, \ldots, y_n)$ is the cycle index polynomial of G acting on the set S.

Taking $u(c) = 1$ for all $c \in C$, we get the number of orbits in X under G. Taking $u(c_i) = t_i$ for $i = 1, \ldots, m$, we get the pattern inventory of orbits in X under G. We record this result in the following theorem.

THEOREM 5.6.5 Let G be a finite group acting on a finite set S having n elements, and let $Z_G(y_1, \ldots, y_n)$ be the cycle index polynomial of G acting on S. Let C be a finite set of m elements, and let $X = C^S$ be the set of all mappings from S to C. Then

(a) The number k of orbits in X under G is given by

$$k = Z_G(m, m, \ldots, m)$$

(b) The pattern inventory of orbits in X under G is given by

$$P_{X/G}(t_1, \ldots, t_m) = Z_G(t_1 + \cdots + t_m, t_1^2 + \cdots + t_m^2, \ldots, t_1^n + \cdots + t_m^n)$$

Note that the pattern inventory $P_{X/G}(t_1, \ldots, t_m)$ is obtained by simply replacing y_i with $t_1^i + \cdots + t_m^i$ for each $i = 1, \ldots, n$ in the cycle index polynomial $Z_G(y_1, \ldots, y_n)$.

In the following example we answer the question that we posed at the beginning of this section.

Example 5.6.6 Let G be the group of symmetries of a rectangle acting on the set S of napkins as shown in the diagram of Example 5.5.2. Find the cycle index polynomial of G acting on the set S. Hence find the number of distinct color patterns with one yellow, two red, and three green napkins.

Solution. As we already know, G is the dihedral group

$$G = D_2 = \{e, \alpha, \beta, \alpha\beta\}$$

In Example 5.5.6 we found the cyclic decompositions of the permutations induced by the elements of G on the set of napkins $\{1, 2, 3, 4, 5, 6\}$. From these, we obtain their cycle indexes:

g	σ_g	Cycle Index
e	$(1)(2)(3)(4)(5)(6)$	$(6, 0, 0, 0, 0, 0)$
α	$(1\,4)(2\,5)(3\,6)$	$(0, 3, 0, 0, 0, 0)$
β	$(1\,5)(2\,4)(3)(6)$	$(2, 2, 0, 0, 0, 0)$
$\alpha\beta$	$(1\,2)(3\,6)(4\,5)$	$(0, 3, 0, 0, 0, 0)$

Hence the cycle index polynomial of G acting on S is

$$Z_G(y_1, y_2, y_3, y_4, y_5, y_6) = \frac{1}{4}\left(y_1^6 + y_2^3 + y_1^2 y_2^2 + y_2^3\right)$$

$$= \frac{1}{4}\left(y_1^6 + y_1^2 y_2^2 + 2y_2^3\right)$$

To find the pattern inventory polynomial with three colors, we replace y_i in Z_G with $(t_1^i + t_2^i + t_3^i)$, so

$$P(t_1, t_2, t_3) = \frac{1}{4}\Big\{(t_1 + t_2 + t_3)^6 + (t_1 + t_2 + t_3)^2\left(t_1^2 + t_2^2 + t_3^2\right)^2 \\ + 2\left(t_1^2 + t_2^2 + t_3^2\right)^3\Big\}$$

The number of patterns with one yellow, two red, and three green napkins is equal to the coefficient of the monomial $t_1 t_2^2 t_3^3$ in the polynomial $P(t_1, t_2, t_3)$. After some computation, we find this coefficient to be 16.

EXERCISES 5.6

1. Find the cycle index polynomial of S_4.

2. Find the cycle index polynomial of S_5.

3. Find the cycle index polynomial of A_4.

4. Let G be the group of symmetries of a square. Find the cycle index polynomial of G acting on the set of vertices of the square.

5. Repeat Exercise 4 for a rectangle.

6. Repeat Exercise 4 for regular hexagon.

7. Let G be the group of symmetries of a regular polygon of n sides. Find the cycle index polynomial of G acting on the set of vertices of the polygon.

8. Let G be the group of rotational symmetries of a regular tetrahedron. Find the cycle index polynomial of G acting on the set of four vertices of the tetrahedron.

9. Let G be the group of rotational symmetries of a regular tetrahedron. Find the cycle index polynomial of G acting on the set of six edges of the tetrahedron.

10. Let G be the group of rotational symmetries of a regular tetrahedron. Find the cycle index polynomial of G acting on the set of four faces of the tetrahedron.

11. Let G be the group of rotational symmetries of a cube. Find the cycle index polynomial of G acting on the set of eight vertices of the cube.

12. Let G be the group of rotational symmetries of a cube. Find the cycle index polynomial of G acting on the set of 12 edges of the cube.

13. Let G be the group of rotational symmetries of a cube. Find the cycle index polynomial of G acting on the set of six faces of the cube.

14. Each side of a regular hexagon is painted red, white, or blue. Find the number of patterns in which each color is used twice.

15. Eight persons are to be seated around a rectangular dining table, with three persons along each longer side. A colored napkin is placed for each person. Find the number of patterns with five green and three red napkins.

16. Repeat Exercise 15 with three green, three red, and two yellow napkins.

5.7 GENERATING FUNCTIONS FOR NONISOMORPHIC GRAPHS

In this section we consider an application of Polya's theorem in graph theory. Informally, a graph consists of a set of vertices of which some pairs (possibly all or none) are joined by line segments or arcs. A formal definition is given below.

For any set V, we write $V^{(2)}$ to denote the set of all pairs of elements in V; that is;

$$V^{(2)} = \{\{x, y\} \mid x, y \in V, \ x \neq y\}$$

Definition 5.7.1 A *graph* G is an ordered pair $G = (V, E)$, where V is a finite nonempty set and E is a subset of $V^{(2)}$. The elements of V are called *vertices* of the graph G, and the elements of E are called its *edges*.

Two vertices a, b in a graph are said to be *adjacent* if the pair $\{a, b\}$ is an edge.

A graph G is commonly represented by a diagram in which the vertices are shown as points or small circles, and two vertices a and b are joined by a segment or an arc if and only if $\{a, b\}$ is an edge. The positions of the vertices in the diagram and the shapes of the arcs joining the vertices are immaterial. For example, the three diagrams below, though they look quite different from one another, represent the same graph $G = (V, E)$ with

$$V = \{a, b, c, d\}, \quad E = \{\{a, b\}, \{a, c\}, \{a, d\}, \{b, c\}, \{c, d\}\}$$

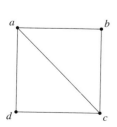
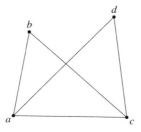
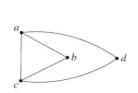

A less trivial and more interesting example is provided by the two diagrams below, which represent the same graph, known as the *Petersen graph*.

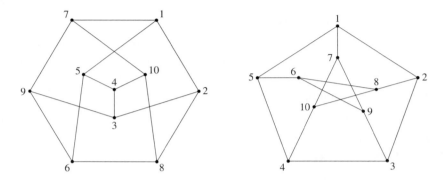

What is the number of all possible graphs on a given set V of n vertices? The set $V^{(2)}$ has $N = \binom{n}{2} = n(n-1)/2$ elements, so there are 2^N distinct subsets of $V^{(2)}$. Hence this is the number of distinct graphs on a given set of n vertices. But among these, there are several cases of graphs that are essentially alike and can be obtained from one another by permuting the vertices. Such graphs are said to be *isomorphic*. More generally, we have the following definition.

Definition 5.7.2 Two graphs $G = (V, E)$ and $G' = (V', E')$ are said to be *isomorphic* if there exists a bijective mapping $f : V \rightarrow V'$ such that for all $\{a, b\} \in V^{(2)}$,

$$\{a, b\} \in E \iff \{f(a), f(b)\} \in E'$$

For example, the graphs represented by the two diagrams below are seen to be isomorphic on taking the mapping $f : \{1, 2, 3, 4\} \rightarrow \{a, b, c, d\}$ given by

$$1 \mapsto a, \ 2 \mapsto c, \ 3 \mapsto b, \ 4 \mapsto d$$

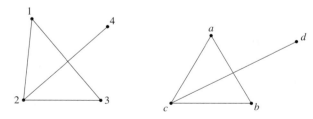

It follows from the definition that two graphs are isomorphic if and only if their respective diagrams can be obtained from each other by relabeling the vertices.

Clearly, then, all graphs that are isomorphic to one another can be represented by the same diagram without any labels attached to the vertices.

Let us now consider graphs that are not isomorphic. An interesting problem is to find the maximum number of nonisomorphic graphs with a given number n of vertices. For small values of n, one can find this number by drawing all possible unlabeled diagrams with n vertices. For example, we can easily see that there are just four different diagrams of a graph with three vertices, as shown below. Hence there are only four nonisomorphic graphs with three vertices. But this method is not practical for large values of n.

As we shall see shortly, this problem is closely related to that of counting color patterns and so can be tackled by applying the theory that we developed in the foregoing sections. Let X be the set of all graphs on a given set V of n vertices, and let $K = (V, V^{(2)})$ be the graph in which every pair $\{a, b\} \in V^{(2)}$ is an edge. Given a graph $G = (V, E)$, let us color the edges in the diagram of the graph K with two colors, say black and white, as follows: If $\{a, b\} \in E$, we color the edge $\{a, b\}$ in K with black, otherwise with white. Thus each graph G on the vertex set V determines a coloring of the edges of K with two colors. It is clear that this gives a one-to-one correspondence between the set X of graphs on V and the set of all 2-colorings of the edges in the diagram of K. In other words, we can identify X with the set of all two-color assignments of the set $V^{(2)}$.

Without loss of generality, we write $V = \{1, \ldots, n\}$ and let S_n be the group of all permutations of V. The natural action of S_n on V induces an action on $V^{(2)}$ by the rule

$$\sigma\{a, b\} = \{\sigma(a), \sigma(b)\}$$

for all $\sigma \in S_n$ and $\{a, b\} \in V^{(2)}$. Two graphs $G = (V, E)$ and $G' = (V, E')$ are isomorphic if and only if the corresponding 2-colorings of $V^{(2)}$ are in the same orbit under the action of the group S_n. So the nonisomorphic graphs on the vertex set V correspond to the orbits in X under S_n. It follows that the number of nonisomorphic graphs on V is equal to the number k of orbits in X under S_n. By Theorem 5.5.5,

$$k = \frac{1}{n!} \sum_{\sigma \in S_n} 2^{\lambda(\bar{\sigma})}$$

where $\lambda(\bar{\sigma})$ is the number of cycles (including cycles of length 1) in the cycle decomposition of the permutation $\bar{\sigma}$ of $V^{(2)}$ induced by σ.

Further, by using Polya's theorem (5.6.2), we can find the number $g(n, m)$ of nonisomorphic graphs on V having m edges, $m = 0, 1, \ldots, N$, where $N = \binom{n}{2}$. Let $Z(y_1, \ldots, y_N)$ be the cycle index polynomial of the group S_n acting on the set $V^{(2)}$. Then the pattern inventory of the orbits in X under S_n is the polynomial

$$P(t_1, t_2) = Z\left(t_1 + t_2, t_1^2 + t_2^2, \ldots, t_1^N + t_2^N\right)$$

The coefficient of the monomial $t^m t^{N-m}$ in this polynomial gives the number $g(n, m)$ of nonisomorphic graphs having m edges. Since $P(t_1, t_2)$ is symmetric in t_1, t_2, it follows that $g(n, m) = g(n, N - m)$.

Putting $t_1 = 1$, $t_2 = x$, we obtain the function

$$f_n(x) = P(1, x) = Z(1 + x, 1 + x^2, \ldots, 1 + x^N)$$

The coefficient of x^m in the polynomial $f_n(x)$ gives the number $g(n, m)$ of nonisomorphic graphs on n vertices having m edges, so

$$f_n(x) = \sum_{m=0}^{N} g(n, m)x^m$$

The polynomial $f_n(x)$ is called the *generating function* for the nonisomorphic graphs on n vertices. Since $g(n, m) = g(n, N - m)$, it follows that $f_n(x)$ is a reciprocal polynomial.

We record this result in the following theorem.

THEOREM 5.7.3 The generating function $f_n(x)$ for the nonisomorphic graphs on n vertices is given by

$$f_n(x) = Z(1 + x, 1 + x^2, \ldots, 1 + x^N)$$

where $N = \binom{n}{2}$ and $Z(y_1, \ldots, y_N)$ is the cycle index polynomial of the group S_n acting on the set $\{1, \ldots, n\}^{(2)}$.

To compute the cycle index polynomial $Z(y_1, \ldots, y_N)$ of S_n acting on $\{1, \ldots, n\}^{(2)}$, it is not necessary to consider every σ in S_n. Suppose σ, σ' have the same cycle structure. Then they are conjugate elements in the group S_n. Therefore the permutations $\bar{\sigma}, \bar{\sigma}'$ induced by them on the set $V^{(2)}$ are also conjugate and have the same cycle structure. So we need to consider just one typical permutation σ in S_n corresponding to each possible cycle structure, and take into account the number of permutations that have the same cycle structure as σ.

Example 5.7.4 Find the generating function $f_4(x)$ for the nonisomorphic graphs on four vertices.

Solution. Here $V = \{1, 2, 3, 4\}$ and therefore

$$V^{(2)} = \{\{1, 2\}, \{1, 3\}, \{1, 4\}, \{2, 3\}, \{2, 4\}, \{3, 4\}\}$$

For the sake of convenience, let us write the pair $\{a, b\}$ as ab. Then we can write $V^{(2)}$ as

$$V^{(2)} = \{12, 13, 14, 23, 24, 34\}$$

The following table shows a typical permuation σ in S_4 for each possible cycle structure, the number $\#(\sigma)$ of permutations that have the same cycle structure as σ, the permutation $\bar{\sigma}$ induced by σ on $V^{(2)}$, the number of cycles in the decomposition of $\bar{\sigma}$, and the monomial contributed by $\bar{\sigma}$ to the cycle index poynomial.

σ	$\#(\sigma)$	$\bar{\sigma}$	$\lambda(\bar{\sigma})$	$\prod y_i^{\lambda_i(\bar{\sigma})}$
$e = (1)$	1	$e = (12)$	6	y_1^6
$(1\,2)$	6	$(13\,23)(14\,24)$	4	$y_1^2 y_2^2$
$(1\,2)(3\,4)$	3	$(13\,24)(14\,23)$	4	$y_1^2 y_2^2$
$(1\,2\,3)$	8	$(12\,23\,13)(14\,24\,34)$	2	y_3^2
$(1\,2\,3\,4)$	6	$(12\,23\,34\,14)(13\,24)$	2	$y_2 y_4$

Hence the cycle index polynomial of S_4 acting on the set $V^{(2)}$ is

$$Z(y_1, y_2, y_3, y_4, y_5, y_6) = \frac{1}{24}\left(y_1^6 + 9y_1^2 y_2^2 + 8y_3^2 + 6y_2 y_4\right)$$

To obtain the generating function $f_4(x)$, we replace y_i with $1 + x^i$ in $Z(y_1, y_2, y_3, y_4, y_5, y_6)$, so

$$f_4(x) = \frac{1}{24}\{(1+x)^6 + 9(1+x)^2(1+x^2)^2 + 8(1+x^3)^2 + 6(1+x^2)(1+x^4)\}$$

Opening brackets and simplifying, we obtain

$$f_4(x) = 1 + x + 2x^2 + 3x^3 + 2x^4 + x^5 + x^6$$

The total number of nonisomorphic graphs is $f_4(1) = 11$.

If we wanted to find just the total number of nonisomorphic graphs, we could obtain it more easily by using Theorem 5.5.5. Then

$$k = \frac{1}{n!}\sum_{\sigma \in S_n} 2^{\lambda(\bar{\sigma})}$$

$$= \frac{1}{24}(2^6 + 6 \cdot 2^4 + 3 \cdot 2^4 + 8 \cdot 2^2 + 6 \cdot 2^2)$$

$$= 11$$

Here are the diagrams of the 11 nonisomorphic graphs on four vertices.

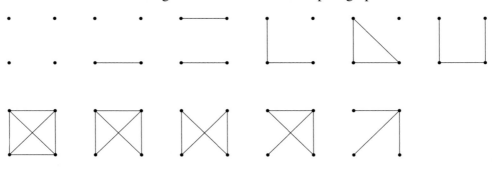

EXERCISES 5.7

1. Find the generating function for the nonisomorphic graphs on five vertices and draw their diagrams.

2. Find the number of nonisomorphic graphs on six vertices.

CHAPTER

6

WALLPAPER PATTERN GROUPS

A wallpaper pattern is obtained by repeating a basic motif. We assume that the basic motif is contained in a parallelogram that is repeated endlessly so as to cover the entire plane. It is clear that such a pattern has infinitely many symmetries. We refer to the symmetry groups of such patterns as *wallpaper pattern groups*. In this chapter we develop a theory of classification of these groups.

A wallpaper pattern is a two-dimensional analogue of a crystal. A crystal is formed by stacking together identical copies of a cluster of atoms in the shape of a parallelopiped. The theory of classification of wallpaper pattern groups presented here can be extended to the classification of symmetry groups of crystals.

6.1 GROUP OF SYMMETRIES OF A PLANE

Let us consider the symmetries of a plane. If we have rectangular coordinate axes in the plane with origin O, then any point in the plane can be represented by its coordinates (x_1, x_2). The entire plane is represented by the set

$$\mathbb{R}^2 = \{(x_1, x_2) \mid x_1, x_2 \in \mathbb{R}\}$$

Given any vector $a = (a_1, a_2) \in \mathbb{R}^2$, let T_a denote the mapping $T_a : \mathbb{R}^2 \to \mathbb{R}^2$ given by

$$T_a(x_1, x_2) = (x_1 + a_1, x_2 + a_2)$$

If we write $x = (x_1, x_2)$, the mapping T_a is given by $T_a(x) = x + a$ for all $x \in \mathbb{R}^2$. It is clear that T_a is bijective and preserves distances; hence it is a symmetry of the plane. Geometrically, T_a shifts every point through the vector a and is therefore called a *translation* through the vector a.

Let σ be an arbitrary symmetry of the plane, and suppose σ takes the origin to the point $a = (a_1, a_2)$—that is, $\sigma(0) = a$. Then $T_{-a}\sigma(0) = T_{-a}(a) = a - a = 0$. Hence the origin is fixed by the mapping $T_{-a}\sigma$. Since σ and T_{-a} are both symmetries, their product $T_{-a}\sigma$ is also a symmetry of the plane. Writing $d = T_{-a}\sigma$, we have $\sigma = T_a d$. Thus we see that every symmetry of the plane can be expressed as a product $T_a d$, where T_a is a translation and d is a symmetry that fixes the origin O.

If a symmetry d fixes the origin O, then d must map any circle around O onto itself. So let us consider the group of symmetries of a circle with center O. When we take a fixed point A on the circle, the position of any point P on the circle is given by the angle θ that OP makes with OA. Let σ be any symmetry of the circle. Then σ preserves the angular distance between any two points on the circle. Suppose σ takes A to A' and OA' makes an angle α with OA. Consider a point B on the circle, and suppose σ takes B to B'. If OB makes an angle β with OA, then OB' must make an angle β or $-\beta$ with OA'—that is, $\sigma(\beta) = \alpha + \beta$ or $\alpha - \beta$. If $\sigma(\beta) = \alpha + \beta$, then $\sigma(\theta) = \alpha + \theta$ for all θ, so σ represents a rotation of the circle about its center through angle α. If, on the other hand, $\sigma(\beta) = \alpha - \beta$, then $\sigma(\theta) = \alpha - \theta$ for all θ. Hence $\sigma(\alpha/2 - \theta) = \alpha/2 + \theta$ for all θ, so σ represents a reflection in the diameter making an angle $\alpha/2$ with OA.

Let R_α denote the rotation of the circle about O through angle α, and let F denote the reflection in the fixed diameter OA. So $R_\alpha(\theta) = \alpha + \theta$ and $F(\theta) = -\theta$. Hence $R_\alpha F(\theta) = \alpha - \theta$ for all θ, so $R_\alpha F$ represents reflection in the diameter making an angle $\alpha/2$ with the fixed line OA. The symmetries of the circle are rotations R_α and reflections $R_\alpha F$, where $0 \le \alpha < 2\pi$. Thus the group D of the symmetries of the cirlce is given by

$$D = \{R_\alpha F^i \mid 0 \le \alpha < 2\pi, \ i = 0, 1\}$$

The identity of the group is $I = R_0 = F^2$. Moreover $FR_\alpha(\theta) = F(\alpha + \theta) = -\alpha - \theta = R_{-\alpha}F(\theta)$. Thus the group D has these defining relations:

$$R_\alpha R_\beta = R_{\alpha+\beta}, \qquad R_{2\pi} = I = F^2, \qquad FR_\alpha = R_{-\alpha}F$$

Let E denote the group of all symmetries of the plane. From the foregoing result and the description of the group D given above, it follows that

$$E = \{T_a R_\alpha F^i \mid a \in \mathbb{R}^2, \ 0 \le \alpha < 2\pi, \ i = 0, 1\}$$

This is a geometric description of the group E. We shall now obtain an algebraic representation for its elements. We have already seen that the translation T_a is given by the mapping $x \mapsto x + a$. Consider the effect of a rotation about the origin on the coordinates of a point. Suppose a point $P(x_1, x_2)$ goes to $P'(x_1', x_2')$ under rotation through angle α. Let OP make an angle θ with the X_1-axis. Then P' makes angle $\theta + \alpha$ with OX_1. Therefore, writing $r = OP = OP'$, we have

$$x_1 = r \cos\theta, \qquad x_2 = r \sin\theta$$
$$x_1' = r \cos(\theta + \alpha), \qquad x_2' = r \sin(\theta + \alpha)$$

From these relations we obtain

$$x_1' = x_1 \cos\alpha - x_2 \sin\alpha, \quad x_2' = x_1 \sin\alpha + x_2 \cos\alpha$$

These equations can be rewritten in the form of a matrix equation.

$$\begin{bmatrix} x_1' \\ x_2' \end{bmatrix} = \begin{bmatrix} \cos\alpha & -\sin\alpha \\ \sin\alpha & \cos\alpha \end{bmatrix} \begin{bmatrix} x_1 \\ x_2 \end{bmatrix}$$

When we write the vector $x = (x_1, x_2)$ as the column $\begin{bmatrix} x_1 \\ x_2 \end{bmatrix}$, the rotation R_α is represented by the mapping $x \mapsto Ax$, where

$$A = \begin{bmatrix} \cos\alpha & -\sin\alpha \\ \sin\alpha & \cos\alpha \end{bmatrix}$$

Further, if F denotes the reflection in the X_1-axis, then F takes the point (x_1, x_2) to $(x_1, -x_2)$. Hence $R_\alpha F$ is represented by the mapping $x \mapsto Ax$, where

$$A = \begin{bmatrix} \cos\alpha & \sin\alpha \\ \sin\alpha & -\cos\alpha \end{bmatrix}$$

Combining these results, we see that the general element $T_a R_\alpha F^i$ in the group E is represented by the mapping $x \mapsto Ax + a$, where $a \in \mathbb{R}^2$ and A is a matrix of the form

$$A = \begin{bmatrix} \cos\alpha & -\sin\alpha \\ \sin\alpha & \cos\alpha \end{bmatrix} \quad \text{or} \quad A = \begin{bmatrix} \cos\alpha & \sin\alpha \\ \sin\alpha & -\cos\alpha \end{bmatrix} \tag{1}$$

In the first case, $\det A = 1$ and the mapping $x \mapsto Ax$ represents a rotation through angle α. In the second case, $\det A = -1$ and the mapping $x \mapsto Ax$ represents a reflection in the line $\theta = \alpha/2$. Further, in both cases

$$AA^T = I = A^T A$$

where I is the identity matrix and A^T denotes the transpose of A.

A square matrix A with real entries is said to be *orthogonal* if $AA^T = I = A^T A$. We thus see that every matrix A of type (1) is orthogonal. We can conversely show that any 2×2 orthogonal matrix is of one of the two forms given in (1). Suppose

$$A = \begin{bmatrix} a & c \\ b & d \end{bmatrix}$$

is an orthogonal matrix. Then

$$\begin{bmatrix} a & b \\ c & d \end{bmatrix} \begin{bmatrix} a & c \\ b & d \end{bmatrix} = A^T A = \begin{bmatrix} 1 & 0 \\ 0 & 1 \end{bmatrix}$$

from which we obtain

$$a^2 + b^2 = 1, \quad c^2 + d^2 = 1, \quad ac + bd = 0$$

From the first two relations it follows that there exist α, β such that

$$a = \cos\alpha, \; b = \sin\alpha \quad \text{and} \quad c = \cos\beta, \; d = \sin\beta$$

Then, using the third relation, we get $\cos(\alpha - \beta) = 0$; hence $\beta = \alpha \pm \pi/2$. This proves

$$A = \begin{bmatrix} \cos\alpha & \mp\sin\alpha \\ \sin\alpha & \pm\cos\alpha \end{bmatrix}$$

Thus we see that the group E consists of all mappings $x \mapsto Ax + a$, where $a \in \mathbb{R}^2$ and A is an orthogonal 2×2 matrix. Let O_2 denote the set of all orthogonal 2×2 matrices. (It is easily seen that O_2 is a group under multiplication.) Further, let (A, a) denote the mapping $x \mapsto Ax + a$. Then the group E can be described as

$$E = \{(A, a) \mid A \in O_2, \; a \in \mathbb{R}^2\}$$

The identity of the group E is $(I, 0)$. For any vector $a \in \mathbb{R}^2$, (I, a) represents a translation through the vector a. For any $A \in O_2$, $(A, 0)$ represents a rotation around the origin O or a reflection in a line through O, according to whether $\det A = 1$ or -1. The product of two elements (A, a) and (B, b) in E is obtained as follows: (B, b) takes x to $Bx + b$, and then (A, a) takes it to $A(Bx + b) + a = ABx + Ab + a$. Hence

$$(A, a)(B, b) = (AB, Ab + a)$$

In particular, it follows that $(I, a)(I, b) = (I, a + b)$ and $(A, 0)(B, 0) = (AB, 0)$. Further, the inverse of (A, a) is given by

$$(A, a)^{-1} = (A^{-1}, -A^{-1}a)$$

We record this result in the following theorem.

THEOREM 6.1.1 The group E of symmetries of a plane is given by

$$E = \{(A, a) \mid A \in O_2, \; a \in \mathbb{R}^2\}$$

where O_2 is the group of orthogonal 2×2 matrices and (A, a) denotes the mapping $x \mapsto Ax + a$. The product of two elements in E is given by the rule

$$(A, a)(B, b) = (AB, Ab + a)$$

Consequently, the inverse of an element is given by

$$(A, a)^{-1} = (A^{-1}, -A^{-1}a)$$

What follows is the geometric interpretation of the mapping $(A, 0)$—that is, $x \mapsto Ax$ in some special cases.

Matrix A	Mapping $(A, \mathbf{0})$
$\begin{bmatrix} 1 & 0 \\ 0 & 1 \end{bmatrix}$	Identity mapping
$\begin{bmatrix} 0 & -1 \\ 1 & 0 \end{bmatrix}$	Rotation through $\pi/2$
$\begin{bmatrix} -1 & 0 \\ 0 & -1 \end{bmatrix}$	Rotation through π
$\begin{bmatrix} 0 & 1 \\ -1 & 0 \end{bmatrix}$	Rotation through $3\pi/2$
$\begin{bmatrix} 1 & 0 \\ 0 & -1 \end{bmatrix}$	Reflection in X-axis
$\begin{bmatrix} -1 & 0 \\ 0 & 1 \end{bmatrix}$	Reflection in Y-axis
$\begin{bmatrix} 0 & 1 \\ 1 & 0 \end{bmatrix}$	Reflection in line $x = y$
$\begin{bmatrix} 0 & -1 \\ -1 & 0 \end{bmatrix}$	Reflection in line $x = -y$

6.2 WALLPAPER PATTERN GROUPS

As we stated earlier, a wallpaper pattern is obtained by repeating a fixed parallelogram pattern infinitely many times so as to cover the whole plane. The vertices of the parallelograms in a wallpaper pattern form a regular array called a *lattice*. More precisely, we define a lattice of points as follows.

Definition 6.2.1 A set L of points in the plane $\mathbb{R}^2$ is called a *lattice* if there exist two linearly independent (that is, nonparallel nonzero) vectors $b_1, b_2 \in \mathbb{R}^2$ such that

$$L = \{mb_1 + nb_2 \mid m, n \in \mathbb{Z}\}$$

The ordered pair (b_1, b_2) is called a *basis* of L.

A lattice may have more than one basis. For example, let $L = \{(m, n) \mid m, n \in \mathbb{Z}\}$. Then the vectors $b_1 = (1, 0)$ and $b_2 = (0, 1)$ form a basis of L. Now for all $m, n \in \mathbb{Z}$, $(m, n) = (m - n)(1, 0) + n(1, 1)$. Hence the vectors $c_1 = (1, 0)$ and $c_2 = (1, 1)$ also form a basis of the lattice L.

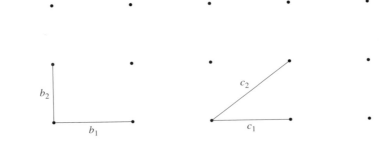

It is clear that a lattice L is a group under addition. If $u, v \in L$, $u = mb_1 + nb_2$, and $v = m'b_1 + n'b_2$, then $u - v = (m - m')b_1 + (n - n')b_2 \in L$. So L is a subgroup of $\mathbb{R}^2$.

Let L be the lattice formed by the vertices of the parallelograms in a given wallpaper pattern P. Then it is clear that for every $a \in L$, the translation $T_a : x \rightarrow x + a$ is a symmetry of P. Thus the group of symmetries of any wallpaper pattern must contain all translations given by some lattice. This motivates the following definition.

Definition 6.2.2 Let E be the group of symmetries of a plane. A subgroup G of E is called a *wallpaper pattern group* (in brief, a *WP group*) if there exists a lattice L such that $(I, a) \in G$ if and only if $a \in L$. L is called the *translation lattice* of G.

The following result follows directly from Definition 6.2.2.

THEOREM 6.2.3 Let G be a WP group with translation lattice L. Let $(A, a) \in G$. Then $(A, b) \in G$ if and only if $b = a + t$ for some $t \in L$.

Proof: If $b = a + t$, where $t \in L$, then $(A, b) = (A, a + t) = (A, a)(I, t) \in G$. Conversely, suppose $(A, b) \in G$. Then $(A, b)(A, a)^{-1} \in G$. But $(A, b)(A, a)^{-1} = (A, b)(A^{-1}, -A^{-1}a) = (I, b - a)$. Hence $b - a \in L$, so $b = a + t$ for some $t \in L$. ∎

Let G be a WP group with translation lattice L, and let T be the set of all translations in G; that is,

$$T = \{(I, a) \mid a \in L\}$$

For all $a, b \in L$, $(I, a)(I, b) = (I, a + b) \in T$ and $(I, a)^{-1} = (I, -a) \in T$. Hence T is a subgroup of G. We call T the *translation subgroup* of G. It is clear that the mapping $a \mapsto (I, a)$ is an isomorphism from L to T. Hence the group T is isomorphic with the group L.

Given a WP group G, let

$$G_0 = \{A \in O_2 \mid (A, a) \in G \text{ for some } a \in \mathbb{R}^2\}$$

where O_2 is the group of orthogonal 2×2 matrices. We claim that G_0 is a subgroup of O_2. Let $A, B \in G_0$. Then $(A, a), (B, b) \in G$ for some $a, b \in \mathbb{R}^2$. Hence $(A, a)(B, b) = (AB, Ab + a) \in G$ and $(A, a)^{-1} = (A^{-1}, -A^{-1}a) \in G$. Therefore AB, $A^{-1} \in G_0$. Since $(I, 0) \in G$, therefore $I \in G_0$. This proves that G_0 is a subgroup of O_2.

Definition 6.2.4 Let G be a WP group. The *point group* of G is defined to be the group

$$G_0 = \{A \in O_2 \mid (A, a) \in G \text{ for some } a \in \mathbb{R}^2\}$$

The following theorem gives the relation between the point group G_0 and the translation subgroup T of G.

THEOREM 6.2.5 Let G be a WP group with translation subgroup T, and let G_0 be the point group of G. Then T is a normal subgroup of G and $G_0 \simeq G/T$.

Proof: Define $f : G \to G_0$ by the rule $(A, a) \mapsto A$. Then

$$f((A, a)(B, b)) = f(AB, Ab + a) = AB$$

Hence f is a homomorphism. By definition, if $A \in G_0$, then $(A, a) \in G$ for some a; hence f is surjective. If $f(A, a) = I$, then $A = I$. But $(I, a) \in G$ if and only if $a \in L$, where L is the translation lattice of G. Hence $\ker f = \{(I, a) \mid a \in L\} = T$. Therefore $T \triangleleft G$ and, by the fundamental theorem of group homomorphisms, $G_0 \simeq G/T$. ∎

Consider the coset decomposition of G by T. The isomorphism $G_0 \simeq G/T$ implies that each element of G_0 determines a unique coset of T. Given $A \in G_0$, there exists $a \in \mathbb{R}^2$ such that $(A, a) \in G$. For every $t \in L$, $(I, t)(A, a) = (A, a + t)$. So the coset determined by (A, a) is $T(A, a) = \{(A, a + t) \mid t \in L\}$. This coset is independent of the choice of the vector a. Let b be any vector in $\mathbb{R}^2$ such that $(A, b) \in G$. Then, by Theorem 6.2.3, $b = a + t$ for some $t \in L$, so $b \in T(A, a)$. Hence G is a disjoint union of the cosets $T(A, a)$, $A \in G_0$; that is, $G = \bigcup_{A \in G_0} T(A, a)$.

We now prove some important properties of the point group that will enable us to determine its structure.

THEOREM 6.2.6 Let G be a WP group, and let L be the translation lattice of G. Let $A \in G_0$. Then for all $t \in L$, $At \in L$. Hence the mapping $(A, 0)$ maps L onto L.

Proof: If $A \in G_0$, then, by definition, $(A, a) \in G$ for some $a \in \mathbb{R}^2$. Moreover $(I, t) \in G$ for all $t \in L$. Hence $(A, a)(I, t)(A, a)^{-1} \in G$. But $(A, a)(I, t)(A, a)^{-1} =$

$(A, At + a)(A^{-1}, -A^{-1}a) = (I, At)$. Hence $At \in L$. Now A is an orthogonal matrix, so A is invertible. Therefore the mapping $t \mapsto At$ is a bijective mapping from L to L. Hence $(A, 0)$ maps L onto L.   ∎

THEOREM 6.2.7 G_0 is a finite group.

Proof: Let L be the translation lattice of G, and let (b_1, b_2) be a basis of L. Consider a circle C around the origin, containing b_1, b_2, and let S be the set of points of L within C. Then S consists of only a finite number of points, say n. Since $0, \pm b_1, \pm b_2 \in S$, it follows that $n \geq 5$. Let $A \in G_0$. By Theorem 6.2.7, $(A, 0)$ maps L onto L. Moreover, the mapping $x \mapsto Ax$ is distance preserving. Hence, for every $t \in S$, $At \in S$. Thus the mapping $(A, 0)$ maps S onto S, so A induces a permutation of S. Now A is completely determined by its effect on b_1, b_2, and hence two distinct elements of G_0 induce distinct permutations of S. Since S has a finite number of permutations, it follows that G_0 has a finite number of elements.   ∎

THEOREM 6.2.8 The point group G_0 is either a cyclic or a dihedral group.

Proof: Let K be the set of those elements A in G_0 for which $\det A = 1$; that is, A is a matrix of the form

$$A = R_\theta = \begin{bmatrix} \cos\theta & -\sin\theta \\ \sin\theta & \cos\theta \end{bmatrix}, \qquad 0 \leq \theta < 2\pi$$

Geometrically, the mapping $(A, 0)$ represents a rotation of the plane through angle θ. It is clear that K is a subgroup of G_0, and for any $R_\theta, R_{\theta'} \in K$, $R_\theta R_{\theta'} = R_{\theta+\theta'}$. We claim that K is cyclic.

If $K = \{I\}$, then K is trivially cyclic. Suppose K is of order greater than 1. Since G_0 is finite, there exists a least positive angle α such that $R_\alpha \in K$. Let R_θ be any element in K, $\theta > 0$. Now $m\alpha \leq \theta < (m+1)\alpha$ for some integer $m > 0$. Since $R_\alpha, R_\theta \in K$, it follows that $R_{\theta-m\alpha} = R_\theta(R_\alpha)^{-m} \in K$. But $0 \leq \theta - m\alpha < \alpha$, and hence $\theta - m\alpha$ must be equal to zero. Therefore $R_\theta = (R_\alpha)^m$. This proves that K is cyclic and generated by R_α. Moreover, if n is the order of the element R_α, then $(R_\alpha)^n = I$, and hence $\alpha = 2\pi/n$. Thus $K = \{I, A, \ldots, A^{n-1}\}$, where

$$A = R_{2\pi/n} = \begin{bmatrix} \cos\frac{2\pi}{n} & -\sin\frac{2\pi}{n} \\ \sin\frac{2\pi}{n} & \cos\frac{2\pi}{n} \end{bmatrix}$$

Now if $K = G_0$, then G_0 is cyclic. Otherwise, suppose G_0 has an element B with $\det B = -1$. Consider the mapping $f : G_0 \to \{1, -1\}$ given by $f(P) = \det P$ for all $P \in G_0$. It is clear that f is a surjective homomorphism, and $\ker f = \{P \in G_0 \mid \det P = 1\} = K$. Hence, by the fundamental theorem of homomorphisms, $G_0/K \simeq \{1, -1\}$. Therefore K is a subgroup of index 2, and $G_0 = K \cup KB$. Thus we have $G_0 = \{I, A, \ldots, A^{n-1}, B, AB, \ldots, A^{n-1}B\}$.

Since B is an orthogonal matrix with $\det B = -1$, it follows that B is of the form

$$B = \begin{bmatrix} \cos\theta & \sin\theta \\ \sin\theta & -\cos\theta \end{bmatrix}$$

It can be easily checked that $B^2 = I = A^n$ and $BA = A^{n-1}B$. This proves that G_0 is a dihedral group.

∎

Let C_n denote the abstract cyclic group of order n, and D_n the dihedral group of degree n. Theorem 6.2.8 has shown that every point group G_0 is isomorphic with C_n or D_n for some integer n. In the next section we shall find the values that n can take.

6.3 CHANGE OF BASIS IN $\mathbb{R}^2$

In our discussion of WP groups so far we have represented a point in the plane by its coordinates with respect to a given system of rectangular axes. But we can choose any two linearly independent (that is, nonparallel) vectors b_1, b_2 in $\mathbb{R}^2$ as a basis, and express the coordinates of a point with respect to these vectors. Any vector $a \in \mathbb{R}^2$ can be expressed uniquely as a linear combination of b_1, b_2; that is, we can express a as $a = a_1 b_1 + a_2 b_2$, where $a_1, a_2 \in \mathbb{R}$. We call a_1, a_2 the *coordinates* (or *components*) of the vector a with respect to the basis (b_1, b_2). In fact, the usual coordinates of a point with respect to the given rectangular axes are its coordinates with respect to the *standard* basis (e_1, e_2), where $e_1 = (1, 0)$ and $e_2 = (0, 1)$.

Now suppose we have two such bases, (b_1, b_2) and (c_1, c_2). Then every vector in $\mathbb{R}^2$ can be expressed as a linear combination of b_1, b_2 and also as a linear combination of c_1, c_2. In particular, the basis vectors b_1, b_2 themselves can be expressed as linear combinations of c_1, c_2. Suppose

$$b_1 = p_1 c_1 + q_1 c_2$$
$$b_2 = p_2 c_1 + q_2 c_2$$

where $p_1, p_2, q_1, q_2 \in \mathbb{R}$. Now, given a vector $v \in \mathbb{R}^2$, let (x_1, x_2) be the coordinates of v with respect to the basis (b_1, b_2). Then

$$v = x_1 b_1 + x_2 b_2$$
$$= x_1(p_1 c_1 + q_1 c_2) + x_2(p_2 c_1 + q_2 c_2)$$
$$= (p_1 x_1 + p_2 x_2)c_1 + (q_1 x_1 + q_2 x_2)c_2$$

Thus the coordinates of v with respect to the basis (c_1, c_2) are (x_1', x_2'), where

$$x_1' = (p_1 x_1 + p_2 x_2), \quad x_2' = (q_1 x_1 + q_2 x_2)$$

These relations can be expressed as a single matrix equation:

$$\begin{bmatrix} x_1' \\ x_2' \end{bmatrix} = \begin{bmatrix} p_1 & p_2 \\ q_1 & q_2 \end{bmatrix} \begin{bmatrix} x_1 \\ x_2 \end{bmatrix}$$

We can rewrite this equation as $x' = Px$, where $x' = \begin{bmatrix} x_1' \\ x_2' \end{bmatrix}$, $x = \begin{bmatrix} x_1 \\ x_2 \end{bmatrix}$, and $P = \begin{bmatrix} p_1 & p_2 \\ q_1 & q_2 \end{bmatrix}$. The matrix P is called the *matrix of transformation* of coordinates from the basis (b_1, b_2) to the basis (c_1, c_2). It is easily seen that the matrix P must be invertible. Let Q be the matrix of transformation from (c_1, c_2) to (b_1, b_2). Then $x = Qx' = Q(Px)$, so $(QP - I)x = 0$. Since this holds for every x, it follows that $QP = I$. Hence P is invertible and $Q = P^{-1}$.

Now suppose (b_1, b_2) and (c_1, c_2) are both bases of a lattice L. Then, by definition, the coordinates of any vector in L with respect to either basis are integers. Let $t \in L$, and suppose the coordinates of t are (m, n) with respect to the basis (b_1, b_2), and (m', n') with respect to (c_1, c_2). Then

$$\begin{bmatrix} m' \\ n' \end{bmatrix} = P \begin{bmatrix} m \\ n \end{bmatrix}$$

where P is the matrix of transformation from the basis (b_1, b_2) to the basis (c_1, c_2). Since this holds for all $t \in L$, and m, n, m', n' are all integers, it follows that the entries in the matrix P must be integers. Likewise, the entries in P^{-1} must be integers. Hence $\det P$, $\det P^{-1}$ are both integers. Now $\det P \det P^{-1} = \det I = 1$ and hence $\det P = \det P^{-1} = \pm 1$.

An invertible matrix P is called *unimodular* if both P and P^{-1} have integer entries. Equivalently, P is unimodular if it has integer entries and $\det P = \pm 1$. We have thus proved the following result.

THEOREM 6.3.1 Given a lattice L, the matrix of transformation of coordinates from one basis of L to another basis of L is a unimodular 2×2 matrix.

Given a 2×2 matrix A and a basis (b_1, b_2) of $\mathbb{R}^2$, these determine a mapping $f : \mathbb{R}^2 \to \mathbb{R}^2$ as follows: Let (x_1, x_2) be the coordinates of a vector v with respect to the basis (b_1, b_2)—that is, $v = x_1 b_1 + x_2 b_2$. Then we define $f(v) = y_1 b_1 + y_2 b_2$, where

$$\begin{bmatrix} y_1 \\ y_2 \end{bmatrix} = A \begin{bmatrix} x_1 \\ x_2 \end{bmatrix}$$

In brief, the mapping f is given by $x \mapsto Ax$, where x represents the coordinates with respect to the basis (b_1, b_2).

We now wish to find a representation of this mapping f when the coordinates are written with respect to another basis (c_1, c_2) of $\mathbb{R}^2$. Let (x_1', x_2'), (y_1', y_2') be the coordinates of the vectors v, $f(v)$, respectively, with respect to the basis (c_1, c_2). Then

$$\begin{bmatrix} x_1' \\ x_2' \end{bmatrix} = P \begin{bmatrix} x_1 \\ x_2 \end{bmatrix}, \quad \begin{bmatrix} y_1' \\ y_2' \end{bmatrix} = P \begin{bmatrix} y_1 \\ y_2 \end{bmatrix}$$

where P is the matrix of transformation of coordinates from the basis (b_1, b_2) to the basis (c_1, c_2). Hence

$$\begin{bmatrix} y'_1 \\ y'_2 \end{bmatrix} = P \begin{bmatrix} y_1 \\ y_2 \end{bmatrix} = PA \begin{bmatrix} x_1 \\ x_2 \end{bmatrix} = PAP^{-1} \begin{bmatrix} x'_1 \\ x'_2 \end{bmatrix}$$

Thus, in terms of coordinates with respect to the basis (c_1, c_2), the mapping f is given by $x' \mapsto A'x'$, where $A' = PAP^{-1}$.

Let us apply the foregoing result to the case when $A \in G_0$, where G_0 is the point group of a given WP group G. Then A is an orthogonal matrix and $\det A = \pm 1$. Let $f : \mathbb{R}^2 \to \mathbb{R}^2$ be the mapping given by $x \mapsto Ax$, where x represents the coordinates with respect to the standard basis (e_1, e_2)—that is, with respect to the original rectangular axes. Let L be the translation lattice of G, and let (b_1, b_2) be a basis of L. Then, in terms of the coordinates with respect to the basis (b_1, b_2), the mapping f is given by $x' \mapsto A'x'$, where $A' = PAP^{-1}$, and P is the matrix of transformation of coordinates from the standard basis (e_1, e_2) to the basis (b_1, b_2). By Theorem 6.2.6, the mapping f maps L onto L. Hence the components of x' and $A'x'$ are integers, which implies that the entries in A' must be all integers. Moreover, $\det A' = \det(PAP^{-1}) = \det A = \pm 1$. Hence A' is a unimodular matrix.

The foregoing result can be expressed as follows: With respect to the standard basis, the elements of a point group G_0 are orthogonal 2×2 matrices. But with respect to a basis of the translation lattice L, the elements of G_0 are unimodular 2×2 matrices.

Let U_2 denote the set of all unimodular 2×2 matrices. It can be easily shown that U_2 is a group under multiplication. Thus, with respect to a basis of L, G_0 is a subgroup of U_2. We record these results in the following theorem.

THEOREM 6.3.2 Let G_0 be the point group of a WP group G with translation lattice L. Let $A \in G_0$, and let (b_1, b_2) be a basis of L. The mapping $x \mapsto Ax$ with respect to the standard basis (e_1, e_2) transforms into $x' \mapsto A'x'$ with respect to the basis (b_1, b_2), where $A' = PAP^{-1}$, and P is the matrix of transformation of coordinates from (e_1, e_2) to (b_1, b_2). Moreover, the matrix A' is unimodular. Consequently, with respect to the basis (b_1, b_2), G_0 is a subgroup of U_2.

This result enables us to find the values of n for which a point group G_0 can be isomorphic to C_n or D_n.

THEOREM 6.3.3 Let G_0 be the point group of a WP group G. Then G_0 is isomorphic with C_n or D_n, where n is equal to 1, 2, 3, 4, or 6.

Proof: We have shown in Theorem 6.2.8 that G_0 is isomorphic with C_n or D_n, for some integer n. In either case, G_0 contains the element

$$A = \begin{bmatrix} \cos\theta & -\sin\theta \\ \sin\theta & \cos\theta \end{bmatrix}$$

where $\theta = 2\pi/n$. Consider the mapping $f : \mathbb{R}^2 \to \mathbb{R}^2$ given by $x \mapsto Ax$, where x represents the coordinates with respect to the standard basis (e_1, e_2). (Geometrically, f represents a rotation through angle $\theta = 2\pi/n$.) Let (b_1, b_2) be a basis of the translation lattice L of G, and let P be the matrix of transformation from the standard basis (e_1, e_2) to the lattice basis (b_1, b_2). With respect to the basis (b_1, b_2), the mapping f is given by $x' \mapsto A'x'$, where $A' = PAP^{-1}$. By Theorem 6.3.2, the entries in the matrix A' are all integers. Hence the trace of A' is an integer. Now, using the property $tr(AB) = tr(BA)$, we have

$$tr(A') = tr(PAP^{-1}) = tr(AP^{-1}P) = tr(A) = 2\cos\theta$$

Hence $2\cos\theta$ must be an integer, which implies $\cos\theta = 0$, $\pm\dfrac{1}{2}$, ± 1. Therefore $\theta = 2\pi/n$, where $n = 1, 2, 3, 4,$ or 6.   ∎

Theorem 6.3.3 shows that the point group of any WP group is isomorphic with one of the ten groups C_n, D_n, $n = 1, 2, 3, 4, 6$. The groups C_2 and D_1 are isomorphic, but we treat them as separate cases. Geometrically, C_2 contains two rotations, but D_1 consists of a rotation and a reflection. In algebraic terms, $C_2 = \{I, A\}$ with $\det A = 1$, but $D_1 = \{I, B\}$ with $\det B = -1$.

6.4 POINT GROUPS AND LATTICE TYPES

We showed in Section 6.3 that the point group G_0 of any WP group G is of one of 10 types—namely, C_n or D_n, where $n = 1, 2, 3, 4,$ or 6. Theorem 6.2.6 stated that for every $A \in G_0$, the mapping $x \mapsto Ax$ maps the translation lattice L onto L. (In brief, we can say that the point group G_0 maps L onto L.) This property entails a restriction on the type of the translation lattice L that is compatible with a given point group G_0.

Let (b_1, b_2) be a basis of L. We say the lattice L is *rectangular, quadratic,* or *rhombic* according to whether the parallelogram formed by the basis vectors b_1, b_2 is a rectangle, square, or rhombus, respectively. (Evidently, a quadratic lattice is rectangular as well as rhombic.) Moreover, a rhombic lattice is called *hexagonal* if the angles in the rhombus are $\pi/2$, $2\pi/3$. In this case, the shorter diagonal of the rhombus divides it into two equilateral triangles, and six such equilateral triangles with a common vertex form a hexagon (see the diagram below). We refer to L as a *general* lattice if it doesn't necessarily belong to any of these types.

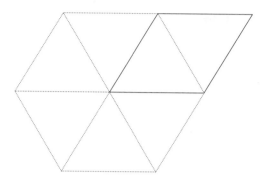

The next two theorems determine the lattice types compatible with the various types of point groups.

THEOREM 6.4.1 Let G_0 be the point group of a WP group G with translation lattice L. Suppose the group G_0 is isomorphic with C_n or D_n for some $n > 2$. Then L is a rhombic lattice.

Moreover, if $n = 4$, then L is a quadratic lattice. If $n = 3$ or 6, then L is hexagonal.

Proof: If G_0 is isomorphic with C_n or D_n, then G_0 contains the element

$$A = \begin{bmatrix} \cos\theta & -\sin\theta \\ \sin\theta & \cos\theta \end{bmatrix}$$

where $\theta = 2\pi/n$. Let t_1 be a nonzero vector of minimal length in L, and let $t_2 = At_1$. By Theorem 6.2.6, $t_2 \in L$. We claim that (t_1, t_2) is a basis of L.

The mapping $x \mapsto Ax$ represents a rotation through an angle $2\pi/n$. Since $n > 2$, it is a rotation through an angle less than π. Hence the vectors t_1, t_2 are not parallel and are of equal length, so every vector can be expressed as a linear combination of t_1, t_2. Let $a \in L$ and suppose

$$a = x_1 t_1 + x_2 t_2$$

where $x_1, x_2 \in \mathbb{R}$. Let

$$b = m_1 t_1 + m_2 t_2$$

where m_1, m_2 are integers closest to x_1, x_2, respectively. Then $b \in L$ and hence $s = a - b \in L$. Now

$$s = y_1 t_1 + y_2 t_2$$

where $y_1 = x_1 - m_1$ and $y_2 = x_2 - m_2$. Hence $|y_1| \leq \dfrac{1}{2}$ and $|y_2| \leq \dfrac{1}{2}$. Since t_1, t_2 are not parallel and have the same length, we have

$$|s| = |y_1 t_1 + y_2 t_2| < |y_1 t_1| + |y_2 t_2| = (|y_1| + |y_2|) |t_1| \leq |t_1|$$

so $|s| < |t_1|$. Since t_1 is a nonzero vector of minimal length in L, it follows that $s = 0$, which implies $y_1 = y_2 = 0$. Hence x_1, x_2 are integers. Thus every vector in L is an integral linear combination of t_1, t_2, which proves that (t_1, t_2) is a basis of the lattice L.

Since the vectors t_1, t_2 are the same length, the parallelogram formed by them is a rhombus, so L is a rhombic lattice. The angle between the vectors t_1, t_2 is $2\pi/n$. If $n = 4$, they are at right angles, so L is a quadratic lattice. If $n = 3$ or 6, the angles in the rhombus are $\pi/3$, $2\pi/3$, and hence the lattice is hexagonal. ∎

THEOREM 6.4.2 Suppose G_0 is isomorphic with D_n for some n. Then the lattice L is either rectangular or rhombic.

Proof: If $G_0 \simeq D_n$, then G_0 contains an element B with $\det B = -1$. The mapping $x \mapsto Bx$ represents a reflection in some line OA. Let a be any nonzero vector in L, neither parallel nor perpendicular to the line OA. Let $b = Ba$. By Theorem 6.2.6, $b \in L$, and hence $a + b, a - b \in L$. Now b is the reflection of a in the line OA. Hence $a + b$ and $a - b$ are nonzero vectors in L, parallel and perpendicular to the line OA, respectively. So there must exist nonzero vectors t_1, t_2 of minimal lengths in L, parallel and perpendicular to OA, respectively. We claim that every vector $t \in L$ parallel to OA is an integral multiple of t_1. Suppose $t = xt_1$ and x is not an integer. Let x' be the greatest integer less than x. Then $x't_1 \in L$ and hence $(x - x')t_1 = t - x't_1 \in L$, a contradiction since $0 < x - x' < 1$. So x must be an integer. Likewise, any vector in L perpendicular to OA is an integral multiple of t_2.

Again, let a be any vector in L and let $b = Ba$. Then $a + b$ and $a - b$ are either zero vectors or parallel and perpendicular to the line OA, respectively. Hence $a + b = mt_1$ and $a - b = nt_2$, where m, n are integers. From these relations, we obtain

$$a = \frac{m}{2}t_1 + \frac{n}{2}t_2$$

We claim that m and n are either both even or both odd. Suppose m is even and n is odd. Then

$$\frac{1}{2}t_2 = a - \left(\frac{m}{2}t_1 + \frac{n-1}{2}t_2\right) \in L$$

which contradicts the fact that t_2 is a nonzero vector of minimal length perpendicular to OA. A similar contradiction occurs if we assume m is odd and n is even.

Now two cases arise: either $\dfrac{t_1 + t_2}{2} \in L$ or $\dfrac{t_1 + t_2}{2} \notin L$.

Case 1. If $\dfrac{t_1 + t_2}{2} \in L$, then $\dfrac{t_1 - t_2}{2} = t_1 - \dfrac{t_1 + t_2}{2} \in L$. We claim that $\left(\dfrac{t_1 + t_2}{2}, \dfrac{t_1 - t_2}{2}\right)$ is a basis of L. We have seen that any $a \in L$ can be expressed as $a = \dfrac{m}{2}t_1 + \dfrac{n}{2}t_2$, where m, n are integers that are both even or both odd. So

$$
\begin{aligned}
a &= \frac{m}{2}t_1 + \frac{n}{2}t_2 \\
&= \left(\frac{m+n}{2}\right)\left(\frac{t_1+t_2}{2}\right) + \left(\frac{m-n}{2}\right)\left(\frac{t_1-t_2}{2}\right)
\end{aligned}
$$

Because m and n are both even or both odd, it follows that $\dfrac{m+n}{2}$ and $\dfrac{m-n}{2}$ are both integers. Thus every vector in L is an integral linear combination of $\dfrac{t_1+t_2}{2}$ and $\dfrac{t_1 - t_2}{2}$. Hence $\left(\dfrac{t_1+t_2}{2}, \dfrac{t_1-t_2}{2}\right)$ is a basis of L. The vectors $\dfrac{t_1+t_2}{2}$ and $\dfrac{t_1-t_2}{2}$ are of equal length, and hence L is a rhombic lattice.

Case 2. Suppose $\dfrac{t_1 + t_2}{2} \notin L$. As before, given any $a \in L$, a can be expressed as

$$
a = \frac{m}{2}t_1 + \frac{n}{2}t_2
$$

where m and n are both even or both odd. We claim that if $\dfrac{t_1+t_2}{2} \notin L$, then m and n are both even. Suppose m and n are odd for some a. Then

$$
\frac{t_1 + t_2}{2} = a - \left(\frac{m-1}{2}t_1 + \frac{n-1}{2}t_2\right) \in L
$$

which contradicts our hypothesis. So m and n are even for every $a \in L$. Thus every vector in L is an integral linear combination of t_1, t_2, which proves that (t_1, t_2) is a basis of L. The basis vectors t_1, t_2 are at right angles; hence L is a rectangular lattice. This proves that the lattice L is either rectangular or rhombic. ∎

The foregoing two theorems determine the lattice type compatible with a point group G_0 in all cases except where G_0 is isomorphic with C_1 or C_2. But in this case it is clear that the translation lattice L is of general type because the identity mapping and a rotation through angle π both map any lattice L onto L.

The following table gives the types of translation lattices that correspond to the various types of point groups.

Type of Point Group G_0	Type of Translation Lattice L
C_1, C_2	General
D_1, D_2	Rectangular or rhombic
C_4, D_4	Quadratic
C_3, C_6, D_3, D_6	Hexagonal

6.5 EQUIVALENCE OF WP GROUPS

Our aim in this chapter, as stated earlier, is to obtain a scheme for classifying WP groups. But what is the criterion for this classification? When do we consider two WP groups to be equivalent or belonging to the same class? Ordinarily, we consider two groups to be equivalent, or abstractly the same, if they are isomorphic. But this criterion is not enough in the case of WP groups. A WP group G is by definition linked with a translation lattice L, and L determines a subgroup of G—namely, the translation subgroup T. Therefore it seems natural to stipulate not only that two equivalent WP groups G, G' should be isomorphic but also that their respective translation subgroups T, T' should correspond to each other under the isomorphism between G and G'. So we have the following definition.

Definition 6.5.1 Let G, G' be WP groups with translation subgroups T, T', respectively. Then G, G' are said to be *equivalent* WP groups if there exists an isomorphism $\phi : G \to G'$ such that $\phi(T) = T'$.

It is clear from the definition that if G, G' are equivalent WP groups under an isomorphism $\phi : G \to G'$, then ϕ induces an isomorphism between T and T', given by the restriction of ϕ to T. We show in the next theorem that ϕ also induces two other isomorphisms—namely, an isomorphism between their translation lattices L, L', and an isomorphism between their point groups G_0, G_0'.

Recall that an element (A, a) in G denotes the mapping $x \mapsto Ax + a$. The product of two elements in G is given by the rule $(A, a)\,(B, b) = (AB + Ab + a)$, and the inverse is given by $(A, a)^{-1} = (A^{-1}, -A^{-1}a)$.

THEOREM 6.5.2 Let G, G' be equivalent WP groups, and let $\phi : G \to G'$ be an isomorphism such that $\phi(T) = T'$, where T, T' are the translation subgroups of G, G', respectively. Let L, L' be the translation lattices of G, G', and G_0, G_0' their point groups, respectively. Then ϕ induces an isomorphism $\phi_1 : L \to L'$ and an isomorphism $\phi_2 : G_0 \to G_0'$ such that

(a)   For all $t \in L$,

$$\phi(I, t) = (I, \phi_1\,(t))$$

(b)   For all $A \in G_0$ and all $a \in \mathbb{R}^2$ such that $(A, a) \in G$,

$$\phi(A, a) = (\phi_2(A), a') \qquad \text{for some } a' \in \mathbb{R}^2$$

(c)   For all $A \in G_0$, $t \in L$,

$$\phi_1(At) = \phi_2(A)\phi_1(t)$$

Proof: **(a)** For any $t \in L$, $(I, t) \in T$ and hence $\phi(I, t) \in T'$. So there exists a unique $t' \in L'$ such that $\phi(I, t) = (I, t')$. This defines a mapping $\phi_1 : L \to L'$ such that

$\phi(I, t) = (I, \phi_1(t))$. Since ϕ is bijective and maps T onto T', it follows that ϕ_1 is bijective. Let $a, b \in L$. Then

$$\phi(I, a + b) = \phi((I, a)(I, b)) = \phi(I, a)\phi(I, b)$$

so

$$(I, \phi_1(a + b)) = (I, \phi_1(a))(I, \phi_1(b)) = (I, \phi_1(a) + \phi_1(b))$$

Hence $\phi_1(a + b) = \phi_1(a) + \phi_1(b)$. This proves that ϕ_1 is an isomorphism.

(b) Let $A \in G_0$. Then $(A, a) \in G$ for some $a \in \mathbb{R}^2$. Let $\phi(A, a) = (A', a')$. We claim that A' is uniquely determined by A, irrespective of the choice of a. Suppose $b \in \mathbb{R}^2$ such that $(A, b) \in G$. Then, by Theorem 6.2.3, $b = a + t$ for some $t \in L$, so $(A, b) = (A, a + t) = (I, t)(A, a)$. Hence $\phi(A, b) = \phi(I, t)\phi(A, a) = (I, \phi_1(t))$ $(A', a') = (A', a' + \phi_1(t))$. Thus ϕ determines a unique element A' in G_0' corresponding to A, irrespective of the choice of a. This defines a mapping $\phi_2 : G_0 \to G_0'$ such that $\phi(A, a) = (\phi_2(A), a')$ whenever $(A, a) \in G$.

Let $A, B \in G_0$. Then $(A, a), (B, b) \in G$ for some a, b. Now $(A, a)(B, b) = (AB, Ab + a)$. Hence $\phi(A, a)\phi(B, b) = \phi(AB, Ab + a)$, which yields

$$(\phi_2(A), a')(\phi_2(B), b') = (\phi_2(AB), c')$$

for some a', b', c'. Thus it follows that $\phi_2(A)\phi_2(B) = \phi_2(AB)$. So ϕ_2 is a homomorphism.

Now consider $\ker \phi_2$. Let $A \in G_0$ such that $\phi_2(A) = I$. Then $\phi(A, a) = (I, a')$ for some a'. Now $(I, a') \in T'$; therefore $(A, a) \in T$ and hence $A = I$. So $\ker \phi_2 = \{I\}$; hence ϕ_2 is injective. Clearly, since ϕ is surjective, it follows that ϕ_2 is surjective. This proves that ϕ_2 is an isomorphism.

(c) Let $A \in G_0$, $t \in L$. Then $(A, a) \in G$ for some a, and also $(I, t) \in G$. Hence $(A, a)(I, t)(A, a)^{-1} = (I, At) \in G$. Therefore

$$\begin{aligned} (I, \phi_1(At)) &= \phi(I, At) \\ &= \phi(A, a)\phi(I, t)(\phi(A, a))^{-1} \\ &= (\phi_2(A), a')(I, \phi_1(t))(\phi_2(A), a')^{-1} \\ &= (I, \phi_2(A)\phi_1(t)) \end{aligned}$$

Hence $\phi_1(At) = \phi_2(A)\phi_1(t)$.

This completes the proof of the theorem. ∎

Theorem 6.5.2 has shown that the point groups of two equivalent WP groups are isomorphic. The next theorem shows that there is in fact a stronger relation between them. Recall from Theorem 6.3.2 that the elements of the point group G_0, when expressed with respect to a basis of the translation lattice L, are unimodular matrices.

THEOREM 6.5.3 Let G, G' be equivalent WP groups, and let $\phi : G \rightarrow G'$ be an isomorphism with $\phi(T) = T'$, where T, T' are the translation subgroups of G, G', respectively. Let L, L' be their translation lattices, respectively, and suppose that the elements of G, G' are expressed with respect to given bases of L, L', respectively. Let G_0, G_0' be their point groups. Then there exists a unimodular matrix P such that

(a)   For all $t \in L$, $\phi(I, t) = (I, Pt)$.
(b)   For all $(A, a) \in G$, $\phi(A, a) = (PAP^{-1}, a')$ for some a'.
(c) $G_0' = PG_0 P^{-1}$.

Proof: **(a)** By Theorem 6.5.2, ϕ induces an isomorphism $\phi_1 : L \rightarrow L'$ and also an isomorphism $\phi_2 : G_0 \rightarrow G_0'$. Let (b_1, b_2), (c_1, c_2) be bases in L, L', respectively. Let $\phi_1(b_1) = b_1'$ and $\phi_1(b_2) = b_2'$. Then (b_1', b_2') is also a basis in L'. For, given $t' \in L'$, there is a unique $t \in L$ such that $\phi_1(t) = t'$. Now $t = mb_1 + nb_2$ for some integers m, n. Hence $t' = \phi_1(t) = m\phi_1(b_1) + n\phi_1(b_2) = mb_1' + nb_2'$. This proves that (b_1', b_2') is a basis of L'.

Let P be the matrix of transformation from the basis (b_1', b_2') to (c_1, c_2). By Theorem 6.3.1, P is a unimodular matrix. Given $t = mb_1 + nb_2 \in L$, we have $\phi_1(t) = mb_1' + nb_2' \in L'$. With respect to the basis (c_1, c_2), let $\phi_1(t) = m'c_1 + n'c_2$. Then $\begin{bmatrix} m' \\ n' \end{bmatrix} = P \begin{bmatrix} m \\ n \end{bmatrix}$. Thus, on identifying the vectors t, $\phi_1(t)$ with their coordinates with respect to the bases (b_1, b_2), (c_1, c_2), respectively, we have the result $\phi_1(t) = Pt$ for all $t \in L$. Hence, by Theorem 6.5.2, $\phi(I, t) = (I, \phi_1(t)) = (I, Pt)$ for all $t \in L$.

(b) Let $(A, a) \in G$, $t \in L$. Then $A \in G_0$, $At \in L$. Hence, by Theorem 6.5.2,

$$PAt = \phi_1(At) = \phi_2(A)\phi_1(t) = \phi_2(A)Pt$$

Since this holds for all $t \in L$, it follows that $PA = \phi_2(A)P$, so $\phi_2(A) = PAP^{-1}$. Hence, by Theorem 6.5.2, $\phi(A, a) = (\phi_2(A, a')) = (PAP^{-1}, a')$ for some a'.

(c) The relation $\phi_2(A) = PAP^{-1}$ holds for all $A \in G_0$, which proves that $G_0' = PG_0 P^{-1}$.   ∎

The relation between the point groups of two equivalent WP groups obtained in the foregoing theorem motivates the following definition.

Definition 6.5.4 Let G_0, G_0' be the point groups of two WP groups with translation lattices L, L', respectively, and suppose that the elements of G_0, G_0' are expressed with respect to given bases of L, L', respectively. If there exists a unimodular matrix P such that $G_0' = PG_0 P^{-1}$, then G_0, G_0' are said to be *equivalent* point groups.

Theorem 6.5.3 implies that the point groups of two equivalent WP groups are equivalent. The converse is not true, however. As we shall see subsequently, two nonequivalent WP groups can have equivalent point groups.

6.6 CLASSIFICATION OF POINT GROUPS

Before we take up the classification of WP groups, which is our goal in this chapter, we consider the classification of point groups. It is clear from the definitions of equivalence of WP groups and point groups, given in Section 6.5, that if the point groups of two WP groups G, G' are nonequivalent, then G, G' themselves must be nonequivalent. So we first find all point groups up to equivalence.

Let us recapitulate some important properties of point groups that we proved in earlier sections. By definition, the point group G_0 of a WP group G consists of matrices A such that $(A, a) \in G$ for some vector $a \in \mathbb{R}^2$, where (A, a) denotes the mapping $x \mapsto Ax + a$. Each A in G_0 is a matrix of the form

$$A = \begin{bmatrix} \cos\theta & -\sin\theta \\ \sin\theta & \cos\theta \end{bmatrix} \quad \text{or} \quad A = \begin{bmatrix} \cos\theta & \sin\theta \\ \sin\theta & -\cos\theta \end{bmatrix}$$

In the first case, the mapping $x \mapsto Ax$ represents a rotation; in the second case, a reflection. In brief, we can say that each element in G_0 represents a rotation or a reflection. In both cases, the matrix A is orthogonal. Thus G_0 is a subgroup of the group O_2 of all orthogonal 2×2 matrices.

However, this description of G_0 holds when the coordinates of a point in the plane are written with respect to rectangular axes or, equivalently, the components of a vector are written with respect to the standard basis (e_1, e_2). If we chose an arbitrary basis (b_1, b_2) for writing the components of a vector, then the mapping $x \mapsto Ax$ transforms into $x' \mapsto A'x'$, where $A' = PAP^{-1}$, and P is the matrix of transformation of coordinates from the basis (e_1, e_2) to the basis (b_1, b_2). Moreover, if (b_1, b_2) is a basis of the translation lattice L of G, then A' is a unimodular matrix. Thus, with respect to a basis of L, each element in G_0 is a unimodular matrix, so G_0 is a subgroup of the group U_2 of all unimodular 2×2 matrices.

In the sequel, we assume that the point group G_0 is expressed with respect to a basis of the translation lattice L. But L can have several bases. Which particular basis do we choose? Suppose (b_1, b_2) and (c_1, c_2) are both bases of L. Let G_0 and G_0' be the representations of the point group of G with respect to these bases. Given $A \in G_0$, let A' be the corresponding element in G_0'. Then $A' = PAP^{-1}$, where P is the matrix of transformation of coordinates from the basis (b_1, b_2) to the basis (c_1, c_2). So $G_0' = PG_0P^{-1}$. By Theorem 6.3.1, the matrix P is unimodular, and hence G_0 and G_0' are equivalent point groups. Therefore, since we are interested in finding all point groups up to equivalence, it is immaterial which particular basis of the translation lattice L is chosen.

For any WP group G, its point group G_0 is isomorphic with one of the groups C_n, D_n, where $n = 1, 2, 3, 4, 5$. The translation lattice L of G is of one of the following five types: general, rectangular, rhombic, quadratic, and hexagonal. The lattice type depends on the point group type—that is, on the group C_n, D_n with which G_0 is isomorphic. If $G_0 \simeq D_1$ or D_2, then L may be rectangular or rhombus. In all other cases, the lattice type is uniquely determined by the point group type. So we have to consider all 12 cases of (G_0, L) combinations. In each of these cases, we wish to find a representation of the point group G_0 with respect to a basis (b_1, b_2) of the translation lattice L.

If $G_0 \simeq C_n$, then $G_0 = \{I, A, \ldots, A^{n-1}\}$, where A represents a rotation through angle $2\pi/n$. The matrix A is completely determined by its effect on the basis vectors b_1, b_2. Suppose $A = \begin{bmatrix} p_1 & p_2 \\ q_1 & q_2 \end{bmatrix}$. Then

$$Ab_1 = \begin{bmatrix} p_1 & p_2 \\ q_1 & q_2 \end{bmatrix} \begin{bmatrix} 1 \\ 0 \end{bmatrix} = \begin{bmatrix} p_1 \\ q_1 \end{bmatrix}$$

$$Ab_2 = \begin{bmatrix} p_1 & p_2 \\ q_1 & q_2 \end{bmatrix} \begin{bmatrix} 0 \\ 1 \end{bmatrix} = \begin{bmatrix} p_2 \\ q_2 \end{bmatrix}$$

Thus the columns of A are the vectors to which the basis vectors b_1, b_2 go under the mapping $x \mapsto Ax$. When the matrix A has been found, the other elements of G_0 are obtained by computing the powers of A.

If $G_0 \simeq D_n$, then $G_0 = \{I, A, \ldots, A^{n-1}, B, AB, \ldots, A^{n-1}B\}$, where A represents a rotation through angle $2\pi/n$, and B represents a reflection in some axis of symmetry of L. As before, the matrices A and B are completely determined by their effects on the basis vectors b_1, b_2. When these matrices A, B have been found, the point group G_0 can be easily written down.

We will now find all point groups up to equivalence.

1. $G_0 \simeq C_1$, L general

Here, obviously, $G_0 = \left\{ \begin{bmatrix} 1 & 0 \\ 0 & 1 \end{bmatrix} \right\}$.

2. $G_0 \simeq C_2$, L general

Here $G_0 = \{I, A\}$, where A represents a rotation through an angle π. So A takes the basis vectors b_1, b_2 to $-b_1, -b_2$, respectively. Hence $A = \begin{bmatrix} -1 & 0 \\ 0 & -1 \end{bmatrix}$. Thus

$$G_0 = \left\{ \begin{bmatrix} 1 & 0 \\ 0 & 1 \end{bmatrix}, \begin{bmatrix} -1 & 0 \\ 0 & -1 \end{bmatrix} \right\}$$

3. $G_0 \simeq C_3$, L hexagonal

Here $G_0 = \{I, A, A^2\}$, where A represents a rotation through an angle $2\pi/3$. Let the basis vectors b_1, b_2 include an angle $2\pi/3$. Then A takes b_1 to b_2, and b_2

to $-(b_1 + b_2)$. Hence $A = \begin{bmatrix} 0 & -1 \\ 1 & -1 \end{bmatrix}$. Therefore

$$G_0 = \left\{ \begin{bmatrix} 1 & 0 \\ 0 & 1 \end{bmatrix}, \begin{bmatrix} 0 & -1 \\ 1 & -1 \end{bmatrix}, \begin{bmatrix} -1 & 1 \\ -1 & 0 \end{bmatrix} \right\}$$

If, however, we take the basis vectors to be c_1, c_2 that include an angle $\pi/3$, we obtain a different representation of the point group. Now A takes c_1 to $-c_1 + c_2$, and c_2 to $-c_1$. Hence $A = \begin{bmatrix} -1 & -1 \\ 1 & 0 \end{bmatrix}$. Thus, with respect to the basis c_1, c_2, the point group is given by

$$G_0' = \left\{ \begin{bmatrix} 1 & 0 \\ 0 & 1 \end{bmatrix}, \begin{bmatrix} -1 & -1 \\ 1 & 0 \end{bmatrix}, \begin{bmatrix} 0 & 1 \\ -1 & -1 \end{bmatrix} \right\}$$

As explained above, the point groups G_0, G_0' must be equivalent, and $G_0' = PG_0P^{-1}$, where P is the matrix of transformation from the basis (b_1, b_2) to the basis (c_1, c_2). In fact, we can write the relations between the two bases as $b_1 = c_1$ and $b_2 = -c_1 + c_2$. Hence the matrix of transformation from the basis (b_1, b_2) to the basis (c_1, c_2) is $P = \begin{bmatrix} 1 & -1 \\ 0 & 1 \end{bmatrix}$. It can be directly checked that $G_0' = PG_0P^{-1}$.

4. $G_0 \simeq C_4$, L quadratic

Here $G_0 = \{I, A, A^2, A^3\}$, where A represents a rotation through an angle $\pi/2$. The basis vectors b_1, b_2 include an angle $\pi/2$. So A takes b_1 to b_2, and b_2 to $-b_1$. Hence $A = \begin{bmatrix} 0 & -1 \\ 1 & 0 \end{bmatrix}$. Therefore

$$G_0 = \left\{ \begin{bmatrix} 1 & 0 \\ 0 & 1 \end{bmatrix}, \begin{bmatrix} 0 & -1 \\ 1 & 0 \end{bmatrix}, \begin{bmatrix} -1 & 0 \\ 0 & -1 \end{bmatrix}, \begin{bmatrix} 0 & 1 \\ -1 & 0 \end{bmatrix} \right\}$$

5. $G_0 \simeq C_6$, L hexagonal

Here $G_0 = \{I, A, A^2, A^3, A^4, A^5\}$, where A represents a rotation through an angle $\pi/3$. If the basis vectors b_1, b_2 include an angle $\pi/3$, then A takes b_1 to b_2, and b_2 to $-b_1 + b_2$. Hence $A = \begin{bmatrix} 0 & -1 \\ 1 & 1 \end{bmatrix}$. Therefore

$$G_0 = \left\{ \begin{bmatrix} 1 & 0 \\ 0 & 1 \end{bmatrix}, \begin{bmatrix} 0 & -1 \\ 1 & 1 \end{bmatrix}, \begin{bmatrix} -1 & -1 \\ 1 & 0 \end{bmatrix}, \begin{bmatrix} -1 & 0 \\ 0 & -1 \end{bmatrix}, \begin{bmatrix} 0 & 1 \\ -1 & -1 \end{bmatrix}, \begin{bmatrix} 1 & 1 \\ -1 & 0 \end{bmatrix} \right\}$$

If we take the basis vectors to include an angle $2\pi/3$, then $A = \begin{bmatrix} 1 & -1 \\ 1 & 0 \end{bmatrix}$. This yields an equivalent representation of the point group, given by

$$G_0' = \left\{ \begin{bmatrix} 1 & 0 \\ 0 & 1 \end{bmatrix}, \begin{bmatrix} 1 & -1 \\ 1 & 0 \end{bmatrix}, \begin{bmatrix} 0 & -1 \\ 1 & -1 \end{bmatrix}, \begin{bmatrix} -1 & 0 \\ 0 & -1 \end{bmatrix}, \begin{bmatrix} -1 & 1 \\ -1 & 0 \end{bmatrix}, \begin{bmatrix} 0 & 1 \\ -1 & 1 \end{bmatrix} \right\}$$

The matrix of transformation from the first basis to the second is $P = \begin{bmatrix} 1 & 1 \\ 0 & 1 \end{bmatrix}$. Hence $G'_0 = P G_0 P^{-1}$.

6. $G_0 \simeq D_1$, L rectangular

Here $G_0 = \{I, B\}$, where B represents a reflection. The basis vectors b_1, b_2 include an angle $\pi/2$. If B represents a reflection in the line along the vector b_1, then B fixes b_1 and takes b_2 to $-b_2$. Hence $B = \begin{bmatrix} 1 & 0 \\ 0 & -1 \end{bmatrix}$. Thus

$$G_0 = \left\{ \begin{bmatrix} 1 & 0 \\ 0 & 1 \end{bmatrix}, \begin{bmatrix} 1 & 0 \\ 0 & -1 \end{bmatrix} \right\}$$

If we take B to be reflection in the line along b_2, then $B = \begin{bmatrix} -1 & 0 \\ 0 & 1 \end{bmatrix}$, which yields an equivalent representation $G'_0 = P G_0 P^{-1}$ of the point group, where $P = \begin{bmatrix} 0 & 1 \\ 1 & 0 \end{bmatrix}$. In fact, P is the matrix of transformation from the basis (b_1, b_2) to (c_1, c_2), where $b_1 = c_2$, $b_2 = c_1$. It is easy to show that this point group G_0 (or G'_0) and the point group $\{I, A\}$ obtained earlier in case 2 above are not equivalent even though they are isomorphic groups. For any P, $\det(P A P^{-1}) = \det A$. Now $\det A = \det \begin{bmatrix} -1 & 0 \\ 0 & -1 \end{bmatrix} = 1$, but $\det B = \det \begin{bmatrix} 1 & 0 \\ 0 & -1 \end{bmatrix} = -1$. Hence there is no P such that $P A P^{-1} = B$.

7. $G_0 \simeq D_1$, L rhombic

Here again $G_0 = \{I, B\}$, where B represents a reflection. If B represents a reflection in the diagonal of the rhombus bisecting the angle between the basis vectors b_1, b_2, then B takes b_1 to b_2 and b_2 to b_1. Hence $B = \begin{bmatrix} 0 & 1 \\ 1 & 0 \end{bmatrix}$. Thus

$$G_0 = \left\{ \begin{bmatrix} 1 & 0 \\ 0 & 1 \end{bmatrix}, \begin{bmatrix} 0 & 1 \\ 1 & 0 \end{bmatrix} \right\}$$

If we take B to be a reflection in the other diagonal, then $B = \begin{bmatrix} 0 & -1 \\ -1 & 0 \end{bmatrix}$. This gives an equivalent representation G'_0 that is related to G_0 by $G'_0 = P G_0 P^{-1}$, where $P = \begin{bmatrix} 1 & 0 \\ 0 & -1 \end{bmatrix}$. In fact, P is the matrix of transformation from the basis (b_1, b_2) to $(b_1, -b_2)$. Using the same argument as in the previous case, we see that this point group is not equivalent to the point group $\{I, A\}$ obtained in case 2 above. Further, we show that it is not equivalent to the point group in case 6, where the lattice L is rectangular. Suppose there exists a unimodular matrix P such that $G'_0 = P G_0 P^{-1}$, where

$$G_0 = \left\{ \begin{bmatrix} 1 & 0 \\ 0 & 1 \end{bmatrix}, \begin{bmatrix} 1 & 0 \\ 0 & -1 \end{bmatrix} \right\}, \quad G_0' = \left\{ \begin{bmatrix} 1 & 0 \\ 0 & 1 \end{bmatrix}, \begin{bmatrix} 0 & 1 \\ 1 & 0 \end{bmatrix} \right\}$$

Then $P \begin{bmatrix} 1 & 0 \\ 0 & -1 \end{bmatrix} P^{-1} = \begin{bmatrix} 0 & 1 \\ 1 & 0 \end{bmatrix}$, so $P \begin{bmatrix} 1 & 0 \\ 0 & -1 \end{bmatrix} = \begin{bmatrix} 0 & 1 \\ 1 & 0 \end{bmatrix} P$. Let $P = \begin{bmatrix} a & b \\ c & d \end{bmatrix}$.

We have

$$\begin{bmatrix} a & b \\ c & d \end{bmatrix} \begin{bmatrix} 1 & 0 \\ 0 & -1 \end{bmatrix} = \begin{bmatrix} 0 & 1 \\ 1 & 0 \end{bmatrix} \begin{bmatrix} a & b \\ c & d \end{bmatrix}$$

which yields $a = c$, $b = -d$, so $\det P = ad - bc = 2ad$. This contradicts the fact that P is unimodular. Hence there does not exist a unimodular matrix P such that $G_0' = P G_0 P^{-1}$, which proves that G_0, G_0' are nonequivalent point groups.

8. $G_0 \simeq D_2$, L rectangular

Here $G_0 = \{I, A, B, AB\}$, where A represents a rotation through angle π and B represents a reflection. As already shown in cases 2 and 6, $A = \begin{bmatrix} -1 & 0 \\ 0 & -1 \end{bmatrix}$ and $B = \begin{bmatrix} 1 & 0 \\ 0 & -1 \end{bmatrix}$. Hence

$$G_0 = \left\{ \begin{bmatrix} 1 & 0 \\ 0 & 1 \end{bmatrix}, \begin{bmatrix} -1 & 0 \\ 0 & -1 \end{bmatrix}, \begin{bmatrix} 1 & 0 \\ 0 & -1 \end{bmatrix}, \begin{bmatrix} -1 & 0 \\ 0 & 1 \end{bmatrix} \right\}$$

The last two elements show that G_0 contains a reflection in each side of the rectangle.

9. $G_0 \simeq D_2$, L rhombic

Here again $G_0 = \{I, A, B, AB\}$, where A represents a rotation through angle π and B represents a reflection in a diagonal of the rhombus. As already shown in cases 2 and 7, $A = \begin{bmatrix} -1 & 0 \\ 0 & -1 \end{bmatrix}$ and $B = \begin{bmatrix} 0 & 1 \\ 1 & 0 \end{bmatrix}$. Hence

$$G_0 = \left\{ \begin{bmatrix} 1 & 0 \\ 0 & 1 \end{bmatrix}, \begin{bmatrix} -1 & 0 \\ 0 & -1 \end{bmatrix}, \begin{bmatrix} 0 & 1 \\ 1 & 0 \end{bmatrix}, \begin{bmatrix} 0 & -1 \\ -1 & 0 \end{bmatrix} \right\}$$

The last two elements of G_0 show that there is symmetry about each diagonal of the rhombus. We claim that the point groups in cases 8 and 9, both isomorphic with D_2, are not equivalent. Suppose there exists a unimodular matrix $P = \begin{bmatrix} a & b \\ c & d \end{bmatrix}$ such that $G_0' = P G_0 P^{-1}$, where G_0, G_0' are the point groups in cases 8 and 9, respectively. Now $\begin{bmatrix} 1 & 0 \\ 0 & -1 \end{bmatrix} \in G$. Hence $P \begin{bmatrix} 1 & 0 \\ 0 & -1 \end{bmatrix} P^{-1}$ must be equal to some element in G_0'. Since $\det \begin{bmatrix} 1 & 0 \\ 0 & -1 \end{bmatrix} = -1$, we know that $P \begin{bmatrix} 1 & 0 \\ 0 & -1 \end{bmatrix} P^{-1}$ must be equal to

$$\begin{bmatrix} 0 & 1 \\ 1 & 0 \end{bmatrix} \text{ or } \begin{bmatrix} 0 & -1 \\ -1 & 0 \end{bmatrix}.$$ As shown earlier, this leads to the result $\det P = 2ad$, which contradicts the hypothesis that P is unimodular.

10. $G_0 \simeq D_3$, L hexagonal

Here $G_0 = \{I, A, A^2, B, AB, A^2B\}$, where A represents a rotation through angle $2\pi/3$ and B represents a reflection in a diagonal of the rhombus. Let the basis vectors b_1, b_2 include an angle $2\pi/3$. Then, as shown in 3, $A = \begin{bmatrix} 0 & -1 \\ 1 & -1 \end{bmatrix}$. If B is a reflection in the shorter diagonal, bisecting the angle between the vectors b_1, b_2, then $B = \begin{bmatrix} 0 & 1 \\ 1 & 0 \end{bmatrix}$. Hence

$$G_0 = \left\{ \begin{bmatrix} 1 & 0 \\ 0 & 1 \end{bmatrix}, \begin{bmatrix} 0 & -1 \\ 1 & -1 \end{bmatrix}, \begin{bmatrix} -1 & 1 \\ -1 & 0 \end{bmatrix}, \begin{bmatrix} 0 & 1 \\ 1 & 0 \end{bmatrix}, \begin{bmatrix} -1 & 0 \\ -1 & 1 \end{bmatrix}, \begin{bmatrix} 1 & -1 \\ 0 & -1 \end{bmatrix} \right\}$$

If, however, B represents a reflection in the longer diagonal, then $B = \begin{bmatrix} 0 & -1 \\ -1 & 0 \end{bmatrix}$. So we get the point group

$$G_0' = \left\{ \begin{bmatrix} 1 & 0 \\ 0 & 1 \end{bmatrix}, \begin{bmatrix} 0 & -1 \\ 1 & -1 \end{bmatrix}, \begin{bmatrix} -1 & 1 \\ -1 & 0 \end{bmatrix}, \begin{bmatrix} 0 & -1 \\ -1 & 0 \end{bmatrix}, \begin{bmatrix} 1 & 0 \\ 1 & -1 \end{bmatrix}, \begin{bmatrix} -1 & 1 \\ 0 & 1 \end{bmatrix} \right\}$$

We claim that these point groups G_0, G_0', both isomorphic with D_3, are not equivalent. Suppose there exists a unimodular matrix $P = \begin{bmatrix} a & b \\ c & d \end{bmatrix}$ such that $G_0' = PG_0P^{-1}$. Then $P \begin{bmatrix} 0 & 1 \\ 1 & 0 \end{bmatrix} P^{-1}$ must be equal to one of the matrices in G_0'.

Now $\det \begin{bmatrix} 0 & 1 \\ 1 & 0 \end{bmatrix} = -1$ and hence $P \begin{bmatrix} 0 & 1 \\ 1 & 0 \end{bmatrix} P^{-1}$ must be equal to $\begin{bmatrix} 0 & -1 \\ -1 & 0 \end{bmatrix}$, $\begin{bmatrix} 1 & 0 \\ 1 & -1 \end{bmatrix}$, or $\begin{bmatrix} -1 & 1 \\ 0 & 1 \end{bmatrix}$. Consider the case $P \begin{bmatrix} 0 & 1 \\ 1 & 0 \end{bmatrix} P^{-1} = \begin{bmatrix} 0 & -1 \\ -1 & 0 \end{bmatrix}$. Then

$$\begin{bmatrix} a & b \\ c & d \end{bmatrix} \begin{bmatrix} 0 & 1 \\ 1 & 0 \end{bmatrix} = \begin{bmatrix} 0 & -1 \\ -1 & 0 \end{bmatrix} \begin{bmatrix} a & b \\ c & d \end{bmatrix}$$

which yields $a = -d$, $b = -c$. Hence $\det P = b^2 - a^2$. Since P is unimodular, we have $b^2 - a^2 = \pm 1$, so either $a = \pm 1, b = 0$ or $a = 0, b = \pm 1$. Hence either $P = \pm \begin{bmatrix} 1 & 0 \\ 0 & -1 \end{bmatrix}$ or $P = \pm \begin{bmatrix} 0 & 1 \\ -1 & 0 \end{bmatrix}$. Now $\begin{bmatrix} 0 & -1 \\ 1 & -1 \end{bmatrix} \in G_0$. Hence $P \begin{bmatrix} 0 & -1 \\ 1 & -1 \end{bmatrix} P^{-1}$

must be an element in G_0'. If we assume $P = \pm \begin{bmatrix} 1 & 0 \\ 0 & -1 \end{bmatrix}$, then

$$P \begin{bmatrix} 0 & -1 \\ 1 & -1 \end{bmatrix} P^{-1} = \begin{bmatrix} 1 & 0 \\ 0 & -1 \end{bmatrix} \begin{bmatrix} 0 & -1 \\ 1 & -1 \end{bmatrix} \begin{bmatrix} 1 & 0 \\ 0 & -1 \end{bmatrix} = \begin{bmatrix} 0 & 1 \\ -1 & -1 \end{bmatrix}$$

If, on the other hand, we assume $P = \pm \begin{bmatrix} 0 & 1 \\ -1 & 0 \end{bmatrix}$, then

$$P \begin{bmatrix} 0 & -1 \\ 1 & -1 \end{bmatrix} P^{-1} = \begin{bmatrix} 0 & 1 \\ -1 & 0 \end{bmatrix} \begin{bmatrix} 0 & -1 \\ 1 & -1 \end{bmatrix} \begin{bmatrix} 0 & -1 \\ 1 & 0 \end{bmatrix} = \begin{bmatrix} -1 & -1 \\ 1 & 0 \end{bmatrix}$$

Thus, in both cases, $P \begin{bmatrix} 0 & -1 \\ 1 & -1 \end{bmatrix} P^{-1} \notin G_0'$. Hence $P \begin{bmatrix} 0 & 1 \\ 1 & 0 \end{bmatrix} P^{-1}$ cannot be equal

to $\begin{bmatrix} 0 & -1 \\ -1 & 0 \end{bmatrix}$. Likewise, if we assume $P \begin{bmatrix} 0 & 1 \\ 1 & 0 \end{bmatrix} P^{-1} = \begin{bmatrix} 1 & 0 \\ 1 & -1 \end{bmatrix}$, we get $a +$

$b = c = d$, so $\det P = a^2 - b^2$. Hence $P = \pm \begin{bmatrix} 1 & 0 \\ 1 & 1 \end{bmatrix}$ or $\pm \begin{bmatrix} 0 & 1 \\ 1 & 1 \end{bmatrix}$. Here again,

$P \begin{bmatrix} 0 & -1 \\ 1 & -1 \end{bmatrix} P^{-1} \notin G_0'$. We arrive at a similar contradiction if we assume that

$P \begin{bmatrix} 0 & 1 \\ 1 & 0 \end{bmatrix} P^{-1} = \begin{bmatrix} -1 & 1 \\ 0 & 1 \end{bmatrix}$. Hence there is no unimodular matrix P such that

$G_0' = P G_0 P^{-1}$. This proves that G_0 and G_0' are nonequivalent point groups.

11. $G_0 \simeq D_4$, L quadratic

Here $G_0 = \{I, A, A^2, A^3, B, AB, A^2 B, A^3 B\}$, where A represents a rotation

through angle $\pi/2$ and B represents a reflection, so $A = \begin{bmatrix} 0 & -1 \\ 1 & 0 \end{bmatrix}$. If we take B

to be a reflection in the side of the square along the vector b_1, then $B = \begin{bmatrix} 1 & 0 \\ 0 & -1 \end{bmatrix}$.

Hence

$$G_0 = \left\{ \begin{bmatrix} 1 & 0 \\ 0 & 1 \end{bmatrix}, \begin{bmatrix} 0 & -1 \\ 1 & 0 \end{bmatrix}, \begin{bmatrix} -1 & 0 \\ 0 & -1 \end{bmatrix}, \begin{bmatrix} 0 & 1 \\ -1 & 0 \end{bmatrix}, \begin{bmatrix} 1 & 0 \\ 0 & -1 \end{bmatrix}, \begin{bmatrix} 0 & 1 \\ 1 & 0 \end{bmatrix}, \right.$$
$$\left. \begin{bmatrix} -1 & 0 \\ 0 & 1 \end{bmatrix}, \begin{bmatrix} 0 & -1 \\ -1 & 0 \end{bmatrix} \right\}$$

G_0 contains reflections in each side and each diagonal of the square.

12. $G_0 \simeq D_6$, L hexagonal

Here $G_0 = \{I, A, A^2, A^3, A^4, A^5, B, AB, A^2 B, A^3 B, A^4 B, A^5 B\}$, where A

represents a rotation through angle $\pi/3$ and B represents a reflection. Let the basis vectors b_1, b_2 include an angle $\pi/3$. Then $A = \begin{bmatrix} 0 & -1 \\ 1 & 1 \end{bmatrix}$. If B is reflection in the longer diagonal, bisecting the angle between the basis vectors b_1, b_2, then $B = \begin{bmatrix} 0 & 1 \\ 1 & 0 \end{bmatrix}$. Hence

$$G_0 = \left\{ \begin{bmatrix} 1 & 0 \\ 0 & 1 \end{bmatrix}, \begin{bmatrix} 0 & -1 \\ 1 & 1 \end{bmatrix}, \begin{bmatrix} -1 & -1 \\ 1 & 0 \end{bmatrix}, \begin{bmatrix} -1 & 0 \\ 0 & -1 \end{bmatrix}, \begin{bmatrix} 0 & 1 \\ -1 & -1 \end{bmatrix}, \begin{bmatrix} 1 & 1 \\ -1 & 0 \end{bmatrix}, \right.$$
$$\left. \begin{bmatrix} 0 & 1 \\ 1 & 0 \end{bmatrix}, \begin{bmatrix} -1 & 0 \\ 1 & 1 \end{bmatrix}, \begin{bmatrix} -1 & -1 \\ 0 & 1 \end{bmatrix}, \begin{bmatrix} 0 & -1 \\ -1 & 0 \end{bmatrix}, \begin{bmatrix} 1 & 0 \\ -1 & -1 \end{bmatrix}, \begin{bmatrix} 1 & 1 \\ 0 & -1 \end{bmatrix} \right\}$$

Since G_0 contains reflections in both the diagonals, we would obtain the same result if we started with the assumption that B represents reflection in the shorter diagonal of the rhombus.

This completes the list of all nonequivalent point groups.

Let us sum up the results of the foregoing discussion. We have shown that there are in total 13 nonequivalent point groups. Seven of these are isomorphic with the groups $C_1, C_2, C_3, C_4, C_6, D_4, D_6$. We will refer to them as point groups of type C_1, C_2, and so on. Two point groups are isomorphic with D_1, whose translation lattices are rectangular and rhombic. We refer to them as point groups of type D_1–R and D_1–Rh, respectively. Likewise, two point groups are isomorphic with D_2, whose translation lattices are rectangular and rhombic. We refer to them as point groups of type D_2–R and D_2–Rh, respectively. Finally, two point groups are isomorphic with D_3, both having a hexagonal lattice. One of them contains a reflection in the shorter diagonal of the rhombus, and the other a reflection in the longer diagonal. We refer to them as point groups of type D_3–S and D_3–L, respectively.

6.7 CLASSIFICATION OF WP GROUPS

In this section, we will find all WP groups (up to equivalence). We showed in Section 6.6 that there are 13 nonequivalent point groups. The following theorem shows (as one would expect) that if G_0 is any of these 13 point groups, there does exist a WP group with G_0 as its point group.

THEOREM 6.7.1 Let G_0 be a point group with associated lattice L. Let

$$G = \{(A, t) \in E \mid A \in G_0, t \in L\}$$

Then G is a WP group with G_0 as its point group.

Proof: Let $(A, t), (B, u) \in G$. Then $(A, t)(B, u) = (AB, t + Au)$. Now $AB \in G_0$ and, by Theorem 6.2.6, $Au \in L$ and hence $(AB, t + Au) \in G$. So the set G is closed under multiplication. Further, $(A, t)^{-1} = (A^{-1}, -A^{-1}t) \in G$. Hence G is a group. Moreover, $(I, a) \in G$ if and only if $a \in L$. This proves that G is a WP group. Clearly, G_0 is the point group of G.

∎

This WP group G is called the *symmorphic group* associated with the point group G_0. We denote it by G_0^S. The symmorphic group is characterized by the property that for each $A \in G_0$, $(A, 0) \in G_0^S$.

If two WP groups are equivalent, then their point groups are also equivalent. Hence the 13 symmorphic groups associated with the 13 nonequivalent point groups are themselves nonequivalent WP groups. The problem now is to find WP groups (if any) that are not equivalent to any of these symmorphic groups. We will discover that there are just four such groups (up to equivalence). So there are in all 17 nonequivalent WP groups.

Let G_0 be a point group with associated translation lattice L. Suppose G is a WP group with G_0 as its point group. By definition, for each $A \in G_0$, there exists a vector $a \in \mathbb{R}^2$ such that $(A, a) \in G$. Moreover, given that $(A, a) \in G$, by Theorem 6.2.3, $(A, a') \in G$ if and only if $a' = a + t$ for some $t \in L$. Let $G_0 = \{A_1, \ldots, A_n\}$. Then there exist vectors $a_1, \ldots, a_n \in \mathbb{R}^2$ such that $(A_i, a_i) \in G$ for each $i = 1, \ldots, n$. When the vectors $a_1, \ldots, a_n$ are known, the group G is completely determined—namely, $G = \{(A_i, a_i + t) \mid t \in L, i = 1, \ldots, n\}$.

So the problem is to find these vectors $a_1, \ldots, a_n$. Theorem 6.7.1 shows that one possible solution is $a_i = 0$ for each $i = 1, \ldots, n$. We shall refer to it as the *trivial solution*. The trivial solution gives us the symmorphic group G_0^S.

To find the general solution for the vectors $a_1, \ldots, a_n$, we start with the observation that these vectors $a_1, \ldots, a_n$ must satisfy certain conditions. Let $A, B, C \in G_0$ and suppose $AB = C$. Let $a, b, c \in \mathbb{R}^2$ such that $(A, a), (B, b), (C, c) \in G$. Then $(A, a)(B, b) = (AB, a + Ab) = (C, a + Ab) \in G$. Hence $a + Ab = c + t$ for some $t \in L$.

Let us define two vectors $u, v \in \mathbb{R}^2$ to be *congruent modulo* L, written $u \equiv v$ (modulo L), if $u - v \in L$. It is obvious that this is an equivalence relation. Moreover, it follows by Theorem 6.2.6, that if $u \equiv v$ (modulo L), then $Au \equiv Av$ (modulo L) for all $A \in G_0$.

Thus we have the following result:

THEOREM 6.7.2 Let G be a WP group with point group G_0 and translation lattice L. Let $A, B, C \in G_0$, and suppose $(A, a), (B, b), (C, c) \in G$. If $AB = C$, then $a + Ab \equiv c$ (modulo L).

In particular, if $A^2 = C$, then $a + Aa \equiv c$ (modulo L).

This simple result has an important consequence. Let $A \in G_0$. Given a vector a such that $(A, a) \in G$, it can be shown easily by repeated application of Theorem 6.7.2 that $(A^i, a + aA + \cdots + A^{i-1}a) \in G$ for any positive integer i. Hence, if G_0 is cyclic and generated by A, then the group G is completely determined if we know a vector a such that $(A, a) \in G$.

If G_0 is isomorphic with $D_n, n > 1$, and generated by elements A, B, then G is completely determined if we can find two vectors a, b such that (A, a), $(B, b) \in G$. The next theorem gives the conditions that the vectors a, b must satisfy.

THEOREM 6.7.3 Let G be a WP group with translation lattice L, and suppose its point group G_0 is isomorphic with D_n, $n > 1$. Let

$$G_0 = \{I, A, \ldots, A^{n-1}, B, AB, \ldots, A^{n-1}B\}$$

and suppose $(A, a), (B, b) \in G$. Then the vectors a, b must satisfy the following conditions:

(a) $(I + A + \cdots + A^{n-1})a \equiv 0$ (modulo L).
(b) $(B + I)b \equiv 0$ (modulo L).
(c) $(AB + I)a + (A - I)b \equiv 0$ (modulo L).

Proof: **(a)** The elements A, B in G_0 satisfy the following defining relations: $A^n = I, B^2 = I, BA = A^{n-1}B$. If $(A, a) \in G$, then, as explained above, it follows from the foregoing theorem that $(A^n, a + aA + \cdots + A^{n-1}a) \in G$. Now $A^n = I$ and $(I, 0) \in G$. Hence $a + aA + \cdots + A^{n-1}a \equiv 0$ (modulo L).

(b) By a similar argument, this follows from the relation $B^2 = I$.

(c) Now $(B, b), (A, a) \in G$; hence $(BA, b + Ba) = (B, b)(A, a) \in G$. On the other hand, $(A^{n-1}, a + aA + \cdots + A^{n-2}a) \in G$, $(B, b) \in G$, so $(A^{n-1}B, a + aA + \cdots + A^{n-2}a + A^{n-1}b) \in G$. Since $BA = A^{n-1}B$, it follows that $b + Ba \equiv a + aA + \cdots + A^{n-2}a + A^{n-1}b$ (modulo L). Using (a), we have $b + Ba \equiv -A^{n-1}a + A^{n-1}b$ (modulo L). Then multiplying both sides of this congruence by A, we obtain $Ab + ABa \equiv -Ia + Ib$ (modulo L). ∎

Thus we see that the problem of finding a WP group G that has a given point group G_0 is reduced to finding vectors a, b that satisfy all three conditions in Theorem 6.7.3. In case G_0 is cyclic, we have to find only one vector a that satisfies the first of these three conditions.

Now suppose that, given a point group G_0, we have found all the solutions for the vectors a, b that satisfy the required conditions and have thereby obtained the corresponding WP groups determined by these vectors. Our task then is to determine whether any of these groups are equivalent. The next theorem gives a sufficient condition for two WP groups to be equivalent.

In the sequel we assume that the elements of a WP group G are expressed with respect to a basis of the translation lattice L associated with it. So the elements of G_0 are unimodular matrices. Every vector $t \in L$ is of the form $t = \begin{bmatrix} m \\ n \end{bmatrix}$, where m, n are integers.

THEOREM 6.7.4 Let G_0 be a point group with associated translation lattice L. Let G, G' be WP groups, both having G_0 as the point group, and suppose the elements of G, G' are expressed with respect to a basis of L. If there exist a unimodular matrix $P \in U_2$ and a vector $s \in \mathbb{R}^2$ such that

$$(P, s)G(P, s)^{-1} = G'$$

then G, G' are equivalent WP groups.

In particular, G, G' are equivalent if there exists $s \in \mathbb{R}^2$ such that

$$(I, s)G(I, s)^{-1} = G'$$

Proof: It is easily seen that the mapping $\phi : G \to G'$ given by $\phi(g) = (P, s)g(P, s)^{-1}$ is an isomorphism. Further, for any $t \in L$, $(P, s)(I, t)(P, s)^{-1} = (I, Pt)$. Since P is unimodular, $Pt \in L$. Hence ϕ maps the translation subgroup T onto T'. This proves that G, G' are equivalent. The second statement follows from the fact that I is unimodular. ∎

The application of this theorem is simplified by the fact that G_0 is generated by two elements A, B (or only one element A if it is cyclic). Every element C in G_0 is of the form A^i or $A^i B$. Let $C = A^i B$ and suppose $(A, a), (B, b) \in G$. Then $(A, a)^i (B, b) = (C, c) \in G$ for some c, So $(C, c + t) = (I, t)(C, c) \in G$ for all $t \in L$. It follows that every element in G is of the form $(I, t)(A, a)^i$ or $(I, t)(A, a)^i (B, b)$. Therefore, if $(P, s)(A, a)(P, s)^{-1}, (P, s)(B, b)(P, s)^{-1} \in G'$, then $(P, s)g(P, s)^{-1} \in G'$ for all $g \in G$ and hence $(P, s)G(P, s)^{-1} = G'$. Thus we see that G, G' are equivalent if there exist P, s such that $(P, s)(A, a)(P, s)^{-1} \in G'$ and $(P, s)(B, b)(P, s)^{-1} \in G'$.

In particular, G, G' are equivalent if there exists s such that $(I, s)(A, a)(I, s)^{-1} \in G'$ and $(I, s)(B, b)(I, s)^{-1} \in G'$. Now $(I, s)(A, a)(I, s)^{-1} = (A, a + s - As)$. Hence G, G' are equivalent if $(A, a + s - As)$ and $(B, b + s - Bs) \in G'$. Thus we have proved the following theorem.

THEOREM 6.7.5 Let G_0 be a point group with associated translation lattice L. Let G, G' be WP groups, both having G_0 as the point group. Let A, B be generators of G_0, and suppose $(A, a), (B, b) \in G$ and $(A, a'), (B, b') \in G'$. If there exists a vector s such that

$$a + s - As \equiv a' \text{ (modulo } L)$$
$$b + s - Bs \equiv b' \text{ (modulo } L)$$

then G, G' are equivalent WP groups.

In particular, if there exists s such that

$$a + s - As \equiv 0 \ (\text{modulo } L)$$
$$b + s - Bs \equiv 0 \ (\text{modulo } L)$$

then G is equivalent to the symmorphic group G_0^S.

If the point group G_0 is cyclic, generated by A, then only the first congruence relation in each case is to be satisfied.

We use this result to prove the following two theorems.

THEOREM 6.7.6 Let G be a WP group with G_0 as its point group. If $G_0 \simeq C_n$, then G is equivalent to the symmorphic group G_0^S.

Proof: The result is obvious if $G_0 = \{I\}$ because then $(I, 0) \in G$ and hence G is symmorphic. Now suppose $G_0 \simeq C_n, n > 1$, so $G_0 = \{I, A, \ldots, A^{n-1}\}$, where the matrix A represents a rotation through angle $2\pi/n$. Hence, for any nonzero vector x, $Ax \neq x$, so $(A - I)x \neq 0$. Therefore $A - I$ is an invertible matrix.

Let $a \in \mathbb{R}^2$ such that $(A, a) \in G$. Set $s = (A - I)^{-1}a$. Then $a + s - As = 0$. Hence, by Theorem 6.7.5, G is equivalent to the symmorphic group G_0^S. ∎

Thus we see that if $G_0 \simeq C_n$, then the symmorphic group associated with G_0 is the unique WP group (up to equivalence) that has G_0 as its point group.

In the next theorem we consider the case where the point group is dihedral.

THEOREM 6.7.7 Let G be a WP group with G_0 as its point group and translation lattice L. Suppose $G_0 \simeq D_n$, $n > 1$, and let

$$G_0 = \{I, A, \ldots, A^{n-1}, B, AB, \ldots, A^{n-1}B\}$$

Then there exists a WP group G', equivalent to G, such that $(A, 0) \in G'$. Moreover, if $(B, b) \in G'$, then the vector b satisfies the following conditions:

$$(A - I)b \equiv 0 \ (\text{modulo } L)$$
$$(B + I)b \equiv 0 \ (\text{modulo } L)$$

Proof: As shown in the proof of Theorem 6.7.6, $A - I$ is an invertible matrix. Let $(A, a) \in G$. Put $s = (A - I)^{-1}a$, and let $G' = (I, s)G(I, s)^{-1}$. Then, by Theorem 6.7.4, G' is equivalent to G. Now $(A, a) \in G$ and hence $(I, s)(A, a)(I, s)^{-1} \in G'$, but $(I, s)(A, a)(I, s)^{-1} = (A, a + s - As) = (A, 0)$. Hence $(A, 0) \in G'$. This proves the first part of the theorem. The second part follows directly by Theorem 6.7.3. ∎

Thus we see that, given $G_0 \simeq D_n$, to solve the problem of finding all WP groups (up to equivalence) that have G_0 as their point group, we may assume

$(A, 0) \in G$. So we have to find only one vector b that satisfies the two congruence relations in the foregoing theorem. (If $G_0 \simeq D_1$, then b has to satisfy only the second of these relations.) Since $(B, b) \in G$ implies $(B, b + t) \in G$ for all $t \in L$, we may assume $b = \begin{bmatrix} x \\ y \end{bmatrix}$, where $0 \le x < 1, 0 \le y < 1$.

Given real numbers x, y, we write $x \equiv y$ (modulo 1) to mean that $x - y$ is an integer. Given vectors u, v, if $u = \begin{bmatrix} u_1 \\ u_2 \end{bmatrix}$, $v = \begin{bmatrix} v_1 \\ v_2 \end{bmatrix}$, and $u \equiv v$ (modulo L), then $u_1 \equiv v_1$ (modulo 1) and $u_2 - v_2$ (modulo 1). In the sequel, we will simply write $x \equiv y$ to mean $x \equiv y$ (modulo 1), and $u \equiv v$ to mean $u \equiv v$ (modulo L).

We will now find the solution for all WP groups (up to equivalence) for each type of point group $G_0 \simeq D_n$. (Refer to Section 6.6 for the description of G_0.)

1. D_1–R: $G_0 \simeq D_1$, L rectangular

Here $G_0 = \{I, B\}$, where $B = \begin{bmatrix} 1 & 0 \\ 0 & -1 \end{bmatrix}$. Suppose $(B, b) \in G$, and let $b = \begin{bmatrix} x \\ y \end{bmatrix}$. Take $s = \begin{bmatrix} 0 \\ -y/2 \end{bmatrix}$. Then $b + s - Bs = \begin{bmatrix} x \\ 0 \end{bmatrix}$, so $\left(B, \begin{bmatrix} x \\ 0 \end{bmatrix} \right) \in G'$, where $G' = (I, s)G(I, s)^{-1}$. The groups G, G' are equivalent, so we may assume $b = \begin{bmatrix} x \\ 0 \end{bmatrix}$. Now the vector b satisfies the condition $(B + I)b \equiv 0$, so

$$ \begin{bmatrix} 2 & 0 \\ 0 & 0 \end{bmatrix} \begin{bmatrix} x \\ 0 \end{bmatrix} = \begin{bmatrix} 2x \\ 0 \end{bmatrix} \equiv \begin{bmatrix} 0 \\ 0 \end{bmatrix} $$

which yields $2x \equiv 0$. Hence $x \equiv 0, \frac{1}{2}$, so we have two solutions (up to congruence) for b; namely,

$$ b_1 = \begin{bmatrix} 0 \\ 0 \end{bmatrix}, \quad b_2 = \begin{bmatrix} \frac{1}{2} \\ 0 \end{bmatrix} $$

Let G_1, G_2 be the corresponding WP groups. (G_1 is in fact the symmorphic group associated with G_0.) We claim that G_1, G_2 are not equivalent.

Suppose on the contrary that G_1, G_2 are equivalent. Then there exists an isomorphism $\phi : G_2 \to G_1$ that maps the translation subgroup T onto T. By Theorem 6.5.3, there exists a unimodular matrix $P \in U_2$ such that $\phi(I, t) = (I, Pt)$ for all $t \in L$ and $PG_0P^{-1} = G_0$, so $PBP^{-1} = B$.

Now $(B, b_2) \in G_2$. By Theorem 6.5.3, $\phi(B, b_2) = (PBP^{-1}, b') = (B, b')$ for some b'. Since $(B, 0) \in G_1$, we have $b' \equiv 0$, so $b' = \begin{bmatrix} u \\ v \end{bmatrix}$, where u, v are integers. Now

$$ (\phi(B, b_2))^2 = (B, b')^2 = (B^2, b' + Bb') = \left(I, \begin{bmatrix} 2u \\ 0 \end{bmatrix} \right) $$

On the other hand, since ϕ is an isomorphism, $(\phi(B, b_2))^2 = \phi((B, b_2)^2)$. Hence

$$(\phi(B, b_2))^2 = \phi\left(B^2, b_2 + Bb_2\right) = \phi\left(I, \begin{bmatrix} 1 \\ 0 \end{bmatrix}\right) = \left(I, P\begin{bmatrix} 1 \\ 0 \end{bmatrix}\right)$$

so $P\begin{bmatrix} 1 \\ 0 \end{bmatrix} = \begin{bmatrix} 2u \\ 0 \end{bmatrix}$. Hence $P = \begin{bmatrix} 2u & m \\ 0 & n \end{bmatrix}$, where u, m, n are integers, and so $\det P = 2un$. This contradicts the hypothesis that P is unimodular. So there is no unimodular matrix P that satisfies the requirement for G_1, G_2 to be equivalent. We thus conclude that if $G_0 \simeq D_1$ and L is rectangular, then there are two nonequivalent WP groups G_1, G_2 that have G_0 as the point group. We shall refer to them as WP groups of type D_1–R-0, D_1–R-1, respectively.

2. D_1–Rh: $G_0 \simeq D_1$, L rhombic

Here we have $G_0 = \{I, B\}$, where $B = \begin{bmatrix} 0 & 1 \\ 1 & 0 \end{bmatrix}$. Suppose $(B, b) \in G$, and let $b = \begin{bmatrix} x \\ y \end{bmatrix}$. Then $(B + I)b \equiv 0$, so

$$\begin{bmatrix} 1 & 1 \\ 1 & 1 \end{bmatrix}\begin{bmatrix} x \\ y \end{bmatrix} = \begin{bmatrix} x + y \\ x + y \end{bmatrix} \equiv \begin{bmatrix} 0 \\ 0 \end{bmatrix}$$

Hence $b = \begin{bmatrix} x \\ -x \end{bmatrix}$. Let $s = \begin{bmatrix} -x \\ 0 \end{bmatrix}$; then $b + s - Bs = 0$. Therefore, by Theorem 6.7.5, the group G is equivalent to the symmorphic group associated with G_0. Thus we conclude that if $G_0 \simeq D_1$ and the lattice L is rhombic, then the symmorphic group is the only WP group (up to equivalence) that has G_0 as its point group.

3. D_2–R: $G_0 \simeq D_2$, L rectangular

Here $G_0 = \{I, A, B, AB\}$, where $A = \begin{bmatrix} -1 & 0 \\ 0 & -1 \end{bmatrix}$ and $B = \begin{bmatrix} 1 & 0 \\ 0 & -1 \end{bmatrix}$. Let $(B, b) \in G$. Then the vector b satisfies the conditions $(A - I)b \equiv 0$ and $(B + I)b \equiv 0$. The first condition yields $2b \equiv 0$. Hence there are four solutions (up to congruence) for b—namely,

$$b_1 = \begin{bmatrix} 0 \\ 0 \end{bmatrix}, \quad b_2 = \begin{bmatrix} \frac{1}{2} \\ 0 \end{bmatrix}, \quad b_3 = \begin{bmatrix} 0 \\ \frac{1}{2} \end{bmatrix}, \quad b_4 = \begin{bmatrix} \frac{1}{2} \\ \frac{1}{2} \end{bmatrix}$$

Clearly, these solutions also satisfy the second condition $(B + I)b \equiv 0$.

Let G_1, G_2, G_3, and G_4 be the corresponding WP groups. Of these, G_1 is the symmorphic group. We claim that only three of these are nonequivalent.

We first prove that G_2 and G_3 are equivalent WP groups. Let $P = \begin{bmatrix} 0 & 1 \\ 1 & 0 \end{bmatrix}$. Then P is unimodular and $P^{-1} = P$. By direct computation, we find that

$$(P, 0)(A, 0)(P, 0)^{-1} = (PAP^{-1}, 0) = (A, 0)$$
$$(P, 0)(B, b_2)(P, 0)^{-1} = (PBP^{-1}, Pb_2) = (AB, b_3)$$

Now $(A, 0), (B, b_3) \in G_3$. Hence $(A, 0)(B, b_3) = (AB, Ab_3) \in G_3$. Now

$$Ab_3 = \begin{bmatrix} -1 & 0 \\ 0 & -1 \end{bmatrix} \begin{bmatrix} 0 \\ \frac{1}{2} \end{bmatrix} = \begin{bmatrix} 0 \\ -\frac{1}{2} \end{bmatrix} \equiv b_3$$

Therefore $(AB, b_3) \in G_3$. Hence

$$(P, 0)G_2(P, 0)^{-1} = G_3$$

which implies that G_2 and G_3 are equivalent.

Now we prove that the groups G_1, G_2, and G_4 are pairwise nonequivalent.

a. G_1 and G_2 *are not equivalent.* Suppose G_1 and G_2 are equivalent. Then there exists an isomorphism $\phi : G_2 \to G_1$ that maps the translation subgroup T onto T. By Theorem 6.5.3, there exists $P \in U_2$ such that $\phi(I, t) = (I, Pt)$ for all $t \in L$, and $PG_0P^{-1} = G_0$. Now $\det(PBP^{-1}) = \det B = -1$; hence PBP^{-1} must be equal to B or AB. Suppose $PBP^{-1} = B$. Then, by Theorem 6.5.3, $\phi(B, b_2) = (PBP^{-1}, b') = (B, b')$ for some b'. Now $(B, 0) \in G_1$; therefore $b' \equiv 0$, so $b' = \begin{bmatrix} u \\ v \end{bmatrix}$, where u, v are integers. Hence

$$(\phi(B, b_2))^2 = (B, b')^2 = (B^2, b' + Bb') = \left(I, \begin{bmatrix} 2u \\ 0 \end{bmatrix}\right)$$

On the other hand, $(\phi(B, b_2))^2 = \phi((B, b_2)^2)$, so

$$(\phi(B, b_2))^2 = \phi(B^2, b_2 + Bb_2) = \phi\left(I, \begin{bmatrix} 1 \\ 0 \end{bmatrix}\right) = \left(I, P\begin{bmatrix} 1 \\ 0 \end{bmatrix}\right)$$

Hence we obtain $P\begin{bmatrix} 1 \\ 0 \end{bmatrix} = \begin{bmatrix} 2u \\ 0 \end{bmatrix}$. Therefore $P = \begin{bmatrix} 2u & m \\ 0 & n \end{bmatrix}$, where u, m, n are integers. So $\det P = 2un$. This contradicts the hypothesis that P is unimodular. A similar contradiction results if we assume $PBP^{-1} = AB$. So there is no P that satisfies the requirement for G_1 and G_2 to be equivalent. This proves that G_1 and G_2 are nonequivalent WP groups.

b. G_1 and G_4 *are not equivalent.* The proof here is similar to the one in the foregoing case. We have $b_4 = \begin{bmatrix} \frac{1}{2} \\ \frac{1}{2} \end{bmatrix}$ and hence $b_4 + Bb_4 = \begin{bmatrix} 1 \\ 0 \end{bmatrix}$. So again, assuming $PBP^{-1} = B$, we get $P\begin{bmatrix} 1 \\ 0 \end{bmatrix} = \begin{bmatrix} 2u \\ 0 \end{bmatrix}$, which, as shown above, contradicts the hypothesis that P is unimodular.

c. G_2 and G_4 *are not equivalent.* Finally we prove that G_2 and G_4 cannot be equivalent. Suppose, on the contrary, that there exists an isomorphism $\phi : G_2 \to G_4$ that maps the translation subgroup T onto T. Then, by Theorem 6.5.3, there exists a unimodular matrix $P \in U_2$ such that $\phi(I, t) = (I, Pt)$ for all $t \in L$, and $PG_0P^{-1} = G_0$.

Since $(A, 0), (B, b_2) \in G_2$, it follows that $(A, 0)(B, b_2) = (AB, Ab_2) = \left(AB, \begin{bmatrix} -\frac{1}{2} \\ 0 \end{bmatrix}\right) \in G_2$. Now $\begin{bmatrix} -\frac{1}{2} \\ 0 \end{bmatrix} \equiv b_2$ and therefore $(AB, b_2) \in G_2$. It can be similarly shown that $(AB, b_4) \in G_4$.

Consider $(\phi(AB, b_2))^2$. Since ϕ is an isomorphism, $(\phi(AB, b_2))^2 = \phi((AB, b_2)^2)$. Now $(AB, b_2)^2 = ((AB)^2, (AB + I)b_2) = (I, 0)$. Hence

$$(\phi(AB, b_2))^2 = \phi(I, 0) = (I, 0)$$

On the other hand, by Theorem 6.5.3, $\phi(AB, b_2) = (PABP^{-1}, b')$ for some b'. Now $P(AB)P^{-1}$ must be equal to B or AB. Suppose $P(AB)P^{-1} = B$. Then $\phi(AB, b_2) = (B, b')$. Now $(B, b_4) \in G_4$ and hence $b' \equiv b_4$. So $b' = \begin{bmatrix} m + \frac{1}{2} \\ n + \frac{1}{2} \end{bmatrix}$, where m, n are integers. Hence

$$(\phi(AB, b_2))^2 = (B, b')^2 = (B^2, (B + I)b') = \left(I, \begin{bmatrix} 2m + 1 \\ 0 \end{bmatrix}\right)$$

Thus we obtain $\begin{bmatrix} 2m + 1 \\ 0 \end{bmatrix} = 0$, which contradicts the fact that m is an integer. A similar contradiction results if we assume $P(AB)P^{-1} = AB$. Hence there is no unimodular matrix P that satisfies the requirement for G_2 and G_4 to be equivalent. This proves that G_2 and G_4 are nonequivalent WP groups.

We thus conclude that if $G_0 \simeq D_2$ and the lattice L is rectangular, there are three nonequivalent WP groups G_1, G_2, and G_4 that have G_0 as the point group. We shall denote them by D_2–R-0, D_2–R-1, D_2–R-2, respectively.

4. D_2–Rh: $G_0 \simeq D_2$, L rhombic

Here $G_0 = \{I, A, B, AB\}$, where $A = \begin{bmatrix} -1 & 0 \\ 0 & -1 \end{bmatrix}$ and $B = \begin{bmatrix} 0 & 1 \\ 1 & 0 \end{bmatrix}$. The vector b satisfies the conditions $(A - I)b \equiv 0$, $(B + I)b \equiv 0$. From these conditions we obtain two solutions for b—namely,

$$b_1 = \begin{bmatrix} 0 \\ 0 \end{bmatrix}, \quad b_2 = \begin{bmatrix} \frac{1}{2} \\ \frac{1}{2} \end{bmatrix}$$

We show that the corresponding WP groups G_1, G_2 are equivalent. Let $s = \begin{bmatrix} -\frac{1}{2} \\ 0 \end{bmatrix}$. Then

$$b_2 + s - Bs = \begin{bmatrix} \frac{1}{2} \\ \frac{1}{2} \end{bmatrix} + \begin{bmatrix} -\frac{1}{2} \\ 0 \end{bmatrix} - \begin{bmatrix} 0 \\ -\frac{1}{2} \end{bmatrix} = \begin{bmatrix} 0 \\ 1 \end{bmatrix} \equiv 0$$

and also

$$0 + s - As = \begin{bmatrix} -1 \\ 0 \end{bmatrix} \equiv 0$$

Hence, by Theorem 6.7.5, G_2 is equivalent to the symmorphic group G_1. Thus we conclude that if $G_0 \simeq D_2$ and the lattice is rhombic, the symmorphic group is the only group (up to equivalence) with point group G_0.

5,6. D_3–S and D_3–L: $G_0 \simeq D_3$, L hexagonal

We prove that in both cases the symmorphic group is the only WP group (up to equivalence) with G_0 as its point group.

First consider the case where G_0 is of type D_3–S. Then $G_0 = \{I, A, A^2, B, AB, A^2B\}$, where $A = \begin{bmatrix} 0 & -1 \\ 1 & -1 \end{bmatrix}$ and $B = \begin{bmatrix} 0 & 1 \\ 1 & 0 \end{bmatrix}$. Let $(B, b) \in G$. Then the vector b satisfies the conditions $(A - I)b \equiv 0$ and $(B + I)b \equiv 0$. Writing $b = \begin{bmatrix} x \\ y \end{bmatrix}$, we have the congruence relations

$$\begin{bmatrix} -x - y \\ x - 2y \end{bmatrix} \equiv \begin{bmatrix} 0 \\ 0 \end{bmatrix}, \quad \begin{bmatrix} x + y \\ x + y \end{bmatrix} \equiv \begin{bmatrix} 0 \\ 0 \end{bmatrix}$$

From these we obtain $x \equiv 2y$ and $3y \equiv 0$. Hence the three solutions for b (up to congruence) are $b = \begin{bmatrix} 0 \\ 0 \end{bmatrix}, \begin{bmatrix} \frac{2}{3} \\ \frac{1}{3} \end{bmatrix}, \begin{bmatrix} \frac{1}{3} \\ \frac{2}{3} \end{bmatrix}$. We claim that the corresponding WP groups are all equivalent. Let b denote any of these solutions. Then $3b \equiv 0$. Hence, if we let $s = b$, the following relations hold:

$$0 + s - As \equiv 0, \quad b + s - Bs \equiv 0$$

This proves that all the groups are equivalent to the symmorphic group.

A similar proof holds for the point group of type D_3–L.

7. D_4: $G_0 \simeq D_4$, L quadratic

Here $G_0 = \{I, A, A^2, A^3, B, AB, A^2B, A^3B\}$, where $A = \begin{bmatrix} 0 & -1 \\ 1 & 0 \end{bmatrix}$ and $B = \begin{bmatrix} 1 & 0 \\ 0 & -1 \end{bmatrix}$. Let $(B, b) \in G$. If we write $b = \begin{bmatrix} x \\ y \end{bmatrix}$, the conditions $(A - I)b \equiv 0$ and

$(B + I)b \equiv 0$ become

$$\begin{bmatrix} -x - y \\ x - y \end{bmatrix} \equiv \begin{bmatrix} 0 \\ 0 \end{bmatrix}, \quad \begin{bmatrix} 2x \\ 0 \end{bmatrix} \equiv \begin{bmatrix} 0 \\ 0 \end{bmatrix}$$

From these relations it follows that there are only two solutions (up to congruence) for b—namely,

$$b_1 = \begin{bmatrix} 0 \\ 0 \end{bmatrix}, \quad b_2 = \begin{bmatrix} \frac{1}{2} \\ \frac{1}{2} \end{bmatrix}$$

Let G_1, G_2 be the corresponding WP groups. We claim that G_1 and G_2 are not equivalent. Suppose G_1 and G_2 are equivalent. Then there exists an isomorphism $\phi : G_2 \to G_1$ that maps the translation subgroup T onto T. By Theorem 6.5.3, there exists a unimodular matrix $P \in U_2$ such that $\phi(I, t) = (I, Pt)$ for all $t \in L$, and $PG_oP^{-1} = G_0$. Now $\det(PBP^{-1}) = \det B = -1$. Hence PBP^{-1} must be equal to one of the matrices $B, AB, A^2 B, A^3 B$. We first show that PBP^{-1} cannot be equal to AB or $A^3 B$.

Suppose $PBP^{-1} = AB$. Then $PB = ABP$. Writing $P = \begin{bmatrix} p & q \\ r & s \end{bmatrix}$, we have

$$\begin{bmatrix} p & q \\ r & s \end{bmatrix} \begin{bmatrix} 1 & 0 \\ 0 & -1 \end{bmatrix} = \begin{bmatrix} 0 & -1 \\ 1 & 0 \end{bmatrix} \begin{bmatrix} 1 & 0 \\ 0 & -1 \end{bmatrix} \begin{bmatrix} p & q \\ r & s \end{bmatrix}$$

which yields

$$\begin{bmatrix} p & -q \\ r & -s \end{bmatrix} = \begin{bmatrix} r & s \\ p & q \end{bmatrix}$$

So $p = r, q = -s$, and hence $\det P = 2ps$. This contradicts the hypothesis that P is unimodular. A similar contradiction results if we assume $PBP^{-1} = A^3 B$.

Now suppose $PBP^{-1} = B$. Then, by Theorem 6.5.3, $\phi(B, b_2) = (PBP^{-1}, b') = (B, b')$ for some b'. Since $(B, 0) \in G_1$, we have $b' \equiv 0$, so $b' = \begin{bmatrix} u \\ v \end{bmatrix}$, where u, v are integers. Hence

$$(\phi(B, b_2))^2 = (B, b')^2 = (B^2, b' + Bb') = \left(I, \begin{bmatrix} 2u \\ 0 \end{bmatrix} \right)$$

On the other hand, since ϕ is an isomomorphism, we have $(\phi(B, b_2))^2 = \phi((B, b_2)^2)$. Hence

$$(\phi(B, b_2))^2 = \phi(B^2, b_2 + Bb_2) = \phi \left(I, \begin{bmatrix} 1 \\ 0 \end{bmatrix} \right) = \left(I, P \begin{bmatrix} 1 \\ 0 \end{bmatrix} \right)$$

so $P \begin{bmatrix} 1 \\ 0 \end{bmatrix} = \begin{bmatrix} 2u \\ 0 \end{bmatrix}$. Hence $P = \begin{bmatrix} 2u & q \\ 0 & s \end{bmatrix}$, where u, q, s are integers. So $\det P = 2us$, which contradicts the hypothesis that P is unimodular.

We arrive at a similar contradiction if we assume $PBP^{-1} = A^2 B$.

So there is no unimodular matrix P that satisfies the requirement for G_1 and G_2 to be equivalent. This proves that G_1 and G_2 are nonequivalent WP groups. We will refer to them as type D_4-0 and D_4-2, respectively.

8. D_6: $G_0 \simeq D_6$, L hexagonal

Here $G_0 = \{I, A, A^2, A^3, A^4, A^5, B, AB, A^2B, A^3B, A^4B, A^5B\}$, where

$$A = \begin{bmatrix} 0 & -1 \\ 1 & 1 \end{bmatrix} \quad \text{and} \quad B = \begin{bmatrix} 0 & 1 \\ 1 & 0 \end{bmatrix}.$$

Let $(B, b) \in G$. Then $(A - I)b \equiv 0$ and $(B + I)b \equiv 0$. Writing $b = \begin{bmatrix} x \\ y \end{bmatrix}$, we have the congruence relations

$$\begin{bmatrix} -x - y \\ x \end{bmatrix} \equiv \begin{bmatrix} 0 \\ 0 \end{bmatrix}, \quad \begin{bmatrix} x + y \\ x + y \end{bmatrix} \equiv \begin{bmatrix} 0 \\ 0 \end{bmatrix}$$

which yield $x \equiv 0$ and $y \equiv 0$; hence $b \equiv 0$. This proves that the symmorphic group is the only group (up to equivalence) with point group G_0.

To sum up the foregoing discussion, we have shown that there are in all 17 WP groups (up to equivalence). Of these, 13 are equivalent to the symmorphic groups associated respectively with the 13 nonequivalent point groups. The remaining four groups that are not equivalent to a symmorphic group are as follows:

D_1–R-1, with point group $G_0 \simeq D_1$, lattice L rectangular, and $b = \begin{bmatrix} \frac{1}{2} \\ 0 \end{bmatrix}$

D_2–R-1, with $G_0 \simeq D_2$, L rectangular, and $b = \begin{bmatrix} \frac{1}{2} \\ 0 \end{bmatrix}$

D_2–R-2, with $G_0 \simeq D_2$, L rectangular, and $b = \begin{bmatrix} \frac{1}{2} \\ \frac{1}{2} \end{bmatrix}$

D_4-2, with $G_0 \simeq D_4$, L quadratic, and $b = \begin{bmatrix} \frac{1}{2} \\ \frac{1}{2} \end{bmatrix}$

What follows is a complete list of all 17 nonequivalent WP groups, with an example of the parallelogram pattern in each case.

1. C_1: $G_0 \simeq C_1$, L general

2. C_2: $G_0 \simeq C_2$, L general

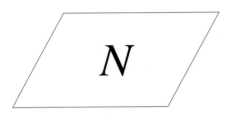

3. C_3: $G_0 \simeq C_3$, L hexagonal

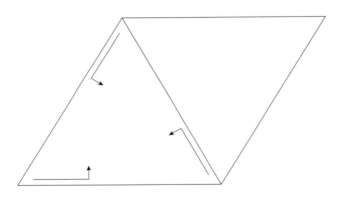

4. C_4: $G_0 \simeq C_4$, L quadratic

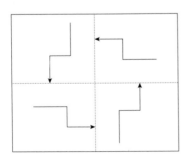

5. C_6: $G_0 \simeq C_6$, L hexagonal

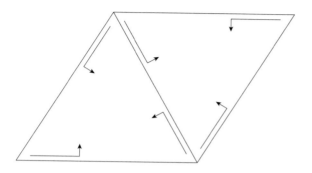

6. D_1–R-0: $G_0 \simeq D_1$, L rectangular

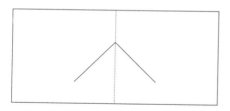

7. D_1–R-1: $G_0 \simeq D_1$, L rectangular

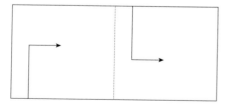

8. D_1–Rh: $G_0 \simeq D_1$, L rhombic

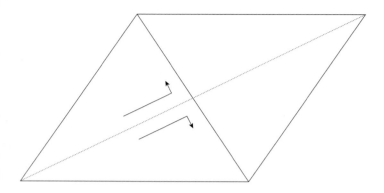

9. D_2–R-0: $G_0 \simeq D_2$, L rectangular

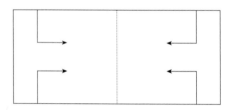

10. D_2–R-1: $G_0 \simeq D_2$, L rectangular

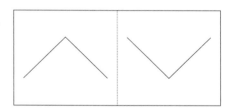

11. D_2–R-2: $G_0 \simeq D_2$, L rectangular

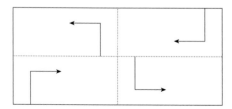

12. D_2–Rh: $G_0 \simeq D_2$, L rhombic

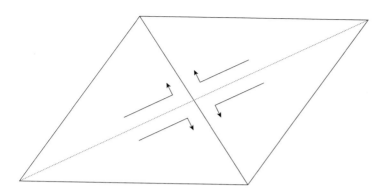

13. D_3–S: $G_0 \simeq D_3$, L hexagonal

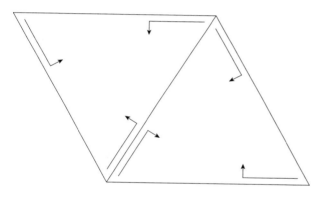

14. D_3–L: $G_0 \simeq D_3$, L hexagonal

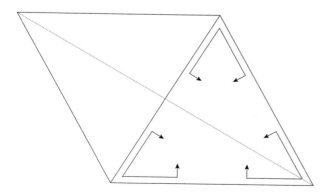

15. D_4–0: $G_0 \simeq D_4$, L quadratic

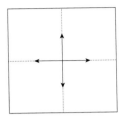

16. D_4–2: $G_0 \simeq D_4$, L quadratic

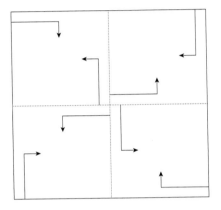

17. D_6: $G_0 \simeq D_6$, L hexagonal

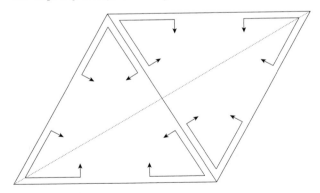

6.8 SAMPLE PATTERNS

EXERCISES 6.8

Identify the symmetry groups of the wallpaper patterns on the following pages, and draw the parallelograms formed by the basis vectors of the lattices.

1.

2.

3.

4.

5.

6.

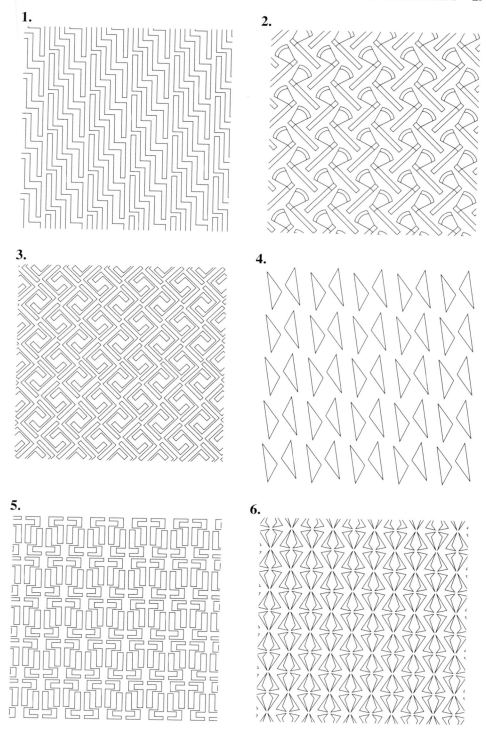

7.

8.

9.

10.

11.

12.

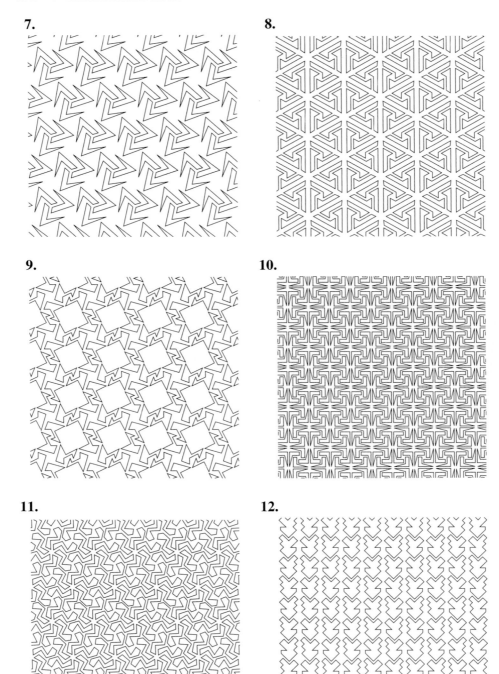

13.

14.

15.

16.

17.

18.

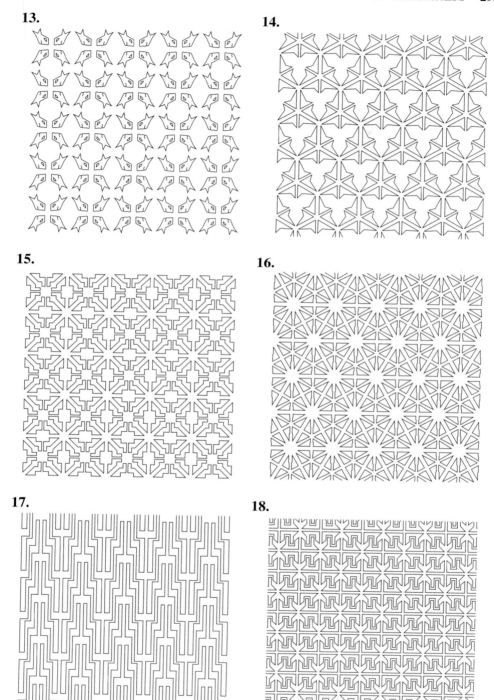

19.

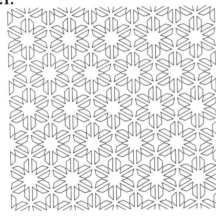

20.

21.

22.

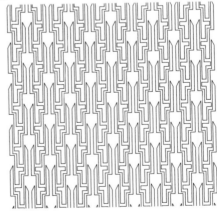

23.

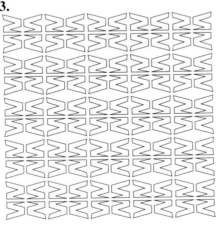

24.

CHAPTER 0: PRELIMINARY ALGEBRAIC CONCEPTS

EXERCISES 0.1

1. $\sum_{r=0}^{n} \binom{n}{r} = (1+1)^n = 2^n$

2. $|A \cup B \cup C| = |A \cup B| + |C| - |(A \cup B) \cap C|$. Further, $|A \cup B| = |A| + |B| - |A \cap B|$, and

$$|(A \cup B) \cap C| = |(A \cap C) \cup (B \cap C)|$$
$$= |A \cap C| + |B \cap C| - |A \cap B \cap C|$$

3. Use induction.

4. Assuming a universal set U, we have

$$
\begin{aligned}
A \triangle B &= (A - B) \cup (B - A) \\
&= (A \cap B') \cup (B \cap A') \\
&= [(A \cap B') \cup B] \cap [(A \cap B') \cup A'] \\
&= [(A \cup B) \cap (B \cup B')] \cap [(A \cup A') \cap (B' \cup A')] \\
&= (A \cup B) \cap (A \cap B)' \\
&= (A \cup B) - (A \cap B)
\end{aligned}
$$

5. $\bigcup \mathcal{F} = (0, \infty); \bigcap \mathcal{F} = \emptyset$

6. The number of injective mappings is $n(n-1) \cdots (n-m+1)$ if $n \geq m$; otherwise 0.

7. If $f(a_1), \ldots, f(a_n)$ are all distinct, then every $b \in B$ must be equal to some $f(a_i)$. And conversely.

8. Draw the graph of $A \times B$ in the xy-plane.

9. $A \times B$ is empty. Conversely, A or B is empty.

10. In $\mathbb{Z}$, 1 divides -1 and -1 divides 1, but $1 \neq -1$.

11. The equivalence class of $(1, 1)$ is $\{(x, x) \mid x \in \mathbb{N}\}$ and that of $(1, 2)$ is $\{(x, x+1) \mid x \in \mathbb{N}\}$.

12. $\mathbb{R}/E = \{\bar{x} \mid 0 \leq x < \alpha\}$, where $\bar{x} = \{x + n\alpha \mid n \in \mathbb{Z}\}$.

13. Congruence modulo 7

14. Let S be a set of $n + 1$ elements. Fix an element a in S, and consider an arbitrary partition of S. Suppose a occurs in a block containing r elements besides a, $0 \leq r \leq n$. These r elements can be chosen in $\binom{n}{r}$ ways. The remaining $n - r$ elements in S can be partitioned in $p(n - r)$ ways. So the total number of possible partitions of S is

$$p(n + 1) = \sum_{r=0}^{n} \binom{n}{r} p(n - r) = \sum_{r=0}^{n} \binom{n}{n - r} p(r) = \sum_{r=0}^{n} \binom{n}{r} p(r)$$

15. (a) No; (b) No; (c) Left distributive

EXERCISES 0.2

1. For all $a, b \in G$, $(ab)^2 = e = ee = a^2 b^2$, so $abab = aabb$. Now use cancellation laws.

2. Consider the product ab. Using the cancellation laws, we see that ab cannot be equal to a, b, or e. Hence $ab = c = ba$, so $c^2 = abba = e$. Further, $bc = bba = a = cb$ and $ca = baa = b = ac$.

3. $\mathbb{Z}_8^* = \{1, 3, 5, 7\}$; $3^2 = 5^2 = 7^2 = 1$

4. By directly checking the product of every two elements in K, we find that it is a Klein's 4-group.

5. Check the product of every two elements.

6. 59

7. Since $\alpha \neq e$ and $n > 2$, there exist distinct $p, q, r \in \{1, \ldots, n\}$ such that $\alpha(p) = q$. Define β to be the transposition $(q\ r)$. Then $\alpha\beta(p) = q$, but $\beta\alpha(p) = r$.

EXERCISES 0.3

1. If $x = a^i$ and $y = a^j$, then $xy = a^{i+j} = yx$.

2. Use induction. Suppose $(ab)^n = a^n b^n$. Then $(ab)^{n+1} = aba^n b^n = aa^n bb^n = a^{n+1} b^{n+1}$.

3. $o(0) = 1$, $o(1) = o(5) = o(7) = o(11) = 12$, $o(2) = o(10) = 6$, $o(3) = o(9) = 4$, $o(4) = o(8) = 3$, $o(6) = 2$

4. Suppose $o(a) = m$. Then $(a^{-1})^m = (a^m)^{-1} = e$. Hence $o(a^{-1})$ divides $m = o(a)$. Similarly, $o(a)$ divides $o(a^{-1})$. Hence $o(a^{-1}) = o(a)$. If a is of infinite order, then a^{-1} must be of infinite order.

5. Suppose $o(ab) = m$. Then $(ba)^m = a^{-1}(ab)^m a = e$. Hence $o(ba)$ divides $m = o(ab)$. Similarly, $o(ab)$ divides $o(ba)$. Hence $o(ba) = o(ab)$. If ab is of infinite order, then ba must be of infinite order. Now $o(aba^{-1}) = o(a^{-1}ab) = o(b)$.

6. The nth roots of unity are $\cos(2\pi i/n) + \sin(2\pi i/n) = (\cos(2\pi/n) + \sin(2\pi/n))^i$, $i = 0, 1, \ldots, n-1$.

7. $\mathbb{Z}_{10}^* = \{1, 3, 7, 9\}$. The generators are 3 and 7.

8. $\mathbb{Z}_9^* = \{1, 2, 4, 5, 7, 8\}$. The generators are 2 and 5.

9. $\mathbb{Z}_{13}^*$ has generators 2, 6, 7, 11. $\mathbb{Z}_{17}^*$ has generators 3, 5, 6, 7, 10, 11, 12, 14.

10. If $o(a) > 2$, then $a \neq a^{-1}$. Now $o(e) = 1$. Hence the number of elements of order 2 is odd.

11. If $a^2 = 1$ in $\mathbb{Z}_p^*$, then $a^2 \equiv 1 \pmod{p}$ in $\mathbb{Z}$. Hence p divides $a^2 - 1$, so $p = a + 1$. Hence $p - 1$ is the only element of order 2.

12. $\mathbb{Z}_2 \times \mathbb{Z}_2 = \{(0,0), (1,0), (0,1), (1,1)\}$. For all $(x, y) \in \mathbb{Z}_2 \times \mathbb{Z}_2$, $2(x, y) = (0, 0)$.

EXERCISES 0.4

1. If G is of order 2, 3, or 5, then, by Theorem 0.4.10, G is cyclic. Suppose G is of order 4. If G has an element of order 4, then it is cyclic. Otherwise, every element other than e is of order 2, so G is a Klein's 4-group.

2. Cyclic subgroups generated by (1), (1 2), (2 3), (3 1), (1 2 3)

3. Cyclic subgroups generated by 0, 1, 2, 3, 4, 6

4. Suppose $x \in m\mathbb{Z} \cap n\mathbb{Z}$. Then m, n both divide x; hence $q \mid x$. So $x \in q\mathbb{Z}$. Conversely, if $q \mid x$, then m, n both divide x.

5. $H \cap K$ is a subgroup of both H and K. By the Lagrange theorem, $|H \cap K|$ divides $|H|$ and $|K|$, so $|H \cap K| = 1$.

6. If a, b are positive real numbers, then a/b is a positive real number. Hence H is a subgroup; $G = H \cup (-1)H$.

7. For any $a \in G$, $a(H \cap K) = aH \cap aK$. Hence every left coset of $H \cap K$ is an intersection of left cosets of H and K. There is a finite number of left cosets of $H \cap K$.

8. Let $a \in G, a \neq e$. Then $[a] = G$, so G is cyclic. If G is infinite, then $[a^2]$ is a proper subgroup of G, a contradiction. Hence G is a finite cyclic group and, by Theorem 0.4.10, its order must be prime.

EXERCISES 0.5

1. Let $h_1, h_2 \in H$ and $n_1, n_2 \in N$. Then $(h_1 n_1)(h_2 n_2)^{-1} = h_1 n_1 n_2^{-1} h_2^{-1} = (h_1 h_2^{-1})$ $(h_2 n_1 n_2^{-1} h_2^{-1}) \in HN$. Hence HN is a subgroup of G. Further, $N \subset HN$ and $(h_1 n_1) n_2 (h_1 n_1)^{-1} = h_1 (n_1 n_2 n_1^{-1}) h_1^{-1} \in N$. Hence $N \triangleleft HN$. If $h \in H$ and $a \in H \cap N$, then hah^{-1} belongs to both H and N. Hence $H \cap N \triangleleft H$.

2. By Exercise 1, HN is a subgroup. Let $g \in G, h \in H$, and $n \in N$. Then $g(hn)g^{-1} = (ghg^{-1})(gng^{-1}) \in HN$. Hence HN is normal.

3. If $\det A = 1$, then $\det(BAB^{-1}) = \det A = 1$.

4. Let P be the diagonal matrix $\text{diag}(-1, 1, \ldots, 1)$. Then $G = N \cup PN$, so $[G : N] = 2$.

5. The only subgroup of order 2 is $\left\{ \begin{bmatrix} 1 & 0 \\ 0 & 1 \end{bmatrix}, \begin{bmatrix} -1 & 0 \\ 0 & -1 \end{bmatrix} \right\}$, which is obviously normal. All subgroups of order 4 have index 2 and are therefore normal.

6. For all $x, y \in \mathbb{Z}$, $m(x + y) = mx + my$. If $mx = my$, then $x = y$.

7. For all $x, y \in G$, $f(xy) = (xy)^m = x^m y^m = f(x)f(y)$.

8. By Exercise 7, f is a homomorphism. If $\gcd(m, n) = 1$, there exist integers s, t such that $sm + tn = 1$. Then for all $x \in G$, $x = x^{sm+tn} = x^{sm}$ because $x^{tn} = (x^t)^n = e$. Hence $x = f(x^s)$, so f is surjective. If $x^m = e$, then $x = (x^m)^s = e$; hence $\ker f = \{e\}$. So f is injective.

9. For all $x, y \in G$, $f(xy) = axya^{-1} = (axa^{-1})(aya^{-1}) = f(x)f(y)$. If $axa^{-1} = aya^{-1}$, then $x = y$. For all x, $x = a(a^{-1}xa)a^{-1} = f(a^{-1}xa)$.

10. Let $o(a) = m$ and $o(f(a)) = k$. Then $(f(a))^m = f(a^m) = f(e) = e$. Hence $k \mid m$. Further, $f(a^k) = (f(a))^k = e$, so $a^k \in \ker f$. If f is injective, then $\ker f = (e)$, so $a^k = e$. Hence $m \mid k$.

11. Suppose G is not cyclic. Then there is no element of order 6 in G. By the Lagrange theorem, every element in G (other than identity e) is of order 2 or 3. Suppose every element is of order 2. Then G is abelian. Let $K = \{e, c\}$, where $c \neq e$. Then the quotient group G/K is of order 3. For all $x \in G$, $(xK)^2 = x^2 K = eK$, which contradicts the Lagrange theorem. Hence G must have an element a of order 3. Let $H = \{e, a, a^2\}$, $b \notin H$. Then $G = H \cup Hb = \{e, a, a^2, b, ab, a^2 b\}$. Now $b^2 \notin Hb$, so $b^2 \in H$. If $b^2 = a$ or a^2, then $b^3 = ab$ or $a^2 b$; hence $o(b) = 6$, a contradiction. Hence $b^2 = e$. Now H, being a subgroup of index 2, is normal. Hence $b^{-1}ab \in H$. If $b^{-1}ab = e$, then $a = e$, which is false. If $b^{-1}ab = a$, then $ab = ba$; hence $o(ab) = 6$, a contradiction. Hence $b^{-1}ab = a^2$. It follows that there is an isomorphism $f : G \to S_3$ such that $f(a) = (1\,2\,3)$ and $f(b) = (1\,2)$.

12. Arguing as in Exercise 11, G has an element a of order 5. Let $H = [a]$. Then $G = H \cup Hb$, where $b \notin H$, $b^2 \in H$. If $b^2 = a, a^2, a^3$, or a^4, then $o(b) = 10$, a contradiction. Hence $b^2 = e$. H is normal; hence $b^{-1}ab \in H$. If $b^{-1}ab = e$, then $a = e$ (false). If $b^{-1}ab = a$, then $ab = ba$, so G is abelian (false). Suppose

$b^{-1}ab = a^2$. Then $b^{-1}a^2b = (b^{-1}ab)(b^{-1}ab) = a^4$. Hence $a^2 = b^{-1}(b^{-1}a^2b)$ $b = b^{-1}a^4b = a^8 = a^3$, a contradiction. Similarly, $b^{-1}ab = a^3$ yields a contradiction. Hence $b^{-1}ab = a^4$.

EXERCISES 0.7

1. Let $a, b \in R$. Then $a + b = (a + b)^2 = a^2 + ab + ba + b^2$, which yields $ab + ba = 0$. In particular, taking $a = b$, we get $a + a = 0$. Hence $ab = ab + (ab + ba) = ba$.

2. See Exercise 4 in Exercises 0.1.

3. 6 is an identity of S.

4. Let $S = \{0, 2, 4, \ldots, 4n\}$. Then S is a subring with identity $2n + 2$.

5. (a) $\{1, 5, 7, 11\}$; (b) $\{a \in \mathbb{Z}_{25} \mid 5 \nmid a\}$; (c) $\{1, 2, \ldots, p - 1\}$

6. Suppose $a^m = b^n = 0$, and let $k = \max(m, n)$. Then $(a + b)^{2k} = \sum_{i=0}^{2k}$ $a^{2k-i}b^i = 0$. For a counter example, let R be the ring of 2×2 matrices over $\mathbb{R}$, and let $a = \begin{bmatrix} 0 & 1 \\ 0 & 0 \end{bmatrix}$ and $b = \begin{bmatrix} 0 & 0 \\ 1 & 0 \end{bmatrix}$. Then $a^2 = b^2 = 0$. But $(a + b)^2 = \begin{bmatrix} 1 & 0 \\ 0 & 1 \end{bmatrix}$, so $a + b$ is not nilpotent.

7. If a, b are nilpotent, then, by Exercise 6, $a - b$ is nilpotent. If $a^m = 0$, then for all $r \in R$, $(ra)^m = r^m a^m = 0$.

8. Let R be a commutative integral domain with n elements, $R = \{a_1, \ldots, a_n\}$. Let b be a nonzero element in R. Then $a_1b, \ldots, a_nb$ are all distinct elements of R, so $R = \{a_1b, \ldots, a_nb\}$. Hence $b = a_qb$ for some q. Write $e = a_q$. Then $e(a_ib) = a_i eb = a_ib$ for all a_i. Hence e is the identity of R. Now $e = a_jb$ for some j. Hence b has an inverse.

9. Let a be a nonzero element in R. Then $A = \{ra \mid r \in R\}$ is a nonzero ideal. So $A = R$; hence $ra = e$ for some r.

10. The only subgroups of the additive group $\mathbb{Z}$ are the cyclic subgroups $n\mathbb{Z} = \{na \mid a \in \mathbb{Z}\}$. These are also ideals in the ring $\mathbb{Z}$. The ideals in the ring $\mathbb{Z}_{15}$ are $\{0\}$, $\{0, 3, 6, 9, 12\}$, $\{0, 5, 10\}$, and $\mathbb{Z}_{15}$.

11. (a) To show that $C[0, 1]$ is not an integral domain, consider the functions f, g defined as follows:

$$f(x) = \begin{cases} 0 & \text{if } 0 \le x \le \frac{1}{2} \\ x - \frac{1}{2} & \text{if } \frac{1}{2} \le x \le 1 \end{cases}$$

$$g(x) = \begin{cases} \frac{1}{2} - x & \text{if } 0 \le x \le \frac{1}{2} \\ 0 & \text{if } \frac{1}{2} \le x \le 1 \end{cases}$$

Then $f(x)g(x) = 0$ for all $x \in [0, 1]$.

(b) If $f, g \in M$, then $f(a) - g(a) = 0$ and $f(a)h(a) = 0$ for all $h \in C[0, 1]$. Hence M is an ideal.

(c) Functions $f, g \in C[0, 1]$ lie in the same coset of M if and only if $f(a) = g(a)$. Hence $C[0, 1]/M = \{f_t + M \mid t \in \mathbb{R}\}$, where f_t denotes the constant function $f(x) = t$. It is clear that the mapping $t \mapsto f_t + M$ is an isomorphism from $\mathbb{R}$ to $C[0, 1]/M$.

12. Let $f : \mathbb{Z} \to \mathbb{Z}$ be a homomorphism. Let $f(1) = a$. Then $a^2 = f(1) f(1) = f(1) = a$. Hence $a = 0$ or 1. If $a = 0$, then $f(n) = 0$ for all $n \in \mathbb{Z}$. If $a = 1$, then $f(n) = n$.

13. Suppose f is a homomorphism with $f(1) = 1$. Then $f(5) = 5$. In $\mathbb{Z}_{10}$, $5 + 5 = 0$; hence $f(5 + 5) = f(0) = 0$. But in $\mathbb{Z}_{15}$, $5 + 5 = 10$. So $f(5 + 5) \neq f(5) + f(5)$, a contradiction.

14. Define $f(a) = a \bmod n$ for all $a \in \mathbb{Z}_m$. Let $a, b \in \mathbb{Z}_m$, and let $a + b = c$ (in $\mathbb{Z}_m$). Then $c \equiv a + b \pmod{m}$. Since n divides m, $c \equiv a + b \pmod{n}$. Hence $f(c) \equiv a + b \pmod{n}$. On the other hand, $f(a) \equiv a \pmod{n}$ and $f(b) \equiv b \pmod{n}$. Hence, in $\mathbb{Z}_n$, $f(a) + f(b) \equiv a + b \pmod{n}$. Thus $f(a + b) = f(a) + f(b)$. Similarly, $f(ab) = f(a)f(b)$.

15. If $f : F \to R$ is a nonzero homomorphism, then $\ker f$ is an ideal in F and $\ker f \neq F$. Hence $\ker f = \{0\}$.

16. It follows from Exercise 15.

17. The mapping $f : \mathbb{Z}_{mn} \to \mathbb{Z}_m \times \mathbb{Z}_n$, given by $f(x) = (x \bmod m, x \bmod n)$, is a homomorphism. By the Chinese remainder theorem, f is bijective.

EXERCISES 0.8

1. Every ideal in $F[x]$ is principal. If $p(x)$ is a polynomial of least degree in an ideal A, then, by Euclidean algorithm, every polynomial in A is a multiple of $p(x)$.

2. $\mathbb{Z}[x]/(x^2 + 1) = \{a + bt \mid a, b \in \mathbb{Z}\}$, where $t = x + (x^2 + 1)$ is the coset of x. The mapping $\phi : \mathbb{Z}[x]/(x^2 + 1) \to \mathbb{Z}[i]$, given by $\phi(a + bt) = a + bi$, is an isomorphism.

3. Neither 0 nor 1 is a root of $x^3 + x + 1$; hence it is irreducible. $F = \{a + bt + ct^2 \mid a, b, c \in \mathbb{Z}_2\} = \{0, 1, t, 1 + t, t^2, 1 + t^2, t + t^2, 1 + t + t^2\}$, where $t^3 + t + 1 = 0$. F^* is a multiplicative group of seven elements. Hence every nonzero element other than 1 is a generator of F^* and so a primitive element. Consequently, the polynomial is primitive.

4. $x^3 + x + 1$ and $x^3 + x^2 + 1$

5. If F is a field of order $2^r, r = 2, 3, 5, 7$, then F^* is a multiplicative group of prime order. Hence every nonzero element other than 1 is primitive.

6. $x^2 + 1$, $x^2 + x + 2(*)$, $x^2 + 2x + 2(*)$, $x^3 + 2x + 1(*)$, $x^3 + 2x + 2$, $x^3 + x^2 + 2$, $x^3 + x^2 + x + 2$, $x^3 + x^2 + 2x + 1(*)$, $x^3 + 2x^2 + 1(*)$, $x^3 + 2x^2 + x + 1(*)$, $x^3 + 2x^2 + 2x + 2$. (Those marked with $*$ are primitive.)

7. Use the polynomial $x^2 + 2x + 2$. $F = \{a + bt \mid a, b \in \mathbb{Z}_3\}$, $t^2 = t + 1$.

8. $F = \{a + bt + ct^2 \mid a, b, c \in \mathbb{Z}_3\}$, $t^2 = t + 2$. The primitive elements are t^i, where i is relatively prime to 26.

CHAPTER 1: BOOLEAN ALGEBRAS AND SWITCHING CIRCUITS

EXERCISES 1.1

1. $a = a \wedge b \Rightarrow a \vee b = (a \wedge b) \vee b = b$. Conversely, $a \vee b = b \Rightarrow a \wedge b = a \wedge (a \vee b) = a$.

2. For all $a \in B$, $a \vee a = a$, so aPa. If aPb and bPa, then $a = b \vee a = a \vee b = b$. If aPb and bPc, then $a \vee b = b$ and $b \vee c = c$. Hence $a \vee c = a(b \vee c) = (a \vee b) \vee c = b \vee c = c$. So aPc.

3. The postulate (B1) holds obviously. For any $a \in B$, $1 \vee a = \mathrm{lcm}(1, a) = a$, and $n \wedge a = \gcd(n, a) = a$. So (B3) holds with 1 and n as the zero and unity elements. Let $a' = n/a$. Then $\gcd(a, a') = 1$ and $\mathrm{lcm}(a, a') = n$. Hence (B4) holds. To prove (B2), write $u = a \wedge (b \vee c)$ and $v = (a \wedge b) \vee (a \wedge c)$. Now $a \wedge b = \gcd(a, b)$ divides both a and b; hence it divides u. Similarly $a \wedge c$ divides u. Hence $v = \mathrm{lcm}(a \wedge b, a \wedge c)$ divides u. Let p be any prime that divides u. Then p divides a, and p divides b or c. Hence p divides $a \wedge b$ or $a \wedge c$, which implies that p divides v. Therefore u divides v, so $u = v$. This proves that $a \wedge (b \vee c) = (a \wedge b) \vee (a \wedge c)$. The second distributive law is proved similarly.

4. It is obvious that (B1) holds. To prove (B2), let $a, b, c \in R$. Then $a \wedge (b \vee c) = a(b + c + bc) = ab + ac + abc = ab + ac + (ab)(ac) = (a \wedge b) \vee (a \wedge c)$. In a Boolean ring, $x + x = 0$ for all x. Hence $(a \vee b) \wedge (a \vee c) = (a + b + ab)(a + c + ac) = a + bc + abc = a \vee (b \wedge c)$. Thus (B2) holds. Let 0 and 1 denote the zero and identity in the ring R. Then $0 \vee a = 0 + a + 0a = a$ and $1 \wedge a = 1a = a$. Hence (B3) holds. Let $a' = 1 - a$. Then $a \vee a' = a + (1 - a) + a(1 - a) = 1$ and $a \wedge a' = a(1 - a) = 0$. Hence (B4) holds. This proves that $(R, \vee, \wedge)$ is a Boolean algebra.

5. See Exercise 2 in Exercises 0.7.

EXERCISES 1.3

1. $(x_2 + x_1)(x_2 + x_3)(x_2 + x_4)$
2. $(x_1 + x_2')(x_1' + x_2)$
3. $(x_1' + x_2)(x_1 + x_3)(x_2 + x_3)$
4. $(x_1 + x_2)(x_1' + x_2')x_3$
5. $(x_1 + x_2)(x_2 + x_3)(x_3 + x_1)$

EXERCISES 1.4

1. $(x_1 + x_2)(x_3 + x_4)$

2. $x'y$

3. $x + y + w'$

4. $xyz' + y'(z + x'f)$

5. $xy + (x + y)z$

6. $xy + z$

7. $xu + (y + xz)(v + uz)$

8. $(x_1 + x_2)x_3$

9. $x_1 x_3'$

11. $x_1 + x_2'$

12. If $x_1 + x_2 + x_3 = x_1 x_2 x_3$, then for each $i = 1, 2, 3$, $x_i = x_i + x_1 x_2 x_3 = x_i + (x_1 + x_2 + x_3) = x_1 + x_2 + x_3$

13. If $x_1 x_2 + x_3 x_2 = 1$, then $x_1 + x_3 = (x_1 + x_3) + (x_1 + x_3)x_2 = (x_1 + x_3) + 1 = 1$

14. $p' = (x'y'z')'(x'y'z)'(xyz')' = (x + y + z)(x + y + z')(x' + y' + z)$

15. (a) $xyz' + x'yz + x'yz' + x'y'z + x'y'z'$;
(b) $(x' + y + z)(x' + y + z')(x' + y' + z')$

16. (a) $xyz + xyz' + xy'z + xy'z' + x'yz$;
(b) $(x + y + z)(x + y + z')(x + y' + z)$

17. (a) $xyzw + xy'z'w + x'y'zw$;
(b) $(x' + y + z' + w')(x' + y' + z + w')(x' + y' + z' + w)(x' + y' + z + w)$ $(x' + y' + z + w)(x' + y + z' + w)(x + y' + z' + w')(x + y' + z' + w)$ $(x + y + z + w')(x + y + z + w)(x + y' + z + w)(x + y + z' + w)$ $(x + y' + z + w')$

18. (a) $x'y'z'$;
(b) $(x + y' + z)(x + y' + z')(x + y + z')(x' + y' + z')(x' + y + z)$ $(x' + y' + z) \, (x' + y' + z')$

19. (a) $xyz + xyyz$;
(b) $(x + y + z)(x + y + z')(x + y' + z)(x + y' + z')(x' + y + z)(x' + y' + z)$

CHAPTER 2: BALANCED INCOMPLETE BLOCK DESIGNS

EXERCISES 2.2

1. Suppose a_i occurs in r_i blocks. Let M be the number of ordered pairs (p, q) such that a_i, a_p lie in the same block B_q. Then $M = (v - 1)\lambda$ and also $M = r_i(k - 1)$. Hence $r_i = (v - 1)\lambda/(k - 1)$ for all i.

2. By theorems, (a) $\Rightarrow$ (b) $\Rightarrow$ (c) $\Rightarrow$ (d). Suppose (a) holds. Then the inner product of every two columns in the incidence matrix A is λ. Hence A^T is the incidence matrix of a BIBD with parameters (b, v, k, r, λ). Now $b \leq v$, so $v = b$.

3. (a) $2r - \lambda$; (b) $2r - 2\lambda$

4. $v - 2k + \lambda$

5. The condition $r(k - 1) = \lambda(v - 1)$ is not satisfied.

6. The condition $r(k - 1) = \lambda(v - 1)$ cannot be satisfied.

7. Since $v \leq b$ and $r < b$, we must have $v = 7$ and $r = 5$. But then the condition $r(k - 1) = \lambda(v - 1)$ cannot be satisfied.

10. Combine two symmetric $(7, 3, 1)$-designs.

11. $(9, 12, 4, 3, 1)$

14. Take the complement of the BIBD in Exercise 11.

15. Combine two copies of the BIBD in Exercise 11.

16. $(\det A)^2 = \det(AA^T) = (k + (v - 1)\lambda)(k - \lambda)^{v-1} = k^2(k - \lambda)^{v-1}$

17. The condition $k(k - 1) = \lambda(v - 1)$ is not satisfied.

18. Use the result of Exercise 16.

19. Use the result of Exercise 16.

20. A symmetric $(6, 5, 4)$-design is to be constructed by using an appropriate fact in the text.

21. 21 teams

22. A symmetric BIBD with parameters $(22, 7, 2)$ is required, which is not possible (Exercise 16).

23. 10; 2

24. Use the BIBD in Exercise 11.

25.

Day 1	1, 2, 9	3, 4, 12	5, 10, 11	6, 14, 15	7, 8, 13
Day 2	1, 3, 10	4, 5, 13	6, 11, 12	7, 15, 2	8, 9, 14
Day 3	1, 4, 11	5, 6, 14	7, 12, 13	8, 2, 3	9, 10, 15
Day 4	1, 5, 12	6, 7, 15	8, 13, 14	9, 3, 4	10, 11, 2
Day 5	1, 6, 13	7, 8, 2	9, 14, 15	10, 4, 5	11, 12, 3
Day 6	1, 7, 14	8, 9, 3	10, 15, 2	11, 5, 6	12, 13, 4
Day 7	1, 8, 15	9, 10, 4	11, 2, 3	12, 6, 7	13, 14, 5

EXERCISES 2.6

1. Consider $S = \{1, 2, \ldots, n - 1\}$. Any nonzero $a \in \mathbb{Z}_n$ can be expressed as a difference of two elements in S in exactly $n - 2$ ways—namely,

$$a = (a + i) - i, \qquad \text{where } 1 \leq i \leq n - 1 \text{ and } i \neq a$$

So S is a difference set. Any other subset of $n - 1$ elements is of the form $k + S$.

2. By direct verification, we find that $\{1, 5, 6, 8\}$ is a $(13, 4, 1)$-difference set.

3. $(15, 7, 3)$

4. Use the difference set in Exercise 2 to obtain a BIBD D with parameters $(13, 4, 1)$. Its complement D' is $(13, 9, 6)$. The BIBD derived from D' has parameters $(9, 12, 8, 6, 5)$.

5. Use the difference set in Exercise 3 to obtain a BIBD with parameters $(15, 7, 3)$. Take its complement.

6. Find the derived BIBD of the $(15, 7, 3)$-design obtained in Exercise 5.

7. Find the derived BIBD of the $(15, 8, 4)$-design obtained in Exercise 5.

8. $a = x - y \Leftrightarrow a = (-y) - (-x)$

9. $Q_{19} = \{x^2 \mid x = 1, \ldots, 9\} = \{1, 4, 5, 6, 7, 9, 11, 16, 17\}$

10. $(19, 9, 4)$

11. $S = \{0, 1, 2, 3, 5, 7, 12, 13, 16\} = -4 + Q_{19}$

12. $\{0, 3, 6, 7, 12, 14, 16, 17, 18\} = -S = 4 - Q_{19}$

13. $(40, 13, 4)$

14. Use the difference set Q_{19} to obtain a $(19, 9, 4)$-design. Find the derived BIBD.

15. Find the complement of the $(9, 18, 8, 4, 3)$-BIBD obtained in Exercise 14.

16. Find the complement of the $(19, 9, 4)$-design obtained in Exercise 14, and find its derived BIBD.

17. Use Q_{23} to obtain a $(23, 11, 5)$-design. Find its derived BIBD.

18. Find the complement of the $(23, 11, 5)$-design obtained in Exercise 15, and find its derived BIBD.

19. Use the difference set in Exercise 13 to obtain a $(40, 13, 4)$-design, and find its derived BIBD.

20. $Q_{31} = \{1, 2, 4, 5, 7, 8, 9, 10, 14, 16, 18, 19, 20, 25, 28\}$

21. Suppose Q_p is a difference set with parameters $\left(p, \dfrac{p-1}{2}, \lambda\right)$.
Then $(p - 1)\lambda = \left(\dfrac{p-1}{2}\right)\left(\dfrac{p-1}{2} - 1\right)$, so $\lambda = (p - 3)/4$, which is not an integer.

22. Suppose $a = x^2$ for some $x \in \mathbb{Z}_p$. Then $a^{(p-1)/2} = x^{p-1} = 1$, so a is not primitive.

23. From $o(a) = 6t$, it follows that $a^{3t} = -1$ and $a^{2t} = a^t - 1$. The differences in S_i are $\pm a^i = a^i, a^{i+3t}$; $\pm a^{i+t} = a^{i+t}, a^{i+4t}$; $\pm a^i(a^t - 1) = a^{i+2t}, a^{i+5t}$. Moreover,
$$F^* = \{a^j \mid j = 0, 1, \ldots, 6t - 1\}$$
$$= \{a^{i+mt} \mid i = 0, \ldots, t - 1; \; m = 0, \ldots, 5\}.$$

24. $\mathbb{Z}_{19}$ is a field of order $6t + 1$ with $t = 3$, and $a = 2$ is a primitive element in $\mathbb{Z}_{19}$. Hence $S_0 = \{0, 1, 8\}$, $S_1 = \{0, 2, 16\}$, $S_2 = \{0, 4, 13\}$.

25. $\mathbb{Z}_{17}$ is a field of order $4t + 1$ with $t = 4$, and $a = 3$ is a primitive element in $\mathbb{Z}_{17}$. Hence $S_i = \{3^i, 3^{4+i}, 3^{8+i}, 3^{12+i}\}$, $i = 0, 1, 2, 3$.

26. $\mathbb{Z}_{31}$ is a field of order $6t + 1$ with $t = 5$, and $a = 3$ is a primitive element in $\mathbb{Z}_{31}$. Hence $S_i = \{0, 3^i, 3^{5+i}\}$, $i = 0, 1, 2, 3, 4$.

27. Let $S_i = \{a_{ij} \mid j = 1, \ldots, k\}$, where $i = 1, \ldots, t$. In the block $g + S_i$, assign the color C_j to the element $g + a_{ij}$. Given any $a \in G$, we have

$$a = (a - a_{ij}) + a_{ij} \in S_i, \qquad i = 1, \ldots, t; \quad j = 1, \ldots, k$$

So a is assigned the color C_j exactly t times.

28. We require a BIBD with parameters $(9, 18, 8, 4, 3)$, which can be obtained from the difference set family $S_1 = \{0, 1, 2, 4\}$, $S_2 = \{0, 3, 4, 7\}$ (see Example 2.6.5). For the color scheme, use the result of Exercise 27.

29. If α is a root of $p(x)$, then $p(x) = (x - \alpha)q(x)$. Conversely, if $p(x) = f(x)g(x)$, then either $f(x)$ or $g(x)$ is of degree 1.

30. $x^2 + x + 1$; $x^3 + x + 1$, $x^3 + x^2 + 1$

31. $x^2 + 1$, $x^2 + x + 2$, $x^2 + 2x + 2$;
$x^3 + 2x + 1$, $x^3 + 2x + 2$, $x^3 + x^2 + 2$, $x^3 + x^2 + x + 2$, $x^3 + x^2 + 2x + 1$, $x^3 + 2x^2 + 1$, $x^3 + 2x^2 + x + 1$, $x^3 + 2x^2 + 2x + 2$

32. $x^2 + 2$, $x^2 + 3$, $x^2 + x + 1$, $x^2 + x + 2$, $x^2 + 2x + 3$, $x^2 + 2x + 4$, $x^2 + 3x + 3$, $x^2 + 3x + 4$, $x^2 + 4x + 1$, $x^2 + 4x + 2$. The fourth, fifth, seventh, and tenth are primitive.

33. $x^2 + x + 3$

34. $25 = 6t + 1$ with $t = 4$. Let a be a root of $x^2 + x + 2$. Then $S_i = \{0, a^i, a^{i+4}\}$, $i = 0, 1, 2, 3$. So $S_0 = \{0, 1, 3a + 2\}$, $S_1 = \{0, a, 4a + 4\}$, $S_2 = \{0, 4a + 3, 2\}$, $S_3 = \{0, 4a + 2, 2a\}$.

35. $S_i = \{a^i, a^{i+6}, a^{i+12}, a^{i+18}\}$, $i = 0, 1, 2, 3, 4, 5$. So
$S_0 = \{1, 2, 3, 4\}$, $S_1 = \{a, 2a, 4a, 3a\}$,
$S_2 = \{4a + 3, 3a + 1, a + 2, 2a + 4\}$, $S_3 = \{4a + 2, 3a + 4, a + 3, 2a + 1\}$,
$S_4 = \{3a + 2, a + 4, 2a + 3, 4a + 1\}$, $S_5 = \{4a + 4, 3a + 3, a + 1, 2a + 2\}$.

CHAPTER 3: ALGEBRAIC CRYPTOGRAPHY

EXERCISES 3.1

1. TPZZPVUPTWVZZPISL

2. MISSION COMPLETED

3. GONE WITH THE WIND

4. LPFRQGVBZABOHVG

5. ETERNAL LIFE

EXERCISES 3.2

1. (a) PCJYTXOWQD; (b) DEMOCRAT

2. (a) PMUGNXKG, WSXG, W; (b) THEORY OF RELATIVITY

3. (a) ARWXNKMJ; (b) ALGEBRA

4. (a) RMBNIL; (b) HELL

5. (a) QTE; (b) OLD

6. (a) KXHKWW; (b) SAD

7. (a) CYBBTI; (b) EQUATIONS

8. If $KK = I$, then $(\det K)^2 = 1$ in $\mathbb{Z}_p$, so p divides $(\det K)^2 - 1$. Hence det $K = 1, p - 1$.

9. $K^{-1} = (\det K)^{-1} adj\, K = \begin{bmatrix} -d & b \\ c & -a \end{bmatrix}$. Hence $K^{-1} = K$ if and only if $a + d = 0$. ($adj\, K$ stands for adjugate of matrix K. See any book on matrix theory for definition of $adj\, K$.)

EXERCISES 3.4

1. $K = (87, 48, 67, 37, 74)$

2. (a) 88 72 12 89 50 34 88 28; (b) FRENCH

3. (a) 73 60 1 186 40 67 14; (b) ANSWER

4. (a) 2519 2286 2079; (b) COMEDY

5. $d = 19$

6. $d = 107$

7. $d = 2273$

8. (a) 522 1529; (b) RUSH

9. (a) 2611 2299 2103; (b) COPPER

CHAPTER 4: CODING THEORY

EXERCISES 4.1

1. The probability that the received vector has fewer than two errors (and hence is decoded correctly) is $P_{corr} = (1 - p)^5 + 5p(1 - p)^4 + 10p^2(1 - p)^3 = 1 - 10p^3 + 15p^4 - 6p^5$. $P_{err} = 1 - P_{corr}$.

2. If $d(a, x) = t + 1$ and $d(a, y) \leq t$, then $d(x, y) \leq d(a, x) + d(a, y) = 2t + 1$, a contradiction.

3. If the minimum distance is 2, then the ordered pairs obtained by deleting the third component in each codeword must be distinct. The total number of such ordered pairs is 9; hence $M \leq 9$. $C = \{000, 011, 022, 101, 112, 120, 202, 210, 221\}$ is a ternary $(3, 9, 2)$-code.

4. The first part is proved as in Exercise 3. To obtain a ternary $(3, q^2, 2)$-code, take $C = \{(x, y, (x + y) \bmod q) \mid x, y \in \mathbb{Z}_q\}$.

5. Suppose C is a binary (n, M, d)-code. Let C_0, C_1 be the subsets of C that contain codewords in which the last component is 0, 1, respectively. Then one of them has at least $M/2$ codewords. Deleting the last component in the codewords, we get a binary $(n - 1, M', d)$-code with $M' \geq M/2$.

6. Suppose C is a binary (n, M, d)-code. Extend every code word $x_1 x_2 \ldots x_n$ in C to $x_1 x_2 \ldots x_n x_{n+1}$, where $x_{n+1} = (x_1 + x_2 + \cdots + x_n) \bmod 2$. If the distance between codewords $x_1 x_2 \ldots x_n$ and $y_1 y_2 \ldots y_n$ is d (odd), then $\sum_{i=1}^{n}(x_i + y_i) = d$, so $\sum_{i=1}^{n} x_i \neq \sum_{i=1}^{n} y_i \pmod 2$. Hence the distance between codewords $x_1 x_2 \ldots x_n x_{n+1}$ and $y_1 y_2 \ldots y_n y_{n+1}$ is $d + 1$. Therefore the new code is a binary $(n + 1, M, d + 1)$-code. Conversely, suppose D is a binary $(n + 1, M, d + 1)$-code. There exist $x, y \in D$ such that $d(x, y) = d + 1$. Suppose x, y differ in the ith component. Delete the ith component from all codewords. The result is a binary (n, M, d)-code.

7. Let A be an alphabet of q symbols. Then $C = A^n$ is a q-ary $(n, q^n, 1)$-code. If D is a q-ary (n, M, n)-code, then the distance between any two codewords of D is n. Hence no two codewords have the same first component, so $M \leq q$. The q-ary repetition code has q codewords; hence $A_q(n, n) = q$.

8. Follows from Exercise 4.

9. Follows from Exercise 5.

10. Here $M = 2$, $n = 2t + 1 = d$, $q = 2$. Hence

$$M \sum_{m=0}^{t} \binom{n}{m}(q - 1)^m = 2 \sum_{m=0}^{t} \binom{2t + 1}{m} = \sum_{m=0}^{2t+1} \binom{2t + 1}{m} = 2^n$$

11. $2048 \sum_{m=0}^{1} \binom{15}{m} = 2^{11}(1 + 15) = 2^{15}$

12. $2^{2^r - r - 1} \sum_{m=0}^{1} \binom{2^r - 1}{m} = 2^{2^r - r - 1}(1 + 2^r - 1) = 2^{2^r - 1}$

13. $4096 \sum_{m=0}^{3} \binom{23}{m} = 2^{12}(1 + 23 + 253 + 1771) = 2^{23}$

14. $9\sum_{m=0}^{1}\binom{4}{m}2^m = 9(1+8) = 3^4$

15. $59049\sum_{m=0}^{1}\binom{13}{m}2^m = 3^{10}(1+26) = 3^{13}$

16. $729\sum_{m=0}^{2}\binom{11}{m}2^m = 3^6(1+22+220) = 3^{11}$

EXERCISES 4.2

1. In a binary linear code, the number of codewords must be a power of 2.

2. $G = \begin{bmatrix} 0 & 1 & 1 \\ 1 & 0 & 1 \end{bmatrix}$

3. By direct computation we find that every codeword is orthogonal to itself and to every other codeword.

4. A binary vector is orthogonal to itself if and only if it has an even number of 1's. If G satisfies the stated conditions, then every two codewords are orthogonal because they can be expressed as linear combinations of rows of G.

5. The first part follows from Exercise 4. The dual code $C^{\perp}$ is generated by a parity-check matrix of C—for example, the canonical parity-check matrix

$$H = \begin{bmatrix} 0 & 1 & 1 & 1 & 0 & 0 & 0 \\ 1 & 0 & 1 & 0 & 1 & 0 & 0 \\ 1 & 1 & 0 & 0 & 0 & 1 & 0 \\ 1 & 1 & 1 & 0 & 0 & 0 & 1 \end{bmatrix}$$

6. The binary repetition code C consists of two codewords, $00\ldots0$ and $11\ldots1$. Its dual consists of all vectors that have an even number of 1's. C has the generator matrix $G = [1 \quad 1 \quad \ldots \quad 1]$. Hence the dual code is generated by

$$H = \begin{bmatrix} 1 & 1 & 0 & \ldots & 0 \\ 1 & 0 & 1 & \ldots & 0 \\ \vdots & \vdots & \vdots & \ddots & \vdots \\ 1 & 0 & 0 & \ldots & 1 \end{bmatrix}$$

7. To prove $(C_1 + C_2)^{\perp} = C_1^{\perp} \cap C_2^{\perp}$, let $y \in C_1^{\perp} \cap C_2^{\perp}$. Then for all $x_1 \in C_1$ and $x_2 \in C_2$, $y \cdot x_1 = 0$, $y \cdot x_2 = 0$, so $y \cdot (x_1 + x_2) = 0$. Hence $y \in (C_1 + C_2)^{\perp}$. Conversely, suppose $y \in (C_1 + C_2)^{\perp}$. Then $y \cdot (x_1 + x_2) = 0$ for all $x_1 \in C_1$ and $x_2 \in C_2$. Hence $y \cdot x_1 = 0$ for all $x_1 \in C_1$, and $y \cdot x_2 = 0$ for all $x_2 \in C_2$. Hence $y \in C_1^{\perp} \cap C_2^{\perp}$.

8. For any $x, y \in \mathbb{F}_2^n$, $w(x + y) = w(x) + w(y) - 2k$, where k is the number of positions where $x_i = y_i$. Hence the sum of two vectors of even weight has even weight.

9. Suppose C has a codeword a of odd weight. Let A and B denote the sets of codewords of even and odd weight in C, respectively. Then $x \mapsto x + a$ is a bijective mapping from A to B.

10. When we perform elementary row operations, G is reduced to the canonical form

$$G^* = \begin{bmatrix} 1 & 0 & 0 & 1 & 1 & 1 & 0 \\ 0 & 1 & 0 & 1 & 0 & 1 & 1 \\ 0 & 0 & 1 & 0 & 0 & 1 & 1 \end{bmatrix}$$

Hence the canonical parity-check matrix is

$$H^* = \begin{bmatrix} 1 & 1 & 0 & 1 & 0 & 0 & 0 \\ 1 & 0 & 0 & 0 & 1 & 0 & 0 \\ 1 & 1 & 1 & 0 & 0 & 1 & 0 \\ 0 & 1 & 1 & 0 & 0 & 0 & 1 \end{bmatrix}$$

11. The canonical form of H is $\begin{bmatrix} 2 & 2 & 1 & 0 \\ 1 & 2 & 0 & 1 \end{bmatrix}$. Hence the canonical generator matrix is $G = \begin{bmatrix} 1 & 0 & 1 & 2 \\ 0 & 1 & 1 & 1 \end{bmatrix}$. Taking all linear combinations of rows of G, we obtain

$$C = \{0000, 1012, 0111, 2021, 0222, 1120, 2102, 1201, 2210\}$$

12. $G = \begin{bmatrix} I_{n-1} & \begin{matrix} 1 \\ 1 \\ \vdots \\ 1 \end{matrix} \end{bmatrix}$

13. $G = \begin{bmatrix} 1 & 0 & 0 & 0 & 0 & 0 & 0 & 0 & 0 & 1 \\ 0 & 1 & 0 & 0 & 0 & 0 & 0 & 0 & 0 & 2 \\ 0 & 0 & 1 & 0 & 0 & 0 & 0 & 0 & 0 & 3 \\ 0 & 0 & 0 & 1 & 0 & 0 & 0 & 0 & 0 & 4 \\ 0 & 0 & 0 & 0 & 1 & 0 & 0 & 0 & 0 & 5 \\ 0 & 0 & 0 & 0 & 0 & 1 & 0 & 0 & 0 & 6 \\ 0 & 0 & 0 & 0 & 0 & 0 & 1 & 0 & 0 & 7 \\ 0 & 0 & 0 & 0 & 0 & 0 & 0 & 1 & 0 & 8 \\ 0 & 0 & 0 & 0 & 0 & 0 & 0 & 0 & 1 & 9 \end{bmatrix}$

14. The canonical form of the given parity-check matrix is

$$\begin{bmatrix} 9 & 8 & 7 & 6 & 5 & 4 & 3 & 2 & 1 & 0 \\ 3 & 4 & 5 & 6 & 7 & 8 & 9 & 10 & 0 & 1 \end{bmatrix}$$

Hence

$$G = \begin{bmatrix} 1 & 0 & 0 & 0 & 0 & 0 & 0 & 0 & 2 & 8 \\ 0 & 1 & 0 & 0 & 0 & 0 & 0 & 0 & 3 & 7 \\ 0 & 0 & 1 & 0 & 0 & 0 & 0 & 0 & 4 & 6 \\ 0 & 0 & 0 & 1 & 0 & 0 & 0 & 0 & 5 & 5 \\ 0 & 0 & 0 & 0 & 1 & 0 & 0 & 0 & 6 & 4 \\ 0 & 0 & 0 & 0 & 0 & 1 & 0 & 0 & 7 & 3 \\ 0 & 0 & 0 & 0 & 0 & 0 & 1 & 0 & 8 & 2 \\ 0 & 0 & 0 & 0 & 0 & 0 & 0 & 1 & 9 & 1 \end{bmatrix}$$

15. No; no; yes

16. $? = 4$

17. Matrices G and G' are row equivalent if and only if $G' = PG$ for some invertible matrix P.

18. $H = \begin{bmatrix} 0 & 0 & 0 & 0 & 0 & 0 & 0 & 1 & 1 & 1 & 1 & 1 & 1 & 1 & 1 \\ 0 & 0 & 0 & 1 & 1 & 1 & 1 & 0 & 0 & 0 & 0 & 1 & 1 & 1 & 1 \\ 0 & 1 & 1 & 0 & 0 & 1 & 1 & 0 & 0 & 1 & 1 & 0 & 0 & 1 & 1 \\ 1 & 0 & 1 & 0 & 1 & 0 & 1 & 0 & 1 & 0 & 1 & 0 & 1 & 0 & 1 \end{bmatrix}$

19. $H = \begin{bmatrix} 0 & 1 & 1 & 1 \\ 1 & 0 & 1 & 2 \end{bmatrix}$. See Exercise 11.

20. $H = \begin{bmatrix} 0 & 0 & 0 & 0 & 1 & 1 & 1 & 1 & 1 & 1 & 1 & 1 & 1 \\ 0 & 1 & 1 & 1 & 0 & 0 & 0 & 1 & 1 & 1 & 2 & 2 & 2 \\ 1 & 0 & 1 & 2 & 0 & 1 & 2 & 0 & 1 & 2 & 0 & 1 & 2 \end{bmatrix}$

EXERCISES 4.3

1. If $f(x) = f_0 + f_1 x + \cdots + f_{n-1} x^{n-1}$, then $f(x) * u(x) = (f_0 + f_1 + \cdots + f_{n-1}) u(x)$.

2. Over $\mathbb{F}_4$, $x^4 - 1 = (x + 1)^4$.

3. $x^5 - 1 = (x - 1)(x^4 + x^3 + x^2 + x + 1)$ (irreducible factors)

4. $x^6 - 1 = (x^3 + 1)^2 = (x + 1)^2 (x^2 + x + 1)^2$

5. $x^7 - 1 = (x + 1)(x^3 + x + 1)(x^3 + x^2 + 1)$

6. Over $\mathbb{F}_3$, $x^4 - 1$ is factored into irreducible polynomials as

$$x^4 - 1 = (x - 1)(x + 1)(x^2 + 1)$$

8. Of the ternary cyclic codes of length 4, only two are of dimension 2—namely, those with generator polynomials $x^2 - 1$ and $x^2 + 1$. Their generator matrices are $\begin{bmatrix} -1 & 0 & 1 & 0 \\ 0 & -1 & 0 & 1 \end{bmatrix}$ and $\begin{bmatrix} 1 & 0 & 1 & 0 \\ 0 & 1 & 0 & 1 \end{bmatrix}$, respectively. The corresponding parity-check matrices are $\begin{bmatrix} 1 & 0 & 1 & 0 \\ 0 & 1 & 0 & 1 \end{bmatrix}$ and $\begin{bmatrix} -1 & 0 & 1 & 0 \\ 0 & -1 & 0 & 1 \end{bmatrix}$. Neither of these is a parity-check matrix of Ham(2, 3).

9. Interchange the third and fourth components.

10. $x^{10} - 1$ can be factored into irreducible polynomials as

$$x^{10} - 1 = (x + 1)^2(x^4 + x^3 + x^2 + x + 1)^2$$

So there are nine polynomials that divide $x^{10} - 1$ (including trivial cases).

EXERCISES 4.4

1. $g(x) = (x^4 + x + 1)(x^4 + x^3 + x^2 + x + 1) = (x^8 + x^7 + x^6 + x^4 + 1)$.
   Since $g(x)$ has 5 nonzero terms, the minimum distance of the code is 5.

2. $g(x) = (x^5 + x^2 + 1)(x^5 + x^4 + x^3 + x^2 + 1) = (x^{10} + x^9 + x^8 + x^6 + x^5 + x^3 + 1)$

3.

$$\begin{aligned} g(x) &= (x^5 + x^2 + 1)(x^5 + x^4 + x^3 + x^2 + 1)(x^5 + x^4 + x^2 + x + 1) \\ &= x^{15} + x^{11} + x^{10} + x^9 + x^8 + x^7 + x^5 + x^3 + x^2 + x + 1 \end{aligned}$$

4.

$$\begin{aligned} g(x) &= (x^5 + x^2 + 1)(x^5 + x^4 + x^3 + x^2 + 1)(x^5 + x^4 + x^2 + x + 1) \\ &\quad \times (x^5 + x^3 + x^2 + x + 1) \\ &= x^{20} + x^{18} + x^{17} + x^{13} + x^{10} + x^9 + x^7 + x^6 + x^4 + x^2 + 1 \end{aligned}$$

Since $g(x)$ has 11 nonzero terms, the minimum distance is 11.

5. 21 divides $2^6 - 1$. Let ζ be a root of $x^6 + x + 1$. Then $\xi = \zeta^3$ is a primitive 21st root of unity in $\mathbb{F}_{2^6}$. Let $p(x), q(x)$ be the minimal polynomials of ξ, ξ^3—that is, ζ^3, ζ^9. Then the generator polynomial of the desired BCH code is

$$\begin{aligned} g(x) &= p(x)q(x) \\ &= (x^6 + x^4 + x^2 + x + 1)(x^3 + x^2 + 1) \\ &= x^9 + x^8 + x^7 + x^5 + x^4 + x + 1 \end{aligned}$$

6.

$$\begin{aligned} g(x) &= (x^6 + x + 1)(x^6 + x^4 + x^2 + x + 1)(x^6 + x^5 + x^2 + x + 1) \\ &= x^{18} + x^{17} + x^{16} + x^{15} + x^9 + x^7 + x^6 + x^3 + x^2 + x + 1 \end{aligned}$$

7. $127 = 2^7 - 1$. Let ζ be a primitive element in $\mathbb{F}_{2^7}$. The minimal polynomials of ζ, ζ^3, ζ^5 are all of degree 7. Hence the generator polynomial $g(x) = m_1(x)m_3(x)m_5(x)$ is of degree 21. The dimension of the code is $127 - 21 = 106$.

8. $255 - 24 = 231$

9. $8 = 3^2 - 1$. Let ζ be a root of $x^2 + x + 2 \in \mathbb{F}_3[x]$. Then ζ is a primitive element in $\mathbb{F}_{3^2}$. The generator polynomial of the required code is

$$
\begin{aligned}
g(x) &= m_1(x)m_2(x)m_4(x) \\
&= (x^2 + x + 2)(x^2 + 1)(x + 1) \\
&= x^5 + 2x^4 + x^3 + x^2 + 2
\end{aligned}
$$

Since $g(x)$ contains 5 nonzero terms, the minimum distance is 5.

10. $80 = 3^4 - 1$. Let ζ be a primitive element in $\mathbb{F}_{3^4}$. The minimal polynomials of $\zeta, \zeta^2, \zeta^4, \zeta^5, \zeta^7, \zeta^8$ are all of degree 4. The minimal polynomial of ζ^{10} is of degree 2. Hence the generator polynomial $g(x) = m_1(x)m_2(x)m_4(x)m_5(x)m_7(x)$ $m_8(x)m_{10}(x)$ is of degree 26. The dimension of the code is $80 - 26 = 54$.

11. We write $k = n/d$. Then $x^n - 1 = (x^k)^d - 1 = (x^k - 1)q(x)$, where $q(x) = 1 + x^k + x^{2k} + \cdots + x^{(d-1)k}$. Let ζ be a primitive nth root of unity. Then $\zeta, \zeta^2, \ldots, \zeta^{d-1}$ are not roots of $x^k - 1$, so they must be roots of $q(x)$. Hence the generator polynomial of the code divides $q(x)$, so $q(x)$ is a codeword. Since $q(x)$ contains d nonzero terms, the minimum distance is d.

12. The components of the syndrome vector $S(a)$ are as follows: $S_1 = 1 + \zeta$, $S_2 = 1 + \zeta^2$, $S_3 = 1 + \zeta^2 + \zeta^3$, $S_4 = 1 + \zeta^4$, $S_5 = 0$, $S_3 = \zeta + \zeta^2 + \zeta^3$. Solving the linear system for the coefficients in the error locator polynomial, we obtain $f(x) = \zeta^3 + \zeta^5 x + \zeta^3 x^2 + x^3$. The solutions of this equation are $x = \zeta, \zeta^3, \zeta^{14}$. Hence the error vector is $e(x) = x + x^3 + x^{14}$. The corrected codeword is $v = a - e = 100001110110010$. Dividing $v(x)$ by the generator polynomial $g(x) = x^{10} + x^8 + x^5 + x^4 + x^2 + x + 1$, we get the message word 11010.

13. The error vector is $e(x) = x^{28} + x^{30}$. The corrected codeword is 10000100101010100111011011110000.

CHAPTER 5: SYMMETRY GROUPS AND COLOR PATTERNS

EXERCISES 5.1

1. There are $\binom{n}{r}$ ways in which r elements $a_1, \ldots, a_r$ can be chosen from $\{1, \ldots, n\}$. Keeping a_1 as the first element in the cycle, we can permute the remaining $r - 1$ elements in $(r - 1)!$ ways. Hence the total number of cycles of length r is $\binom{n}{r}(r - 1)!$.

2.

Cycle Structure	Type	Number
(1, 1, 1, 1, 1)	(1) = e	1
(1, 1, 1, 2)	(1 2)	10
(1, 1, 3)	(1 2 3)	20
(1, 4)	(1 2 3 4)	30
(1, 2, 2)	(1 2) (3 4)	15
(2, 3)	(1 2) (3 4 5)	20
(5)	(1 2 3 4 5)	24

3.

Cycle Structure	Type	Number
(1, 1, 1, 1, 1, 1)	(1) = e	1
(1, 1, 1, 1, 2)	(1 2)	15
(1, 1, 1, 3)	(1 2 3)	40
(1, 1, 4)	(1 2 3 4)	90
(1, 5)	(1 2 3 4 5)	144
(1, 1, 2, 2)	(1 2) (3 4)	45
(1, 2, 3)	(1 2) (3 4 5)	120
(3, 3)	(1 2 3) (4 5 6)	40
(2, 4)	(1 2) (3 4 5 6)	90
(2, 2, 2)	(1 2) (3 4) (5 6)	15
(6)	(1 2 3 4 5 6)	120

4. For all $a \in G$, $a = eae^{-1}$. If $b = xax^{-1}$, then $a = x^{-1}b(x^{-1})^{-1}$. If $b = xax^{-1}$ and $c = yby^{-1}$, then $c = (yx)a(yx)^{-1}$.

5. $\alpha = (1\ 7\ 3)(2\ 5\ 4\ 6)$, $\beta = (1\ 4\ 2)(3\ 6\ 7\ 5)$. So α and β have the same cycle structure $(3, 4)$. One possible σ such that $\beta = \sigma\alpha\sigma^{-1}$ is $\sigma = \begin{pmatrix} 1 & 2 & 3 & 4 & 5 & 6 & 7 \\ 1 & 3 & 2 & 7 & 6 & 5 & 4 \end{pmatrix}$.

6. Let A_i $(i = 1, \ldots, n)$ denote the set of all permutations $\sigma \in S_n$ such that $\sigma(i) = i$. Then $k_n = n! - |A_1 \cup \cdots \cup A_n|$. Using the result of Exercise 3 in Exercises 0.1 we have

$$|A_1 \cup \cdots \cup A_n| = \sum_{r=1}^{n}(-1)^{r-1}\binom{n}{r}(n - r)! = n!\sum_{r=1}^{n}\frac{(-1)^{r-1}}{r!}$$

To prove the second part, we use the relation

$$e^{-1} = \sum_{r=0}^{\infty}\frac{(-1)^r}{r!}$$

7. Let $\alpha, \beta \in N$. Then $\alpha^{-1}\beta$ moves a finite number of elements, so N is a subgroup of S_X. To show that it is normal, let $\alpha \in N$, $\sigma \in S_X$. Let $A = \{x \in X \mid \alpha \text{ moves } \sigma(x)\}$. Then A is a finite set. For all $x \notin A$, $\sigma^{-1}\alpha\sigma(x) = x$, so $\sigma^{-1}\alpha\sigma \in N$.

8. Let τ be the transposition $(1\ 2)$. Then $\sigma \mapsto \tau\sigma$ is a bijective mapping from A_n to the set of odd permutations in S_n. Hence $S_n = A_n \cup \tau A_n$, so A_n is of index 2 and hence normal.

9. See Exercise 7 in Exercises 0.2.

EXERCISES 5.2

1. D_3, D_2, D_4, D_2, D_2

2. C_1 : F, G, J, L, P, Q, R; C_2 : N, S, Z; D_1 : A, B, C, D, E, K, M, T, U, V, W, X, Y; D_2 : H, I, O

3. D_2, D_1, D_2

4. D_1, D_2, D_1, D_4, D_3

7. See the table given in the text.

8. The rotational symmetries of the cube correspond to the permutations of the four diagonals.

9. Suppose $a^i \in Z(D_n)$. Then $a^i b = ba^i = (a^{n-1})^i b = a^{-1}b$, so $a^{2i} = e$. Hence $2i = 0$ or n. Therefore $Z(D_n) = \{e\}$ or $\{e, a^{n/2}\}$, according to whether n is odd or even.

EXERCISES 5.5

1. $\dfrac{1}{8}(m^4 + 2m + 3m^2 + 2m^3)$

2. $\dfrac{1}{10}(m^5 + 4m + 5m^3)$

3. $\dfrac{1}{4}(m^4 + 3m^2)$

4. $\dfrac{1}{2p}(n^p + (p-1)n + pn^{(p+1)/2})$

5. $\dfrac{1}{12}(3^6 + 2 \cdot 3 + 2 \cdot 3^2 + 4 \cdot 3^3 + 3 \cdot 3^4) = 92$

6. $\dfrac{1}{2}(3^{11} + 3^6) = 88,938$

7. $\dfrac{1}{6}(2^6 + 2 \cdot 2^2 + 3 \cdot 2^3) = 16$

8. $\dfrac{1}{8}(3^{12} + 2 \cdot 3^3 + 3 \cdot 3^6 + 2 \cdot 3^7) = 67,257$

9. $\dfrac{1}{2n}\left(\sum_{r|n} \varphi(r)m^{qn/r} + N\right)$, where

$$N = \begin{cases} nm^{qn/2} & \text{if } q \text{ is even} \\ nm^{(qn+1)/2} & \text{if } q \text{ is odd and } n \text{ is odd} \\ \frac{n}{2}(m^{qn/2} + m^{(qn+2)/2}) & \text{if } q \text{ is odd and } n \text{ is even} \end{cases}$$

10. $\dfrac{1}{6}(4^3 + 2 \cdot 4 + 3 \cdot 4^2) - 4 = 16$

11. $\dfrac{1}{8}(4^4 + 2 \cdot 4 + 3 \cdot 4^2 + 2 \cdot 4^3) - 4 = 51$

12. $\dfrac{1}{6}(m^6 + 2m^2 + 3m^3)$

13. $\dfrac{1}{4}(m^{12} + 2m^6 + m^8)$

14. $\dfrac{1}{4}(m^{24} + 3m^{12})$

15. $\dfrac{1}{4}(m^{35} + m^{18} + m^{20} + m^{21})$

16.

$$
k = \begin{cases}
\dfrac{1}{4}(m^{pq} + m^{(pq+1)/2} + m^{(pq+p)/2} + m^{(pq+q)/2}) & \text{if } p, q \text{ are both odd} \\[2mm]
\dfrac{1}{4}(m^{pq} + 3m^{pq/2}) & \text{if } p, q \text{ are both even} \\[2mm]
\dfrac{1}{4}(m^{pq} + 2m^{pq/2} + m^{(pq+p)/2}) & \text{if } p \text{ is even and } q \text{ is odd} \\[2mm]
\dfrac{1}{4}(m^{pq} + 2m^{pq/2} + m^{(pq+q)/2}) & \text{if } p \text{ is odd and } q \text{ is even}
\end{cases}
$$

17. $\dfrac{1}{12}(m^4 + 11m^2)$

18. $\dfrac{1}{24}(m^6 + 8m^2 + 12m^3 + 3m^2)$

EXERCISES 5.6

1. $\dfrac{1}{24}(y_1^4 + 6y_1^2 y_2 + 3y_2^2 + 8y_1 y_3 + 6y_4)$

2. $\dfrac{1}{120}(y_1^5 + 10y_1^3 y_2 + 20y_1^2 y_3 + 15y_1 y_2^2 + 20y_2 y_3 + 30y_1 y_4 + 24y_5)$

3. $\dfrac{1}{12}(y_1^4 + 3y_2^2 + 8y_1 y_3)$

4. $\dfrac{1}{8}(y_1^4 + 2y_1^2 y_2 + 3y_2^2 + 2y_4)$

5. $\dfrac{1}{4}(y_1^4 + 3y_2^2)$

6. $\dfrac{1}{12}(y_1^6 + 3y_1^2 y_2^2 + 4y_2^3 + 2y_3^2 + 2y_6)$

7. $\dfrac{1}{2n}\left(\sum_{r|n} \varphi(r) y_r^{n/r} + n y_1 y_2^{(n-1)/2}\right)$ if n is odd;

$\dfrac{1}{2n}\left(\sum_{r|n} \varphi(r) y_r^{n/r} + \frac{n}{2}\left(y_1 y_2^{(n-1)/2} + y_2^{n/2}\right)\right)$ if n is even

8. $\dfrac{1}{12}\left(y_1^4 + 3y_2^2 + 8y_1 y_3\right)$

9. $\dfrac{1}{12}\left(y_1^6 + 3y_1^2 y_2^2 + 8y_3^2\right)$

10. $\dfrac{1}{12}\left(y_1^4 + 3y_2^2 + 8y_1 y_3\right)$

11. $\dfrac{1}{24}\left(y_1^8 + 9y_2^4 + 6y_4^2 + 8y_1^2 y_3^2\right)$

12. $\dfrac{1}{24}\left(y_1^{12} + 3y_2^6 + 6y_1^2 y_2^5 + 6y_4^3 + 8y_3^4\right)$

13. $\dfrac{1}{24}\left(y_1^6 + 6y_1^2 y_4 + 3y_1^2 y_2^2 + 6y_2^3 + 8y_3^2\right)$

14. The cycle index polynomial is $Z = \dfrac{1}{12}\left(y_1^6 + 3y_1^2 y_2^2 + 4y_2^3 + 2y_3^2 + 2y_6\right)$. Hence the pattern inventory with three colors is

$$P(t_1, t_2, t_3) = \frac{1}{12}\Big\{(t_1 + t_2 + t_3)^6 + 3(t_1 + t_2 + t_3)^2\left(t_1^2 + t_2^2 + t_3^2\right)^2$$
$$+ 4\left(t_1^2 + t_2^2 + t_3^2\right)^3 + 2\left(t_1^3 + t_2^3 + t_3^3\right)^2 + 2t_1^6 + 2t_2^6 + 2t_3^6\Big\}$$

The coefficient of $t_1^2 t_2^2 t_3^2$ in $P(t_1, t_2, t_3)$ is 11.

15. The cycle index polynomial is $Z = \dfrac{1}{4}\left(y_1^8 + 2y_1^2 y_2^3 + y_2^4\right)$. Hence the pattern inventory with two colors is

$$P(t_1, t_2) = \frac{1}{4}\Big\{(t_1 + t_2)^8 + 2(t_1 + t_2)^2\left(t_1^2 + t_2^2\right)^3 + \left(t_1^2 + t_2^2\right)^4\Big\}$$

The coefficient of $t_1^5 t_2^3$ in P is 17.

16.

$$P(t_1, t_2, t_3) = \frac{1}{4}\Big\{(t_1 + t_2 + t_3)^8 + 2(t_1 + t_2 + t_3)^2\left(t_1^2 + t_2^2 + t_3^2\right)^3$$
$$+ \left(t_1^2 + t_2^2 + t_3^2\right)^4\Big\}.$$

The coefficient of $t_1^3 t_2^3 t_3^2$ in P is 146.

EXERCISES 5.7

1. The cycle index polynomial of S_5 acting on the edges is

$$Z(y_1, \ldots, y_{10}) = \frac{1}{120}\big(y_1^{10} + 10y_1^4 y_2^3 + 20y_1 y_3^3 + 30y_2 y_4^2 + 15y_1^2 y_2^4$$
$$+ 20y_1 y_3 y_6 + 24y_5^2\big)$$

Hence

$$
\begin{aligned}
f_5(x) &= Z(1+x, 1+x^2, \ldots, 1+x^{10}) \\
&= \frac{1}{120}\{(1+x)^{10} + 10(1+x)^4(1+x^2)^3 + 20(1+x)(1+x^3)^3 \\
&\quad + 30(1+x^2)(1+x^4)^2 + 15(1+x)^2(1+x^2)^4 \\
&\quad + 20(1+x)(1+x^3)(1+x^6) + 24(1+x^5)^2\} \\
&= 1 + x + 2x^2 + 4x^3 + 6x^4 + 6x^5 + 6x^6 + 4x^7 + 2x^8 + x^9 + x^{10}
\end{aligned}
$$

2. 156

CHAPTER 6: WALLPAPER PATTERN GROUPS

EXERCISES 6.8

1. C_2
2. D_1-R-1
3. D_2-R-2
4. C_2
5. D_2-R-1
6. D_2-Rh
7. C_3
8. D_3-S
9. C_4
10. D_2-Rh
11. C_6
12. D_1-Rh
13. D_2-R-0
14. D_3-L
15. D_4-0
16. D_6
17. D_1-R-0
18. D_2-Rh
19. D_3-S
20. D_4-2
21. D_6
22. C_4
23. D_2-R-0
24. D_1-Rh

INDEX